LEICESTER MUSEUM STUDIES
General Editor: Susan M. Pearce

Museums and the Natural Environment

Museums and the Natural Environment

The Role of Natural History Museums in Biological Conservation

Peter Davis

Leicester University Press
London and New York

For Glyn and Helen

LEICESTER UNIVERSITY PRESS
A Cassell Imprint
Wellington House, 125 Strand, London WC2R 0BB, England
215 Park Avenue South, New York, NY 10003, USA

First published in 1996

British Library Cataloguing in Publication Data
A catalogue record for this book is available from the British Library

ISBN 0 7185 1548 X

Library of Congress Cataloging-in-Publication Data

Davis, Peter, 1947 Nov. 15–
Museums and the natural environment : the role of natural history museums in biological conservation / Peter Davis.
p. cm. — (Leicester museum studies)
Includes bibliographical references and index.
ISBN 0-7185-1548-X
1. Natural history museums. 2. Environmentalism. 3. Biological diversity conservation. I. Title. II. Series: Leicester museum studies series.
QH70.A1D38 1996
508'.075—dc20 95–41536
CIP

Typeset by York House Typographic Ltd, London
Printed and bound in Great Britain by Biddles Limited, Guildford and King's Lynn

Contents

General preface to series

Museums are an international growth area. The number of museums in the world is now very large, embracing some 13,500 in Europe, of which 2,300 are in the UK: some 7,000 in North America; 2,800 in Australasia and Asia; and perhaps 2,000 in the rest of the world. The range of museum orientation is correspondingly varied and covers all aspects of the natural and the human heritage. Paralleling the growth in numbers, comes a major development in the opportunities open to museums to play an important part in shaping cultural perceptions within their communities, as people everywhere become more aware of themselves and their surroundings.

Accordingly, museums are now reviewing and rethinking their role as the storehouses of knowledge and as the presenters to people of their relationship to their own environment and past and to those of others. Traditional concepts of what a museum is, and how it should operate, are confronted by contemporary intellectual, social and political concerns which deal with questions like the validity of value judgements, bias in collecting and display, the demystifying of specialized knowledge, the protection of the environment, and the nature of our place in history.

These are all large and important areas and the debate is an international one. The series *Leicester Museum Studies* is designed to make a significant contribution to the development of new theory and practice across the broad range of the museum operation. Individual volumes in the series will consider in depth particular museum areas, defined either by disciplinary field or by function. Many strands of opinion will be represented, but the series as a whole will present a body of discussion and ideas which should help to redress both the present poverty of theory and the absence of a reference collection of substantial published material, which curators everywhere currently see as a fundamental lack. The community, quite rightly, is now asking more of its museums. More must be given, and to achieve this, new directions and new perspectives must be generated. In this project, *Leicester Museum Studies* is designed to play its part.

SUSAN M. PEARCE
Department of Museum Studies
University of Leicester

List of plates

List of figures

Acknowledgements

Much of the early research for this book was carried out during an extended visit to the USA in 1993, when I spent a very productive month based in the Academy of Natural Sciences, Philadelphia. My special thanks are due to Bob Peck, Fellow of the Academy, who not only introduced me to many contacts in that institution and elsewhere, but also provided me with accommodation and the benefit of his considerable knowledge of the history and development of American natural history museums. Other staff of the Academy provided me with an insight into the work of that institution and the broader philosophy which drives museums in America; it is a pleasure to record my gratitude to Sandy Sage, George Davis, and Nancy Darstadter for fielding my questions, and to the unflagging assistance of the Librarian, Carol Spawn and her staff. A brief yet very useful visit to the Smithsonian Institution, Washington, was made possible by the Archivist, Bill Deiss; the staff of the Museum Reference Center of the Smithsonian Institution Libraries kindly gave me the freedom to access the books and journals there. My visit to the USA was made possible by a grant from the Research Committee of the University of Newcastle upon Tyne.

Librarians and curators around the world have helped me with numerous inquiries and acted as a sounding-board for ideas, and it is impossible to mention them all by name. However, special mention is due to the staff of the ICOM Documentation Centre, Paris, who provided me with an initial list of 'environmental' publications, and Gina Douglas, Librarian of the Linnean Society of London, who gave me considerable help in locating older texts and engravings. Joseph Guntavid (Sabah), Peter Lups (Natural History Museum, Berne, Switzerland), Urs Kneubuhl (Swiss Alpine Museum, Berne, Switzerland), Robert de Jong (South Africa), Ann Rhoads (Morris Arboretum, Philadelphia), Stephen Bitgood (Jacksonville State University), K. Elaine Hoagland (Association of Systematics Collections, Washington), were amongst the many people with whom I discussed ideas or who provided information and encouragement. The source for many of the facts presented about museums in the USA were often anonymous voices on the end of a telephone, the museum press and public relations officers who, faced with a strange accent and odd requests, provided me with annual reports, mission statements, budgets and other printed information relating to their institutions; to all of them I am most grateful.

Closer to home, Chris Hill (London) gave me invaluable information on the activities of the ICOM Natural History Committee, Emma Watson (NHM) brought me up to date with the UK Systematics Forum and SA2000, Paul Williams of the Natural History Museum introduced me to *Worldmap*, Elizabeth Goodhew and Mike Graham gave me details of the environmental activities at the Horniman Museum, Forest Hill, London and Towneley Hall Museum, Burnley respectively. Ann Pickering, Curator of the Moor Bank Botanic Garden, University of Newcastle kindly gave me access to her unpublished PhD thesis on the contemporary role of botanic gardens.

Sue Pearce (Department of Museum Studies, University of Leicester) was initially responsible for encouraging me to write this book, and despite the numerous deadlines which came and went, was patient, understanding and entirely supportive. My special thanks are due to her, and to the three individuals who kindly read and commented on the text, Geoff Stansfield, Alec Coles and Alwynne Wheeler. Geoff Stansfield not only provided valuable guidance, but also kindly provided access to his bibliographies relating to museums and the natural sciences, and his extensive library.

Thanks are due to my closest support team, my secretary Joanna Keith, and my colleague Andrew Newman, who effectively dealt with much of my normal work-load, so enabling me to have the time and space to write this book. Finally, special thanks go to my wife Sally for her patience and understanding whilst this work was in progress. I am particularly thankful that every once in a while she insisted on prising me from the computer in order to experience the natural environment.

I am grateful to HarperCollins *Publishers* Limited for permission to quote from *If I Ran the Zoo* by Dr Seuss.

List of abbreviations

AAM	American Association of Museums
AAZPA	American Association of Zoological Parks and Aquaria
ABRC	Advisory Board for the Research Councils
AFA	American Forestry Association
AMNH	American Museum of Natural History
AONB	Area of Outstanding Natural Beauty
ASC	Association of Systematics Collections
BIN21	Biodiversity Information Network
BCG	Biology Curators Group
BCSD	Business Council for Sustainable Development
BRC	Biological Records Centre
BRISC	Biological Recording in Scotland Committee
BSBI	Botanical Society of the British Isles
BTO	British Trust for Ornithology
CAS	California Academy of Sciences
CBI	Confederation of British Industry
CBS	Canadian Biological Survey
CDC	Conservation Data Center
CERES	Coalition for Environmentally Responsible Economies
CHIN	Canadian Heritage Information Network
CIN	Conservation Information Network
CIS	Collections Impact Statement
CITES	Convention on International Trade in Endangered Species
CMN	Canadian Museum of Nature
CoEnCo	Council for Environmental Conservation
CFC	Chlorofluorcarbon
COSHH	Control of Substances Hazardous to Health Regulations
CPC	Centre for Plant Conservation
CPRE	Council for the Protection of Rural England
CRU	Collections Research Unit
ECC	Entomology Collections Committee
ECY	European Conservation Year
EIA	Environmental Impact Assessment
EIS	Environmental Impact Statement
EPA	Environmental Protection Agency
FENSCORE	Federation for Natural Sciences Collections Research

FTC	Federal Trade Commission
GCG	Geological Curators Group
GCR	Geological Conservation Review
GIS	Geographical Information System
HCFC	Hydrochlorofluorocarbon
HFC	Hydrofluorocarbon
IBUNAM	Biological Institute of the National University of Mexico
ICOM	International Council of Museums
INBio	Instituto Nacional de Biodiversidad (Costa Rica)
IOPN	International Office for the Preservation of Nature
ISIS	International Species Inventory System
IUBS	International Union of Biological Sciences
IUCN	International Union for the Conservation of Nature
IUMS	International Union for Moral and Social Action
IUPN	International Union for the Preservation of Nature
JANET	Joint Academic Network
MBG	Missouri Botanic Garden
MOU	Memorandum of Understanding
NBIC	National Biodiversity Information Centre
NBS	National Biological Survey
NEPA	National Environmental Protection Act
NFBR	National Federation for Biological Recording
NHM	Natural History Museum (London)
NIC	National Institute for the Conservation of Cultural Property
NMNH	National Museum of Natural History
NRIC	Nature Reserves Investigation Committee
NSF	National Science Foundation
NSFNET	National Science Foundation Network
OAL	Office of Arts and Libraries
OSHA	Occupational Safety and Health Administration
RIGS	Regionally Important Geological and Geomorphological Sites
RSNC	Royal Society for Nature Conservation
RSPB	Royal Society for the Protection of Birds
RSPCA	Royal Society for the Prevention of Cruelty to Animals
SA2000	Systematics Agenda 2000
SCOPE	Scientific Committee on Problems of the Environment
SMNH	Shanghai Museum of Natural History
SPNHC	Society for the Preservation of Natural History Collections
SPNR	Society for the Promotion of Nature Reserves
SSSI	Site of Special Scientific Interest
UKIC	United Kingdom Institute for Conservation
UNCED	United Nations Conference on Environment and Development
UNEP	United Nations Environment Programme
UNESCO	United Nations Educational, Scientific and Cultural Organization
WCCR	World Council on Collections Resources

WINS	Women into Natural Sciences Program
WWF	World Wide Fund for Nature
WWW	World Wide Web

Introduction

> A very large number of all kinds of corals, and other harder sea plants, a multitude of various sorts of crystals, several head-dresses of different races of men, musical instruments etc. Various stuffed birds and fish, where the birds often stood fast on small bits of board as naturally as if they still lived. Skeletons of various four-footed beasts, among which were particularly noticed that of a young elephant, the stuffed skin of a camel, and an African many-striped ass. Several human skeletons larger and smaller, the head and other parts of a frightfully large whale.

This brief quotation from Per Kalm's account of his visit to Sir Hans Sloane's 'curiosities' in 1748 provides an insight into the world of enlightenment collecting, which focused largely on the natural world. It is from these chaotic and undisciplined beginnings that natural history museums emerged. As efforts to document and order the natural world increased in the nineteenth century individuals and societies began to establish collections on a scale hitherto unknown. Throughout Europe national museums were formed to document and preserve collections which related not only to specific countries, but to their dependencies, whilst provincial towns and cities vied with one another to create museums of significance, demonstrations of civic pride. For the majority of these museums, zoological, botanical and geological collections were of great importance, a reflection of the nineteenth-century fascination with natural history which spread throughout the developing world. As a result, there are few towns which fail to possess a natural history museum, or at the very least a collection of natural history specimens in their local museum.

What is the significance of these collections and institutions at the end of the twentieth century? Sadly, for some people the very word 'museum' still hints at irrelevance, and natural history collections are the nadir – cases of faded, stuffed specimens accompanied by meaningless text. Yet in the past 35 years museums have changed radically, not simply by adopting new techniques of presentation, including information technology, but also by reflecting changes in attitudes, and by adopting new philosophical approaches. Natural history museums have been at the forefront of change since the early twentieth century, as they have gradually shaken off the shackles which the technology, attitudes and objectives that their nineteenth-century predecessors had imposed on them. Perhaps the most significant catalyst for change for natural history museums has been the growing awareness of environmental change – the phenomenon known as environmentalism, which had its roots in the radical political mêlée of the 1960s.

The environmental movement has entered all spheres of our lives. Throughout the world countries have enacted legislation to promote environmental protection, often supported by new or strengthened governmental agencies. International Conventions to protect species endangered by trade or overhunting, and to conserve biological and geological sites of world-wide significance have been agreed and implemented. A growing number of voluntary conservation organizations have added their voice and energies to the environmental debate. Public interest in environmental matters has been aroused by these interest groups, many of which have as a key policy the promotion of environmental awareness.

The media have played an increasingly important role, in the provision of high-quality wildlife programmes, and by reporting information relating to environmental matters. The quality of television programmes devoted to wildlife, conservation issues and scientific discovery has never been greater and the immediacy of world-wide communication through television, radio and the press has meant that information about major environmental disasters is quickly available, usually with thoughtful analysis of key issues. The careful presentation of facts need not necessarily lack emotion – pictures of oiled seabirds or the dead, tuskless bodies of elephants continue to act as a reminder of human stupidity and greed. The provision of wonderful wildlife programmes has made us more aware than ever of the world around us. This informal access to knowledge has been supplemented by formal environmental education, both in the classroom and through field studies, and has played its role in fostering environmental awareness, especially in young people.

The consequent large-scale concern about the environment, in particular the dramatic implications of population growth, global warming, depletion of ozone, overcropping of major biological resources and loss of major world habitats such as tropical rainforest, has encouraged a dramatic shift towards 'green' consumerism. Perhaps even more remarkable is the change in attitude to 'nature' itself – we now recognize nature as fragile, not the perceived savage wilderness of the eighteenth century. We are also aware of our custodial role as guardians of the world's environment, its wild places and wildlife, and the inevitable consequences of ignoring that role.

This book is about how natural history museums have responded to the early conservation movement of the late nineteenth century and to the growing demands of environmentalism; it begins with a brief description of the conservation movement, a backdrop on which the role of museums can be superimposed. What follows describes how museums have made the transition from the essential roles of collecting and naming the natural world, to new objectives which reflect a growing respect and concern for it. It has been a move from collecting for its own sake to collecting with a conservation aim; from capturing specimens to capturing and interpreting data, from working alone to networking with conservation organizations, from providing static taxonomic displays to producing entertaining, informative exhibitions with an environmental message. It is about the change from providing information to providing interpretation with the aims of changing perceptions and attitudes and influencing behaviour.

The dual role of natural history museums in research and education has enabled them to make a real contribution to science and public understanding of environmental issues. Having examined the historical background to the adoption of an environmental philosophy, (Chapter 3), the nature of museum exhibitions and educational activities are explored, both within the museum building itself (Chapter 4) and beyond, in field museums and ecomuseums (Chapter 5). The biodiversity debate is one which is exercising the minds of scientists world-wide, and museums are making a significant input in two ways, first through their taxonomic and systematics expertise, and secondly through their recording activities in the field. The background to the biodiversity debate and the role of systematics are discussed in Chapters 6 and 7, looking at the role of the world's major scientific museums in mainstream taxonomy and the assessment of biodiversity at the global level. All museums can and do contribute to researching the geographical distribution of species, and the history and current practice of biological and geological recording are explored in Chapters 8 and 9. In some circles zoos and botanic gardens are regarded as types of museum, as they share much of the philosophy and objectives – it is only the live nature of the collections which make them different. Zoological and botanical gardens have also changed dramatically, the changes being very comparable to those seen in natural history museums – these are described in Chapters 10 and 11. Finally, environmental auditing in museums is discussed in Chapter 12. Just how environmentally friendly are the world's natural history museums?

The naturalist E. O. Wilson has referred to humankind's love of the natural world as 'biophilia', a condition that can be expressed through the love of, for example, animals, gardens, wilderness and landscape. Most individuals who work in natural history museums will be afflicted with biophilia, often to extreme degrees. Most people find it hard to imagine why anyone could spend their lives investigating the status and distribution of ants, lichens, spiders, fish or sponges, and least of all regard it as an occupation of any relevance. Yet we live in a world where human beings, simply through their demands for space and resources, continue to threaten these organisms and the habitats they live in. Paradoxically, it is only in the face of this threat that we are becoming increasingly aware that an important resource is being lost, and, despite the best efforts of the world's taxonomists, we still do not know how great that resource is. Animals and plants play a vital role in ecological systems and they have utilitarian benefits, not least as foodstuffs and sources of medicines. However, there are important moral reasons for conserving biodiversity too, the aesthetic, the 'right' to exist, the demands of the future. For all these reasons natural history museums have a vital role to play as part of a wider conservation movement; they are well placed to help document biodiversity, and from a sound scientific base are able to convince the public that the natural environment is important. The evidence presented here indicates that museums have adopted a new and vitally important role, with their ultimate aim being to foster the conservation ethic.

Peter Davis
Newcastle upon Tyne
1995

1. *Origins of the conservation movement*

Only in the last moment of human history has the delusion arisen that people can flourish apart from the rest of the living world.

E. O. Wilson, *The Diversity of Life*

Wednesday, 22 April 1970. Earth Day. Throughout the USA, 20 million people took to the streets. Politicians, anxious to be seen to be aware of environmental issues, deserted Congress to attend local rallies and declare their concern and support. The 1960s had made popular protest a powerful vehicle for change; it was also a decade that had witnessed environmental disasters on a grand scale, from huge oil spills to the impact of persistent insecticides. This growing recognition of the effect of humankind on the Earth had widened conservationist concerns; from a love of wilderness, spectacular landscapes and wildlife had sprung a deeper understanding of interrelationships, chemical and biological cycles and the fragility of ecosystems. However, this was not a day of protest, more a celebration for planet Earth. It was a celebration which called for action – new legislation to protect natural areas, new laws to combat pollution, better standards of urban design, and social change. It is difficult now to comprehend the scale of participation in Earth Day, and the depth of feeling in the cities of America. It was an event of such magnitude, and with such political impact, that for many historians of the environmental movement, Earth Day marks the beginning of the phenomenon we refer to now as environmentalism.

As the environment entered the political arena during the 1960s, it was accompanied by growing concerns about world peace, nuclear disarmament and a rapidly growing human population. However, as Nicholson (1986) has remarked:

> The implications of living in harmony with the biosphere and its life-support systems are fully consistent with the objectives of the other great global movements for peace and disarmament, for family planning and reduction of population growth and for development and remedying of mass poverty. They are also consistent with the aspirations of growing numbers, especially among the young, towards a more decent and reasonable way of living, recognising that the earth can with good housekeeping provide enough for everyone's need, although not enough for everyone's greed.

The new movement of environmentalism needed an ethical and moral base, and a range of sophisticated perceptions; only now, in the late twentieth century are these ethics and opinions in place. Conservation of wilderness, species and non-renewable resources is seen as a necessity for survival, and has a growing knowledge backed by strong organization. In many ways the 1960s proved to be a decisive decade for conservation in ensuring that ethical stances were made and mechanisms developed which would allow conservationist objectives to flourish. The advances made in organization of the movement, coupled with strong media involvement in promoting wildlife and environmental issues, harnessed supporters and participants world-wide. In the 1990s there are environmental bodies of all persuasions, from those concerned with global issues to local preservation societies or groups seeking to conserve a particular species. The public in general are now well informed about major environmental issues such as tropical deforestation, global warming and ozone depletion. This knowledge has encouraged, in some countries, and in certain sections of society, a change in behaviour and attitude. A new lifestyle is emerging which requires a consideration of the environmental impact of our actions as individuals. This book is about how museums have helped to promote this new outlook, and how the conservation ethic and environmentalism have influenced the philosophy, development and current activities of museums, especially natural history museums. Before examining these issues, it is important to trace the history of the conservation movement in America and Britain, the ways in which the various building blocks of 'the new conservation' were put in place. An understanding of the social and political context of the development of the conservation ethic, and the rise of environmentalism from the 1960s provides a backdrop on which the activities of museums can be superimposed. How have museums served the environmental movement so far? – and how might they respond to the needs of the future?

The development of environmental awareness in America

Although Earth Day is accepted by some political historians as a benchmark, it was the culmination of a long and celebrated history of concern for the American environment. The New World offered wonderful opportunities – and above all it had what Europe had long lost – real wilderness. For the seventeenth century pioneers however, wild America was there to be tamed and subdued – the simple need to survive was paramount. As Stegner (1990) notes:

> The New World . . . lay there so temptingly, like an unlocked treasure house with the watchman sleeping, that nobody thought of limits, nobody thought of preservation, until generations of living in America and 'breaking' its wilderness had taught us to know it, and knowing it had taught us to love it . . . .

The frontiersmen undoubtedly knew nature, and indeed survived by their knowledge, but were far from being conservationists. The idealized, buckskin-clad figure, epitomized by Cooper's 'Leatherstocking' in *The Pioneers* (1823), a romantic image of man at one with nature, and sympathetic to it, (a guise also adopted by the ornithologist Audubon) is divorced from the reality of wilderness man, who in reality led a brutal existence. In effect it was only following the activities of the pioneers and frontiersmen, and the resultant loss of habitat, especially woodland, that others recognized the need for wise use of the vast natural resources of their new country. Celebrated and respected names such as Emerson, Thoreau, William Penn, George Washington, Benjamin Franklin and Thomas Jefferson were to play important roles in establishing a conservation ethic in a rapidly developing country.

Ralph Waldo Emerson, (1803–1882) as a transcendentalist, believed that a correspondence or parallelism existed between the higher realm of spiritual truth and the lower one of material objects, and as a consequence saw religious significance in every aspect of nature. He was to state in 1836 that 'nature is the symbol of the spirit ... the world is emblematic.' (Emerson, 1883). Emerson's friend and disciple Henry David Thoreau (1817–1862) developed the philosophical approach to nature, turning his back on the driving concerns of the majority of his countrymen – progress, improvement and the accumulation of wealth – and devotedly recorded the wildlife around his home in Concord, Massachusetts. Although he personally rarely visited true wilderness, in his writings he recognizes the need for it; indeed Thoreau anticipates many of the ideas which are now central to the environmental movement, including for example, the need for open space in urban areas. He sensed the consequences of industrialization, and the need for the tranquillity of nature described so elegantly in *Walden, or Life in the Woods* (1854). Thoreau was not an activist however, and neither did he appreciate the lines which thread nature together as a whole. The man to develop these ideas, and describe man's impact on the environment was waiting in the wings.

It was a man sadly and surprisingly little known to European conservationists who shifted the focus of concern for countryside and wildlife from Britain and Europe to North America in the nineteenth century – George Perkins Marsh (1801–1882). Marsh (Plate 1) was a polymath – a linguist, politician, diplomat and architect, a Congressman for Vermont, and a major promoter of the Smithsonian Institution. He was Minister of Turkey, and for 21 years Minister Plenipotentiary to Italy. His years in Europe, where he observed environmental destruction, man-made deserts and the remains of once-great civilizations, undoubtedly influenced the thinking which culminated in the publication which 'was the rudest kick in the face that American initiative, optimism and carelessness had yet received' (Stegner, 1990). His far-sighted book, *Man and Nature; or, Physical Geography as Modified by Human Action*, first published in 1864, retitled more accurately *The Earth as Modified by Human Action* for the second edition of 1874, placed the fundamental issues of environmental destruction and its consequences before the human (and especially the American) conscience. He states the purpose of his book is 'to point out the dangers of imprudence and the necessity of caution in all operations which, on a large scale,

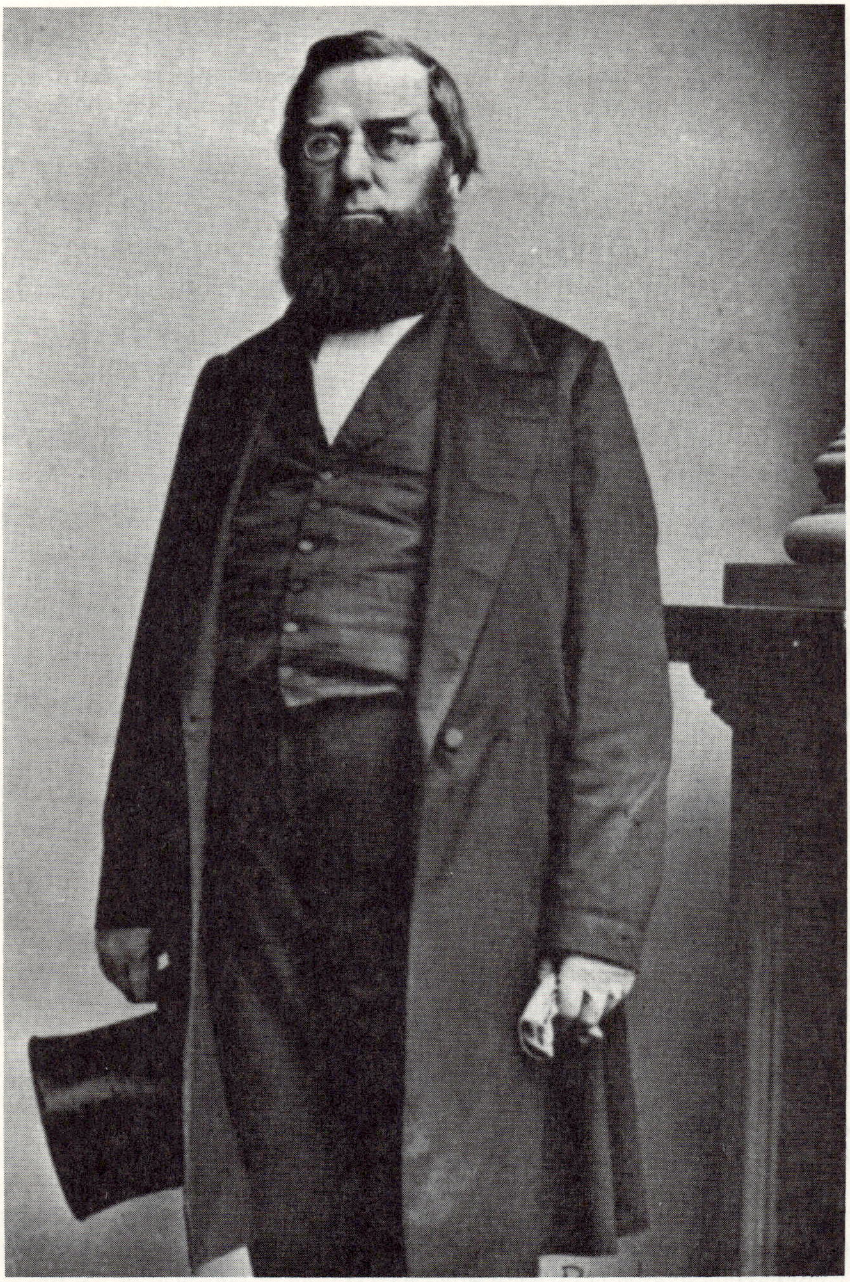

Plate 1 George Perkins Marsh (1801–1882). (Reproduced courtesy of the Library of Congress.)

interfere with the spontaneous arrangements of the organic or the inorganic world.'

Marsh's achievement is more remarkable when we consider that it was published in a country where natural resources were seemingly endless, and the pioneering spirit called for forest destruction on a grand scale – indeed Walt Whitman's 'Song of the Redwood Tree' (1874) demanded that woodland must 'disappear' in order to 'build a grander future' (quoted in Huth, 1990). Predictably, Marsh's warnings went unnoticed, and the deforestation of America proceeded unabated in the late nineteenth century, probably the greatest and swiftest period of forest clearance in history, only exceeded by the current destruction in the Amazonian and Asiatic rainforests.

The devastation of natural habitats in nineteenth-century America ran counter to the growing interest in wilderness, landscape and natural history which had sprung from the early scientific investigation of the continent. John Bartram (1699–1777) had established the first botanical garden in North America, on the outskirts of Philadelphia in 1730, and travelled from Florida to Lake Ontario in his search for plants. Nicholas Collin, a member of the American Philosophical Society, described forests as 'a national treasure' in his lecture to members delivered on 3 April 1789. Charles Willson Peale (1741–1827) founded his natural history museum in Philadelphia in 1785, an establishment which invited visitors to enter 'the great school of nature'. Thomas Jefferson's *Notes on the State of Virginia* (1781) has emotional passages which indicate a new understanding of nature, an aesthetic appreciation of landscape akin to romanticism. St Jean de Crevecoeur's *Letters from an American Farmer* (1782) and William Bartram's *Travels* (1791) continued the lyrical appreciation of nature. There is little doubt that the naturalist William Bartram (1739–1823) was influenced by Pope, Burke and the Lake poets.

Despite widespread interest in the American landscape, and Marsh's published predictions, by 1875 the scale of forest destruction was remarkable. This inspired a group of determined ornithologists to form the American Forestry Association (AFA), a body which was influential in amending the General Revision Act in 1891 to include a clause promoting the establishment of 'forest reserves'. Three Presidents (Harrison, Cleveland and Theodore Roosevelt) took advantage of this to protect 59 million acres of forest from logging interests before angry north-western Congressmen pushed through a Bill which rescinded the President's powers in 1907. As Stegner (1990) notes:

> Nearly every aspect of environmentalism since the founding of the A.F.A. has demonstrated the same pattern: a charismatic and influential individual who discerns a problem and formulates a public concern; a group that forms itself around him or around his ideas, and exerts educational pressure on the public and political pressure on Congress; legislation that creates some new kind of reserve – national park, national forest, national monument, national wildlife sanctuary or wilderness area; and finally, an increasingly specific body of regulatory law for the protection of what has been set aside.

Other areas had been set aside long before the AFA made a case for forest reserves, including the world's first National Park, Yellowstone, in 1872. A group of explorers in Montana were overawed by its natural beauty; aided by paintings and photographs produced by members of the US Geological and Geographical Survey, they convinced Congress to set aside the area for public recreation and enjoyment.

America can justifiably claim to have originated the National Park concept, and once again, one determined individual can take much of the credit for developing the idea of preserving wilderness. As Nash (1982) notes:

> Although the creation of Yellowstone National Park and the Adirondack Forest Preserve [in New York State (1885)] marked a weakening of traditional assumptions about uninhabited land, in each case wilderness preservation was almost accidental and certainly not the result of a national movement.

John Muir (1838–1914), 'the grandest character in national park history' (Mills, 1916) became the champion that natural areas needed, earning a reputation as 'the most magnificent enthusiast about nature in the United States' (Anon, 1912). Born in Dunbar, Scotland, Muir (Plate 2) emigrated to Wisconsin in 1849, an area then still semi-wilderness; Muir never forgot the first impact of the area around Fountain Lake Farm, later recalling 'Nature streaming into us, wooingly teaching her wonderful glowing lessons . . . every wild lesson a love lesson, not whipped but charmed into us. Oh, that glorious Wisconsin wilderness!' (quoted in Wolfe, 1945). Muir was a rebel, and frequently at odds with his stern father; his genius as an inventor provided his means of escape, allowing him to enter the state university at Madison in 1861. Muir's knowledge and horizons grew rapidly, but he also grew restless, eventually leaving for the Canadian wilderness in 1864, thence to Indianapolis from where in 1867 he made his long trek south to the Gulf of Mexico (Muir, 1916). Sailing to California, he reached San Francisco in 1868, soon leaving the city to reach the Sierra Nevada, 'amidst mountains capable of satisfying Muir's enthusiasm, developing his wilderness philosophy and inspiring his most powerful writing' (Nash, 1982). Astounded by their beauty he began his conservation efforts through his writings and political contacts (see Nash, 1982 and Wolfe, 1945) – Yosemite, Sequoia and the Petrified Forest were dedicated as protected areas in 1890. Two years later Muir founded the Sierra Club to look after their interests and enjoy their beauty, an organization which has ever since been central to environmental conservation in the USA. Muir was an extraordinary character, he was a persuasive writer and talker as well as a remarkable field naturalist, and as such was able to influence the great and the good. Often seen as a tough individual, in effect his fight for wild landscapes reflected his sensitive nature and his love for civilization. He was one of the few to recognize Marsh's warning that destruction of wilderness would diminish humankind.

The Sierra Club became engaged in one of the most controversial battles in conservation – San Francisco's efforts to dam the Hetch Hetchy Valley, an area within the Yosemite National Park. This brought into conflict two very

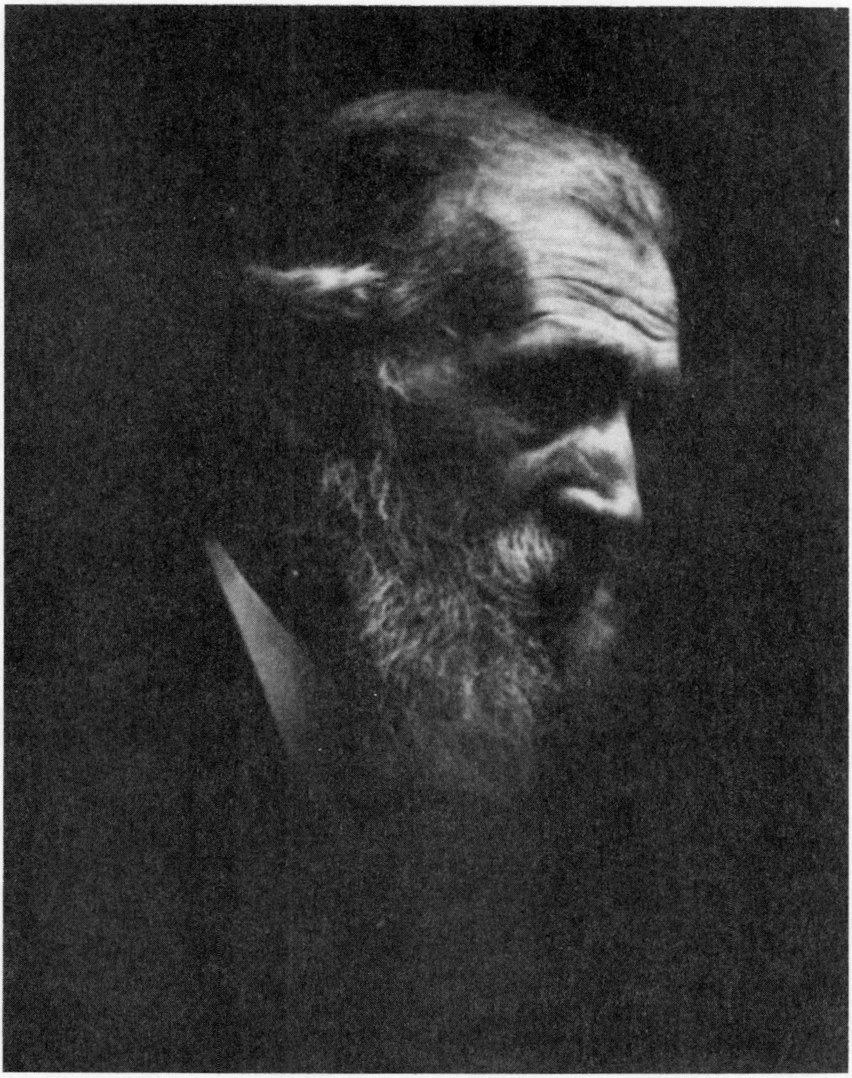

Plate 2 John Muir (1838–1914). (Reproduced courtesy of the Sierra Club Colby Library.)

distinct views on natural areas – the conservationists, (the word was first employed in relation to saving and using spring run-off – see Hays, 1959) who held a 'wise-use' view, and the preservationists, led by Muir and the Sierra Club, who believed that the best natural areas should not be exploited for profit, water, power or any other purpose. The conservationists won that particular battle when Congress approved the dam in 1913. However, such was the strength of feeling about the outcome, that in 1916 when the National Parks Act was passed, it included a 'use without impairment' clause – the very guarantee that Muir had sought. This was a turning-point in conservation history in America; the preservationist ethic gained ground and replaced the conservationist view, so much so that the Sierra Club won the next major battle when they had removed the proposal for a dam in the Dinosaur National Monument from the Colorado Water Storage Project Bill of 1959. Stegner (1990) remarks on this incident:

> Historians have concluded that the Dinosaur controversy marked the coming-of-age of conservation, the first time the diverse single-interest organisations joined together for strength, and proved they had a lot of it. But Dinosaur is not understandable without Hetch Hetchy. The tactics, the publicity campaigns, the political skirmishing, were similar in these instances. The difference between defeat in the first instance and victory in the second is that Muir did not have the National Park Act and the 'use without impairment' clause to fight from, while Brower (leader of the campaign) and his alliance did. And the precedent has been instructive in later struggles.

It is important to distinguish between National Parks and Wilderness Areas – the first conforms to the conservationist 'wise-use' ethic, whereas preservation of wilderness implies agreement with Marsh in denying most kinds of human interference. The concept of wilderness has its critics, particularly because it ring-fences utilitarian reserves; but many more people now appreciate the other human uses implicit in safeguarding such areas, protecting watersheds and genetic resources in particular. Protecting wilderness parallels other major conservation efforts – breeding endangered species for example – by being a potent symbol of our caring attitude to the environment.

One of the greatest supporters of wilderness was Aldo Leopold (1887–1948):

> Blessed with a poet's gift of words, trained in forestry, he brought as much sensitivity as Thoreau to his study of nature, and more science. His book (*A Sand County Almanac*, 1949) comes to a climax in the essay called *The Land Ethic*, in which he argues for the kind of responsibility in our land relations that civilised peoples are supposed to show in their human relations . . . he wanted a portion of the land left wild, for ultimately all human endeavour has to come back to the wilderness for its justification and its new beginnings (Stegner,1990).

The efforts of Leopold (Plate 3), and those of many other committed conservationists, bore fruit in 1964 with the passing of the Wilderness Act and immediate designation of 9.1 million acres of wilderness, defined as 'an area where the earth and its community of life are untrammelled by man, where man himself is a visitor who does not remain.' Today the National Wilderness Preservation System contains 89 million acres (36 million hectares) of designated wilderness at 465 locations, managed by a variety of agencies.

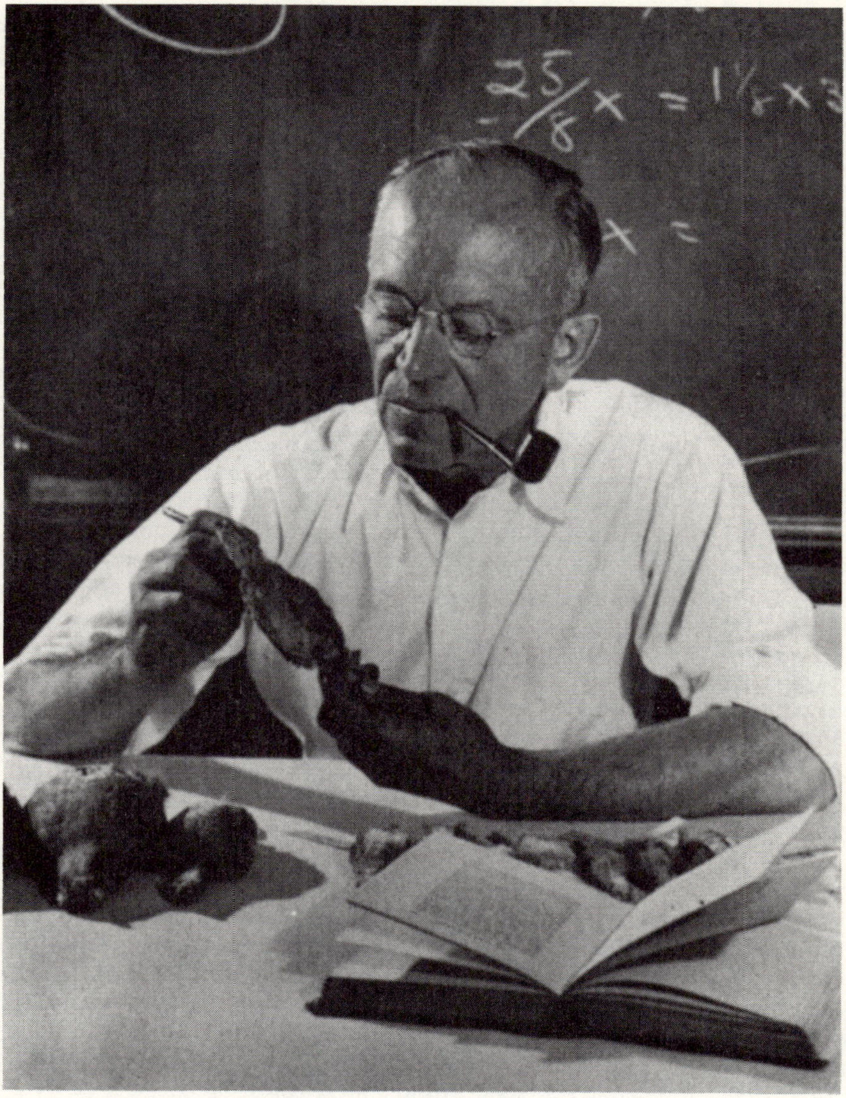

Plate 3 Aldo Leopold (1887–1948). (Photograph by Robert Oetking reproduced courtesy of the University of Wisconsin-Madison Archive.)

The conservation movement in Britain

In contrast to the USA, Britain's environment has long been influenced by people. Even by the beginning of the nineteenth century an agricultural economy had confined nature to pockets of land in a managed landscape; there was no sharp division between land which had been tamed and wilderness, no sudden and dramatic realization of humankind's impact which is central to the story of the conservation movement in America. In contrast, the roots of the conservation ethic in Britain can be traced back to the tradition of field natural history and the humanitarian crusade against cruelty to animals. The two countries also differ in their appreciation of the roles played by conservationists. Whilst the USA reveres the achievements of outstanding individuals and their search for a conservation ethic, Britain's key players are frequently shrouded beneath the cloak of the organizations they helped to create.

The history of Britain's naturalists and their achievements has been described in a classic study by David Allen (1976), who traces the beginnings of the social organization of naturalists back to the end of the seventeenth century and the Enlightenment, a period which saw the first use of the geologist's hammer (1696), the vasculum (1704) and the butterfly net (1711). The seventeenth century had also seen the publication of the first identification aids for naturalists, including John Ray's *Catalogus Plantarum Angliae* (1677) and *Synopsis Methodica Stirpium Brittanicarum* (1690), Martin Lister's *Historiae Animalium Angliae* (1678) and Francis Willoughby's *Ornithologia* (1676). Early natural history studies in Britain focused on botany, a consequence of the interest in medicinal plants and the long tradition of British field natural history evolved from these early 'herbarizings' of the seventeenth century. It is interesting to contemplate that as the interest in nature was beginning to become established, huge changes were taking place in the British countryside as woodland cover was destroyed and wetlands drained; by the late seventeenth century 50 per cent of England and Wales was classified as agricultural land.

The Enlightenment enabled mankind to take a new, more objective approach to the natural world which would eventually lead to a greater understanding and a genuine passion for it by the end of the eighteenth century. Making collections of natural history specimens was an essential element in this new approach to nature, and an important step in promoting the social standing of naturalists. The exchange of specimens between individuals did much to establish natural history as an acceptable pastime, creating links which cut across traditional barriers of age and class. The band of amateur naturalists began to grow, and a more learned view of the natural world came about through careful observation and collection of specimens. The eighteenth century saw the adoption of many advances, including the adoption of the classification systems of the Swede Linnaeus (1707–1778) following the publication of his *Species Plantarum* in 1753, which introduced the consistent use of binomial names in botany, and the formation of societies devoted to the study of natural history, including the Linnean Society of London (1788). Specialist literature in English became more widely available, including Gilbert White's *Natural History of Selborne* (1788), which captures one man's delight in the diversity and ingenuity of nature. Thomas (1983) comments on the availability and success of works on natural history:

> The eighteenth century saw the popularisation of natural history by authors who wrote in the vernacular rather than Latin and who aimed to entertain as well as to instruct. In France library catalogues reveal that the Abbe Pluche's *Spectacle de la*

> *Nature* (1732) and the Comte de Buffon's *Histoire Naturelle* (1749–1804) had a greater vogue than even the writings of Voltaire. In England popular writers of natural history like John Hill, Oliver Goldsmith, Thomas Pennant and William Bingley enjoyed similar success: Peter Collinson thought in 1747 that works on natural history 'sell the best of any books in England'.

Illustrated works of natural history proved especially popular (though often expensive), and artists who specialized in natural history subjects (including Moses Harris, William Lewin and Thomas Bewick) became well known. By 1763 *The Critical Review* could declare that natural history was 'the favourite study of the times'.

It was in the late eighteenth and early nineteenth centuries, when the study of natural history was an amateur pastime, that the seeds of the conservation movement in Britain were sown, and a respect for the natural world began to become engendered in society. Animals, plants, even rocks and fossils, were no longer seen as objects to revile and despise, but became appreciated for their beauty and the intellectual challenges they posed. This applied not just to collections of specimens, but to living animals and plants. Thomas (1983) records several eighteenth-century attempts at practical conservation, including Rowland Hill's attempts in the 1790s to establish a refuge for toads 'a place of retirement . . . called . . . a toadery'. Allen (1976) refers to the wildlife reserve established by the eccentric Charles Waterton in 1813, and Fitter (1963) has claimed that Waterton 'was probably the first man to put up nestboxes purely to attract birds for the pleasure of seeing them.'

This new sensitivity and concern saw a marked change in attitudes not only to wildlife but to wild landscapes, an expression of 'the collective enjoyment of the scenery of nature for its own sake and in its original, unmodified condition' (Allen, 1976), a 'Romantic' attitude which would later be adopted across the Atlantic. The growing understanding of plants, animals and rocks was a prerequisite for the Romantic movement and its humility and respect for landscape and natural beauty (see Plate 4). Romanticism in Europe, and American transcendentalism of the mid-nineteenth century, have contributed much to the philosophy of modern environmentalism. As Pepper (1984) notes, for the Romantics,

> nature had an integrity of its own. Man could relate to it in a very intimate way, but if he were to be removed from the scene nature would nonetheless go on being what it was. It is but a short step from this position to that of the modern ecocentrics . . . the idea that nature has a purpose and meaning by itself.

David Allen (1976) contends that by the early nineteenth century the Romantic movement had collapsed into sentimentality, with numerous publications (especially those dealing with botany) in France and England presented as poetical texts. The increased accessibility of natural history made it a compulsion for many people. For some, nature made the transformation from being not simply wonderful, exciting or pleasing to the eye, but a spiritual experience, to be pursued with evangelical fervour. Lowe (1983) recounts the words of the President of the Cotteswold Naturalists Club 'We seek a healthy, a most fascinating, far more than either, a most

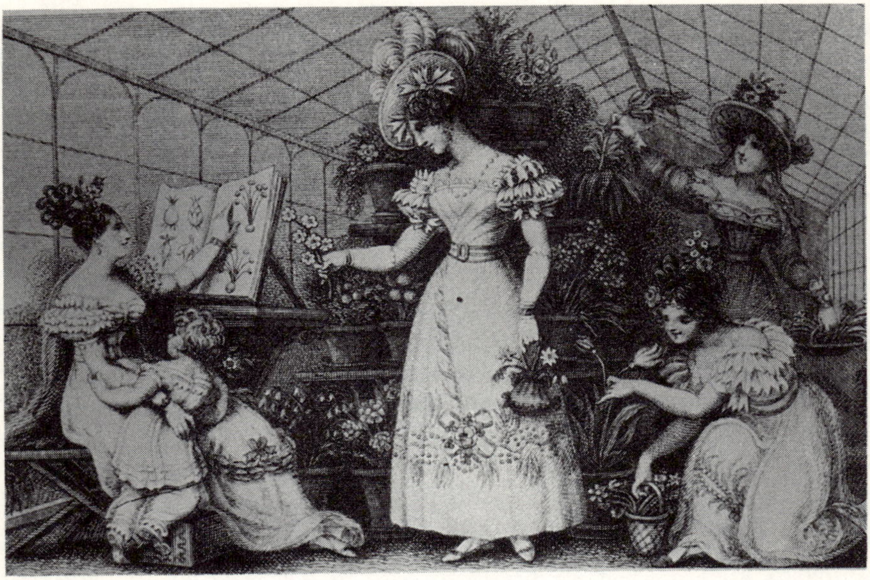

Plate 4 The Romantic view of natural history – 'Ladies botanizing, *c.*1830'. (Reproduced by permission, Mary Evans Picture Library.)

holy study. For what is the study of Natural History, but an approach to the Creator through his works.' The Victorian obsession with self-improvement and travel recruited more and more people to the study of natural history, supported by a wealth of newly published books and specialist journals. Books were often written in a popular style which resulted in huge sales – the Reverend J. G. Wood's *Common Objects of the Country* (1858) sold 100,000 copies in a single week. Collecting became a passion, with the size of the collection, rather than its quality, often paramount. Sheltering behind the maxim 'through Nature up to Nature's God' the greedy collector could justify superfluous collecting activities. The major benefit for today's curators and researchers, apart from the number and size of such collections held in store, is the meticulous care that the Victorian naturalists paid to labelling and organizing their specimens.

The early 1800s saw tremendous energy in data gathering, publishing and the founding of new national societies devoted to various branches of natural history (The Zoological Society 1826, The British Association for the Advancement of Science 1831, The Entomological Society 1833, The Palaeontographical Society, 1847). This period also saw the growth of Literary and Philosophical Societies in a number of urban centres, whose members did much to advance the knowledge of the natural history of their regions and build significant museums. However, as David Allen (1976) remarks:

> There was one other Early Victorian invention which was to have a more far-reaching influence and bring more benefits to more people than all of the rest

put together. This was the masterpiece of social mechanics, the natural history field club.

The Berwickshire Naturalists' Field Club was founded in 1831 by George Johnston (1797–1855), a doctor in Berwick who had experienced the delight of field natural history when a student at Edinburgh University, where he took part in the excursions of the Plinian Society. The Berwickshire Club modelled itself on similar lines, meetings were held outdoors as the local naturalists explored the natural history of the border country together. The meetings were not simply scientific excursions, as considerable time was devoted to socializing – members breakfasted together, rode and botanized together, and enjoyed dinner in the evening where toasts would be drunk; papers of interest were read and discussed afterwards. Johnston (see Plate 5) had made natural history a socially acceptable pastime, and the Berwickshire Club became a highly successful model. Without the demands of expensive premises, costs could be kept low, and it is no surprise that the field club idea spread to all corners of Britain, becoming the ancestor of natural history societies in Britain today. Many field clubs produced journals of distinction, some having an unbroken run until the present day – the Berwickshire Club's *History* continues to publish fascinating accounts of the natural history and archaeology of the Borders. These journals provide modern conservationists with essential information about past distribution of species, the lists and catalogues produced by the nineteenth-century naturalists providing a yardstick against which environmental change can be measured. The collections made by individual field club members, which were frequently the basis of publications, perform the same function.

The success of George Johnston's invention can be measured in the proliferation of field clubs and natural history societies – Lowe (1983) estimates that by the 1880s there were several hundred of them, with a combined membership of around 100,000. The fascination with nature had resulted in a small army of dedicated, knowledgeable naturalists, a substantial literature, well-documented and accurate regional and local lists of plants, animals and geological features, and large collections of specimens held by societies and individual naturalists. It was from this body of men and women that the first concerns for the protection of nature were to be voiced.

The second intellectual movement which fostered conservation in Britain was the major humanitarian preoccupation of the nineteenth century, the crusade against cruelty to animals. The gradual erosion of the idea of the superiority of mankind, the recognition of the moral and social implications of man's cruelty to animals and the sentimentality which characterized Victorian society resulted in a movement which initially focused on the barbarities of animal baiting, blood sports and the treatment of domestic animals, especially cattle and horses. The Society for the Prevention of Cruelty to Animals, founded in 1824 (receiving a Royal charter in 1840) became the vehicle for action through education, individual prosecutions and the promotion of legislation. Before long the Society had turned its

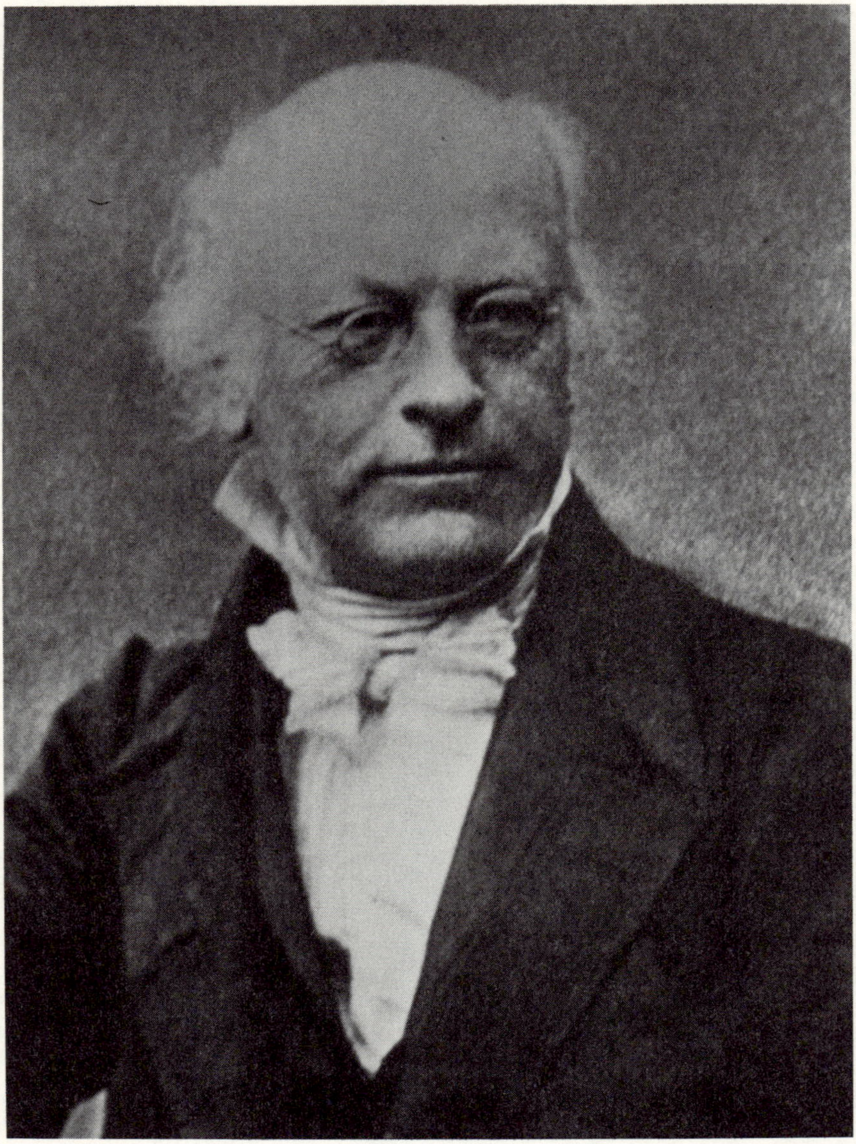

Plate 5 George Johnston (1797–1855) of Berwick-upon-Tweed, founder of the Berwickshire Naturalists' Field Club.

attention to wildlife issues, joining the naturalists to promote the cause of protecting birds from unwarranted levels of persecution.

The shooting of seabirds on their nesting cliffs at Flamborough Head in Yorkshire prompted concerned ornithologists to form in 1868 the East

Riding Association for the Protection of Seabirds, often considered the first wildlife preservation society in the world. The Society, with the considerable support of the RSPCA, lobbied Parliament to pass the Sea Birds Protection Bill in 1869 – the first-ever legislation to protect wildlife. As Lowe (1983) notes:

> Though the reform was prompted by humanitarian motives, the debate which surrounded it also raised the issue of the impact of shooting on bird populations. Much was made, for example, of the birds' role in preserving the balance of nature and in keeping down harmful insects.

As concern grew about the shooting of birds during the breeding season, the British Association, the Association for the Protection of British Birds (founded in 1870) and the RSPCA were instrumental in establishing additional legislation for the protection of wildfowl (1872, 1876) and the Wild Birds Protection Act of 1880.

Emotion and science had combined to enact this first bird protection legislation; it was raw emotion that dominated the fight against the plumage trade. The dictates of fashion in the 1880s, in particular the grotesque adornment of women's hats with bird feathers, resulted in the deaths of millions of birds. A militant, predominantly female, reaction against the plumage trade occurred on both sides of the Atlantic. America's Audubon Society, and in Britain the Selborne League and the Plumage League, formed in 1885 to voice their protest. The two British Leagues joined forces as the Selborne Society; and an amalgamation in 1891 between the quaintly named Fur, Fin and Feather Folk of Croydon (1889) and Manchester's Society for the Protection of Birds (1889) led to the formation of what is one of Britain's most influential conservation bodies, the Royal Society for the Protection of Birds – the Royal charter being granted in 1904. Both groupings sought to change public opinion, but by very different means; the RSPB took a militant stance, the Selborne Society the route of education and persuasion. The latter's decision to work with the plumage trade to devise methods of regulation caused a deep split which effectively delayed the enactment of the Importation of Plumage (Prohibition) Act until 1921. It is of interest to note that the Natural History Museum voiced its protest against the plumage trade in 1914 (Stearn, 1981, pp. 105–6), an early indicator of the growing involvement of museums in conservation.

The history of the early conservation legislation in Britain was a strangely fought battle between those who still perceived wildlife as utilitarian (for hunting, gourmet foods, or for commercial gain) and the humanitarian–naturalists who felt a more emotional attachment. The humanitarian movement which dominated this early period needed the numerical support provided by the army of dedicated naturalists; humanitarian elements can also be recognized in the struggle to preserve the British landscape which began in the late nineteenth century.

The years from 1870–1940 have been christened the 'preservationist' period in British conservation by Lowe (1983), recognized as a time when there was 'a gathering sense of the vulnerability and loss of wildlife, demanding urgent counter measures'. It was also a time of expansion of towns and cities, and marked changes in the rural economy, when appreciation of open space and 'natural' landscapes came to the fore in the face of growing threats to them. The Commons Preservation Society, formed in 1865, was the first influential society formed as a reaction against the loss of public amenity, a reaction driven by the needs of recreation, preservation, sanctuary and wilderness. As Allen (1976) notes, the philosophical concept of wilderness, and its utilitarian benefits:

> made little impact over here, this more sober version of the cluster of attitudes that we have since come to recognise as 'conservation' was a regrettably tardy influence on British thinking and missed inclusion in the ideological mix at this formative period.

The early twentieth century brought a growing awareness of man's destructive power as a result of changes in land management. Lowe (1983) quotes Conwentz's (1909) observations on the growing impact of economic change:

> Anyone travelling in England . . . may see for himself that the constant cultivation of the land and the growth of the industrial undertakings have threatened, and in many places considerably damaged, interesting tracts of country as well as natural monuments . . . it is much to be regretted that so many types of scenery and of the vegetable and animal worlds should pass away irrevocably.

Concern was not simply focused on Britain, as the wider perspective of the Empire brought other evidence of the loss of species and the destruction of habitat. Charles Rothschild, founder of the Society for the Promotion of Nature Reserves (SPNR) in 1912 remarked that 'civilisation and progress are gradually but surely forcing back and destroying the beauties of wild life' (Rothschild, 1914); Ray Lankester stated that 'Once man is present in the neighbourhood, even at long distance, he upsets the "balance of nature" ' (Lankester, 1915). It was sometime after Marsh's activities in the USA that Britain recognized the dangers. With new sensibilities, nature and the countryside were now seen as fragile and under threat and as destruction of habitats continued they became more and more valued as an antidote to the harsh realities of the industrial society. The growing recognition of the threat to 'traditional' countryside (and hence to cultural identity), coupled with a recognition of its aesthetic, recreational and scientific value spurred the preservationists to positive action. The new preservationist societies (including The National Trust (1894) and the Council for the Preservation of Rural England (1926)) began to acquire land and lobby for new legislation to protect surviving wild places. The National Trust, always at the forefront of the preservationist movement, became one of the first organizations to own and manage areas of biological significance, notably Wicken Fen in Cambridgeshire and Blakeney Point in Norfolk.

The formation of the British Ecological Society in 1913 marked a turning-point in the history of biological conservation in Britain. Scientific evaluation and careful selection of potential nature reserves replaced the more opportunistic approach of the preservationists; habitat protection was recognized as the key to saving rare species, and a growing understanding of habitats as dynamic, changing systems made their active management a necessity. This brought the ecologists into conflict with the preservationists, who saw reserves as places of quiet contemplation, which suffered no interference from man. This concept of the nature reserve was questioned – was this really the function of such an area or were they, as the ecologists proclaimed, outdoor laboratories where useful scientific work could be conducted? The ecologists were interested in typical species and discovering how ecosystems worked, and went beyond the purely descriptive approach of their predecessors, the early conservationists, who were preoccupied with taxonomy and rarities, describing new species or producing accounts of the distribution of animal and plant groups. According to Lowe (1976) ecology had arisen as a discipline in conscious reaction to the obsession of systematics with novelty and rarity. This divergence between systematics and ecology is a fascinating phenomenon which has had considerable implications, including the decline in the teaching of systematic biology, and in collection building. There can be little doubt that as ecological science and environmentalism progressed hand in hand, the activities of systematists became undervalued and the status of the traditional collector declined, indeed collecting was seen to be in total contradiction of the conservation ethic. The implications for museums as a repository of collections and systematists, have been considerable, and these are discussed later.

During the 1930s and 1940s the ecologists began to dominate the British conservation movement. The Nature Reserves Investigations Committee (NRIC), appointed under the auspices of SPNR were given the task of making the case for nature conservation and drawing up a short list of potential reserves. The British Ecological Society established its own committee, under the chairmanship of Arthur Tansley, to investigate the *raison d'être* of nature reserves. Official committees chaired by Julian Huxley and James Ritchie analysed the evidence from the BES, NRIC and others; the government acted on their advice to establish the Nature Conservancy as a chartered research council in 1949. As a consequence scientific methodology has been a key element in the British conservation movement, with reserve management closely linked to underlying ecological principles. The scientific nature of conservation practice has in turn lead to its recognition as an academic discipline, and ultimately to conservation as a profession.

Much of the enthusiasm and drive for the creation of a wildlife agency (The Nature Conservancy) had come from the voluntary sector. Similar demands had been made for greater access to the British countryside and the creation of National Parks, spearheaded by the CPRE from 1929; direct action came with mass trespasses on Kinder Scout in the Peak District in April 1932. The two campaigns for the establishment of nature reserves and National Parks ran in parallel, yet divergence of views between the access lobby and that of the wildlife conservationists about their purpose resulted in the failure of the recommendations for the establishment of National Parks made by the government's review body (the Addison Committee) in 1931. Evans (1992) describes the efforts made by the voluntary sector to make National Parks a reality in Britain, efforts lead by the Standing

Committee on National Parks established in 1936. This Committee united voluntary bodies of all persuasions; with John Dower as its visionary leader a National Parks Bill was drafted by 1939. Dower produced a further report in 1945, which advocated quick action to establish National Parks and to give them administrative independence; it has been described by Bonham-Carter (1971) as 'a masterly document . . . now acknowledged as the basic blueprint of recreational planning in the countryside'.

It was not until 1949 that the National Parks and Access to the Countryside Act was passed by Parliament; enshrined within it were the duties of The Nature Conservancy, and the mechanisms to establish National Parks and Areas of Outstanding Natural Beauty (AONBs), and reconstituted the Standing Committee as the National Parks Commission. The Bill was introduced with great political posturing, which Evans (1992) notes:

> only served to heighten the disappointments of the Act's reality. None of our National Parks would qualify for the title by international standards, for they are all too intensively used to be of sufficient amenity value. The label AONB has proved to be nothing more than a paper designation.

Environmental progress

The accounts of progress towards concern for the environment in the USA and Britain support the view that 'The (environmental) movement did not begin in one country and then spread to another; it emerged in different places at different times, and usually for different reasons' (McCormick, 1989). By the 1950s most of the developed countries of the world had established organizations for conserving landscapes and wildlife, and the concept of the National Park and the nature reserve was firmly established. Networks which demanded biological and geological knowledge had been created; natural history museums, with their taxonomic expertise and accumulated data became integral to those networks. The stage was set for the next major development – environmentalism.

References

Allen, D. E. (1976) *The Naturalist in Britain; A Social History*, Allen Lane, London.

Anon. (1912) 'About the Yosemite', *American Review of Reviews*, 45, 766–767.

Bartram, W. (1791) *Travels through North and South Carolina, Georgia, East and West Florida*, Philadelphia.

Bonham-Carter, V. (1971) *The Survival of the English Countryside*, Hodder and Stoughton, London.

Cooper, J. F. (1823) *The Pioneers*, Charles Wiley, New York.

Emerson, R. W. (1883) *Nature, Addresses and Lectures, The Works of Ralph Waldo Emerson*, Standard Library Edition, 14 vols, Boston.

Evans, D. (1992) *A History of Nature Conservation in Britain*, Routledge, London and New York.

Fitter, R. (1963) *Wildlife in Britain*, Penguin, Harmondsworth.

Hays, S. P. (1959) *Conservation and the Gospel of Efficiency: The progressive Conservation Movement, 1890–1920*, Harvard University Press, Cambridge, MA.

Huth, H. (1990) *Nature and the American*, University of Nebraska Press, Lincoln and London.

Lankester, R. (1915) *Diversions of a naturalist*, Methuen, London.

Lowe, P. D. (1976) 'Amateurs and professionals; the institutional emergence of British plant ecology', *Journal of the Society for the Bibliography of Natural History*, 7, 517–535.

Lowe, P. D. (1983) 'Values and Institutions in the History of British Nature Conservation'. In Warren, A. and Goldsmith, F. B. *Conservation in Perspective*, 329–352, John Wiley and Sons Ltd.

McCormick, J. (1989) *The Global Environmental Movement – Reclaiming Paradise*, Belhaven Press, London.

Mills, E. (1916) Untitled. *Sierra Club Bulletin*, X, (January), 25.

Muir, J. (1916) *A thousand mile walk to the Gulf.* Edited by William F. Bade, Houghton Mifflin, Boston.

Nash, R. (1982) *Wilderness and the American Mind*, Yale University Press, New Haven and London, 3rd Edition.

Nicholson, M. (1986) *The New Environmental Age*, Cambridge University Press, Cambridge.

Pepper, D. (1984) *The Roots of Modern Environmentalism*, Croom Helm, London, Dover, NH.

Rothschild, N. C. (1914) 'The preservation of nature', *Countryside*, 416–418.

Stearn, W. T. (1981) *The Natural History Museum at South Kensington*, Heinemann, London.

Stegner, W. (1990) 'It all began with conservation', *Smithsonian*, 21 (1), 35–43.

Thomas, K. (1983) *Man and the Natural World; Changing Attitudes in England 1500–1800*, Allen Lane, London.

Wolfe, L. M. (1945) *Son of the Wilderness – The Life of John Muir*, University of Wisconsin Press, Madison: WI.

2. *The rise of environmentalism*

It's clear the future holds great opportunities. It also holds pitfalls. The trick will be to avoid the pitfalls, seize the opportunities, and get back home by six o'clock.

Woody Allen

Humanistic value systems must be replaced by supra-humanistic values that bring all plant and animal life into the sphere of legal, moral and ethical consideration. And in the long run, whether anyone likes it or not, force will eventually have to be brought to bear against those who continue to desecrate the environment.

Greenpeace Chronicles, 1979

Environmentalism is the new religion of the 'us generation' replacing the 'me generation'.

Courier-Mail, Brisbane, 4 May 1990

Politics and the American environment

The passing of the Wilderness Act in 1964 was an indication that the campaign to conserve remote natural areas in America was succeeding; further battles were won throughout the 1960s as other legislation followed, including the Wild and Scenic Rivers Act (1968) and the Endangered Species Acts (1966, 1969, 1973). New long-range wildlife conservation policies were put in place, and new National Parks and a National Seashore (Cape Cod) designated. President Johnson, speaking to Congress in February 1965 drew attention to these natural areas, and called for a 'New Conservation'.

> Our conservation must not be the classic conservation of protection and development, but a creative conservation of restoration and innovation. Its concern is not with nature alone, but with the total relation between man and the world around him. Its object is not just man's welfare but the dignity of man's spirit.

Aware of the progress made, he informed Congress in March 1968 that 'Our environment can sustain our growth and nourish our future. Or it can overwhelm us. History will say that in the 1960s the Nation began to take action so long delayed' (quoted in Caulfield, 1989).

In truth, the Kennedy and Johnson administrations had achieved much – but it was not enough for a population whose concerns were now moving on to wider environmental issues. The public now had an inkling of what ecology was, and of the delicate balance of nature. Hence the powerful reaction to the publication of Rachel Carson's *Silent Spring* in 1962, a book which elicited a public outcry for direct government action to protect the wild for its own intrinsic value. However, the activists wanted more. Conserving wilderness, plants and animals was certainly progress, but the perception was now of a greater need: to preserve all aspects of the environment, to introduce regulations which would protect the air, rivers and streams from misuse of chemicals, and punish those who polluted them. As a consequence, by the late 1960s the need to combat pollution was firmly established in the political arena. A better-educated population had wider perspectives, people were beginning to look at habitat destruction in other countries, and the Moon landings which preceded Earth Day reinforced this global view. Images of a delicate blue planet captured everyone's imagination. Other concerns were voiced: overpopulation, poverty, urban squalor. There can be little doubt that environmentalism was forged at a time of strife and social change, a time when there was a widespread fear that the fabric of planet Earth was crumbling, and that the very survival of humankind was itself threatened. The period between the publication of *Silent Spring* and Earth Day is for good reason regarded as a time of 'environmental revolution'.

Earth Day was the catalyst that transformed a specialist field of interest into one of popular concern, which in turn put the environment firmly into the American political arena. Gaylord Nelson, then a US Senator from Wisconsin, was the driving force behind Earth Day. Twenty years later, reflecting on its impact he stated:

> Earth Day – as was intended – demonstrated to the Washington Establishment and the public that there *was* an environmental movement. The principal and lasting effect was to make environmental concerns a permanent part of the political dialogue in this country. Obviously not all members of Congress are now what I would call environmentalists, but almost without exception they are sensitive to environmental issues – because they have to address them while campaigning – and they have at least one staff member who is knowledgeable about them (Gilbert, 1990).

Henry Caulfield, in his historical analysis of the environmental movement examines this 'élite theory', the critical role of political leaders operating as environmental policy entrepreneurs. He concludes that 'Both the conservation movement and the environmental movement have been identified by the press and popular literature, as well as by historians and political scientists, as political realities of this century' (Caulfield, 1989).

It was under Richard Nixon's Presidency that Congress passed the National Environmental Protection Act (NEPA) in 1969. This statute recast the government's role from conserver of wilderness to protector of soil, air, water and land, with the intent 'to create and maintain conditions under which man and nature can exist in productive harmony'. Pressing forward his initiative, Nixon proclaimed that the 1970s would be a decade of environmental transformation, and on 10 February 1970 presented the House of Representatives and the Senate with a 37-point message on the environment, along with a request for 4 billion dollars for the improvement of water treatment facilities. He asked for national standards to be set and enforced for air quality and vehicle emissions, sought legislation to prevent waste dumping in the Great Lakes, ordered a clean-up of federal facilities known to be causing pollution, and approved a National Contingency Plan for the treatment of oil spills. Overseeing all these improvement programmes was an autonomous regulatory body, the Environmental Protection Agency (EPA), founded in December 1970, which has proved to be a powerful and committed force in combating pollution and countering the environmental effects of new developments through its demand for Environmental Impact Statements. Unreasonable expectations about what the EPA could and should do has lead the environmental movement to regard it with some scepticism, and even as a scapegoat for the country's environmental ills.

Despite these misgivings, the various environmental and conservation bodies have had a quantifiable impact since 1970 – a 30 per cent decline in air pollution, a substantial reduction in the dumping of domestic sewage into surface waters, the banning of several poisonous chemicals, the designation of ten new National Parks and an increase in populations of formerly threatened species including the bald eagle, alligator, bighorn sheep and peregrine falcon.

Reactions to the changing British environment

Conservationist concerns in early 1950s Britain centred around the pollution of the atmosphere, as thousands of people died in smog-filled industrial cities; as a result in 1956 the Clean Air Act was passed. Fears were also voiced about the state of Britain's rivers, leading to the implementation of legislation which culminated in the Water Resources Act (1963), which demanded that river boards consider wildlife conservation and landscape issues. Greatest concern was raised by the impact of persistent chemicals, especially the organochlorine compounds such as aldrin, dieldrin and DDT; the numbers of top predators dropped alarmingly in the late 1950s and early 1960s as a result of cumulative poisoning. *Silent Spring*, which had had such an impact in America, was published in Britain in 1963, with shattering effect. Nicholson (1970) regards the book as 'probably the greatest and most effective single contribution hitherto towards informing public opinion on the true nature and significance of ecology'. The banning of the use of persistent pesticides was initially partial, sometimes voluntary; a total ban on all uses was not enforced until 1989.

By the 1950s the pattern of nature conservation in Britain had been established, and these new concerns prompted an increased interest in the British countryside from those who recognized the threats to their heritage. The growth of the county trust for the nature conservation movement in the late 1950s and early 1960s was indicative of a new awareness; there was an unprecedented increase in the number of individuals joining conservation organizations such as the National Trust and the RSPB. Specialist conservation bodies were established, including Trusts to aid the survival of otters and birds of prey, both of which had declined in numbers as a result of the use of pesticides. In 1958 The Council for Nature was formed as an umbrella organization for the increasing number of conservation organizations, and launched its National Nature Weeks from 1963 to promote conservation more widely. The public as a whole were more concerned and demanded to be better informed. As Lowe (1983) suggests, 'The elitism of preserving wild plants and animals for the sole amusement of naturalists had to give way to a new emphasis on presenting and interpreting nature to an eager public.' Integral to popularization was the BBC Natural History Unit (founded in 1957), promoting awareness of wildlife through radio and television programmes which have become an established part of popular culture in Britain.

As in America, public awareness of the changing environment was growing, and continued to increase through publicity surrounding the 'Countryside in 1970' conferences held in 1963, 1965 and 1970 under the Presidency of the Duke of Edinburgh. These meetings took nature conservation away from its traditional nature reserve approach, stressing the need for conservation in the 'wider countryside', and recognizing the role of landowners of all persuasions as guardians of landscape and wildlife heritage. Media interest in the Conferences was considerable, arousing public concern; the environment was newsworthy. There was much to report, including the continuing impact of organochlorine pesticides, the wreck of the *Torrey Canyon* (1967) (see Nicholson, 1970) and the mass kill of seabirds in the Irish Sea in the autumn of 1969 which has been linked to contamination by polychlorobiphenyls (PCBs). These biological disasters were instrumental in continuing to foster changes in thinking which had begun in Britain in the early 1960s.

Earth Day went largely unnoticed in Britain, and there were no mass demonstrations calling for protection of the wider environment; political reaction to environmental concerns was limited. Yet awareness about environmental issues, and an interest in countryside and wildlife grew rapidly in the late 1960s and early 1970s. Various indicators have been used to demonstrate the growth of the environmental movement in Britain at that time, including content analysis of newspapers, public opinion and social survey data, analysis of pressure group membership and the advent of new political institutions and environmental legislation (Sandbach, 1980).

Efforts continued to establish laws to protect Britain's environment throughout the 1960s and 1970s, including the Deer Acts (1959, 1963) and the Countryside Acts (1967, 1968), the latter creating specialist organizations for the promotion of recreation and amenity in the countryside. WWF was founded in Britain in 1961; Enterprise Neptune was launched in 1965 by the National Trust to raise funds for purchase of sections of Britain's coastline; yet still a traditional 'preservationist' approach to conservation and the countryside existed; conservation as a positive concept and a co-operative venture had still to be adopted. The need to bring together interested parties and break down barriers was a key objective of European Conservation Year 1970 (ECY); the Environment Council (formerly CoEnCo) continues its bridge-building activities. ECY also did much to promote the growth of conservation organizations such as the RSPB, SPNC and County Trusts; a revitalized Nature Conservancy Council was formed in 1973, and published the long-needed *Nature Conservation Review* (Ratcliffe, 1977). This work, and the RSNC's *Red Data Book on Vascular Plants* (Perring and Farrell, 1977) drew attention once again to the need for legislation which specifically addressed species conservation. As a result the Conservation of Wild Creatures and Wild Plants Act had reached the statute book in 1975, but with severe limitations. CoEnCo, through Wildlife Link, was instrumental in pushing through revised protection, culminating in The Wildlife and Countryside Act 1981, which examined the wider conservation issues of habitats and their management, and which currently (1995) is the basis of wildlife law in Britain. Throughout the 1980s new campaigns aimed at raising environmental awareness were launched, including RSNC's British Wildlife Appeal (1985) and the Countryside Commission's 'Watch over the National Parks' (1985); more specialist groups came into being, including the British Dragonfly Society (1983), the Whale and Dolphin Conservation Society (1987) and the Bat Conservation Trust (1990). As a consequence the voluntary sector of the conservation movement is stronger than ever. However, the statutory body, the Nature Conservancy Council, has been fragmented with the formation of the Countryside Council for Wales and Scottish Natural Heritage, (with a resultant loss of a single government body with nature conservation as its primary function in those countries) a change viewed with concern by many conservationists. It has recently been suggested that a similar amalgamation may occur in England, English Nature (the rump of the Nature Conservancy Council) joining forces with the Countryside Commission; as yet (1995) the two bodies remain discrete.

Current British involvement with wildlife and the countryside has been quantified by Evans (1992), who records that:

> The nature conservation movement as we know it in Britain today ... boasts a combined membership in excess of 4 million. The National Trust ... is the world's largest conservation society, with 226,000 ha of land and a membership of over 2 million. ... Local trusts for nature conservation are supported by 250,000 members and maintain nearly 2,000 reserves between them. A network of over 240 National Nature Reserves extends to 170,000 ha. More than 4,700 Sites of Special Scientific Interest cover nearly 900,000 ha ... Annually, over a quarter of a million volunteers spend in excess of 1.6 million workdays on practical conservation tasks. (See Figure 2.1)

The new conservationists

Vesey-Fitzgerald (1969) remarked that 'The future of the wild life of Britain does not lie in a few reserves, but in man's attitude of mind.' The activities of naturalists and conservation organizations were slowly beginning to alter perceptions, but:

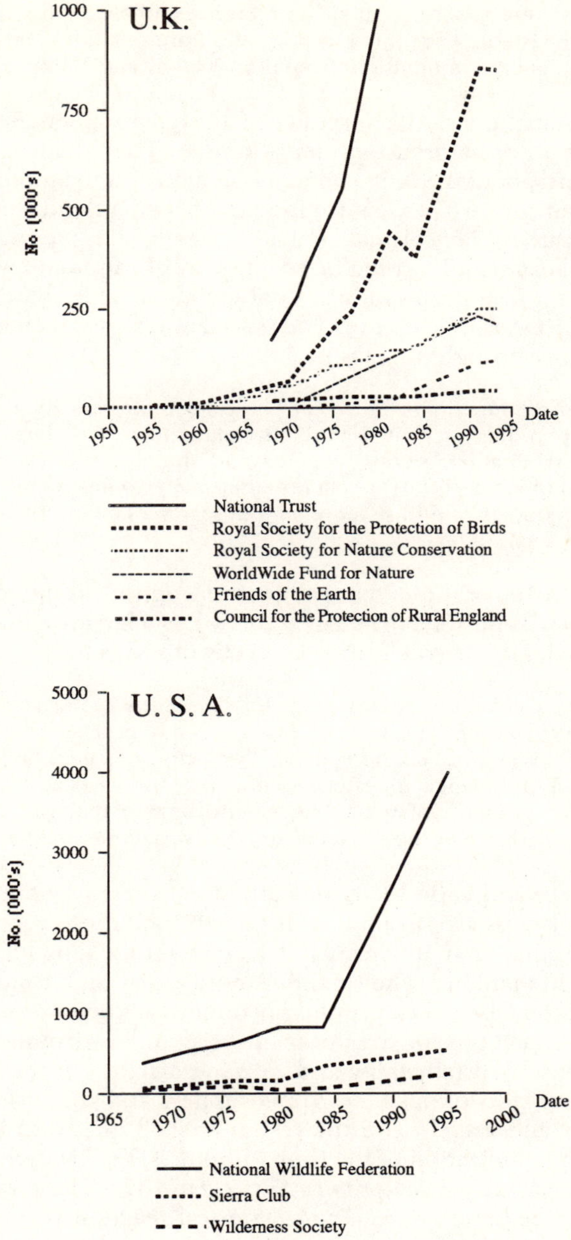

Figure 2.1 The growth of conservation organizations in the United Kingdom and the United States of America.

> the new conservationists, using more extravagant tactics to tackle wider environmental issues, were dissolving public indifference and official inertia in a much more positive fashion. Certainly it is they, and not the nature conservationists, who are remembered in hindsight from the 1960s (Evans, 1992).

The new conservationists, the product of the environmental revolution, differed from the naturalist–preservationists. They had rediscovered Marsh's concern about the impact of man on nature, and amplified and re-emphasised that concern as a fear for the survival of mankind, and they were political activists. As McCormick (1989) notes 'Their message was that environmental catastrophe could be avoided only by fundamental changes in the values and institutions of industrial societies.' This new approach to the environment was indicative of wider social and political change, and in the United States brought together:

> the consumer movement, including the corporate reformers; the movement for scientific responsibility; a revitalised public health movement; birth control and population stabilisation groups; pacifists and those who stress participatory democracy in which decisions are made consensually; young people who emphasize direct action; and a diffuse movement in search of a new focus for politics. (quoted in Pursell, 1973).

In many ways the environmental revolution encapsulated the mood of the 1960s, but why it came about so quickly and with such conviction cannot be easily explained. McCormick (1989) correctly observes that:

> The elements of change were emerging long before the 1960s; when they finally intersected with one another, and with wider socio-political factors, the result was a new force for social and political change. Six factors in particular seem to have played a role in the change: the effects of affluence, the age of atomic testing, the book *Silent Spring*, a series of well-publicised environmental disasters, advances in scientific knowledge, and the influence of other social movements.

In the late 1960s and early 1970s concern about the environment, and in particular the key issues of population growth, pollution, economics and technology, prompted an intellectual debate about the long-term future of the Earth and humankind. The (mainly American) academics at the centre of the debate were labelled as 'prophets of doom', so uncompromising were they in their predictions for the decay of social and environmental conditions. Questions of sustainability and carrying capacity of the Earth have long been the subject of debate for philosophers and scientists, including Malthus, Marx and Engels, and the revival of such questions in the early 1970s led to the publication of the Club of Rome's *The Limits to Growth* and the *Ecologist* magazine's *A Blueprint for Survival* in 1972. These works, which prompted both reasoned debate and hysterical reaction (e.g. Beckmann, 1973), and the 'prophets' themselves, did much to widen the environmental debate and draw attention to major issues which became the subject of political discussion.

Reaction by mainstream political parties in Europe to the needs of the environment in the early 1970s was so ineffective that the evolution of a 'green' political movement was inevitable. In 1983, 27 representatives of *die Grunen* took their seats in the West German Bundestag; 'Green' politicians were represented on seven European governments by 1988. The British Green Party (the first in Europe, 1973) has never had the success of its continental counterparts, an issue examined by Rudig and Lowe (1986). The 1994 European elections have indicated a downturn in the fortunes of the Greens compared to their huge success in 1989. This latter protest vote has had a resounding impact on European politics, reflected at a national level; although the Green Party has had little political success in Britain, there can be few doubts that its efforts have had a significant impact by persuading the other political groupings to take environmental concerns seriously. The efforts of both old and new conservationists have pushed environmental conservation further into the political spotlight in the early 1990s; it is interesting to consider that the nature conservation movement, with its emphasis on wilderness and wildlife could not succeed in achieving this alone. Although the loss of species and habitats would be catastrophic, it is difficult for most individuals (including politicians) to comprehend the mechanisms involved; in contrast the impact of global climate change, the threat of sea level rise (and hence loss of land and homes) or the loss of protective ozone (and an increased risk of skin cancers through exposure to high levels of UV radiation) can be understood at a personal level (see Figure 2.2). The 'new conservation' needed new organizations to relay its message; it also needed to recruit members to participate in direct action, to campaign for change in environmental policy. Two organizations have spearheaded the new approach; Friends of the Earth, founded in 1969 in San Francisco, now has offices world-wide promoting wise use of natural resources; Greenpeace (founded as Don't make a Wave in 1969) initially focused its activities on nuclear pollution issues, but now has a wide environmental remit.

The new radical approach adopted by Greenpeace and Friends of the Earth was refreshing and timely, if at first worryingly unorthodox to the traditional naturalist–conservationist. Positive, often ruthless and sometimes daring campaigns have been waged by these bodies; courting publicity on issues as diverse as whaling, recycling, rainforest destruction, nuclear energy and the dumping of nuclear waste, transport policies, and international trade in wildlife, they have successfully raised environmental awareness across the globe. Shock tactics are regarded as legitimate, and graphic descriptions and photographs are regularly used with dramatic effect to drive home messages – Nash (1989) compares such methods with those of the abolitionists in America. In their publications the environmental lobby continue to address regional, national issues – the summer 1994 edition of *Earth Matters* (the magazine of Friends of the Earth UK) covered topics as diverse as wind farms, import of hardwoods, peat extraction in Ireland, the Cardiff barrage, Britain's road transport programme, the European Union's Habitats Directive and the trade in frogs' legs from Bangladesh; on all these fronts, and in many countries Friends of the Earth continues to play an active role. Bold, aggressive tactics have had a dramatic effect; although initially regarded as an extreme element (and one shunned

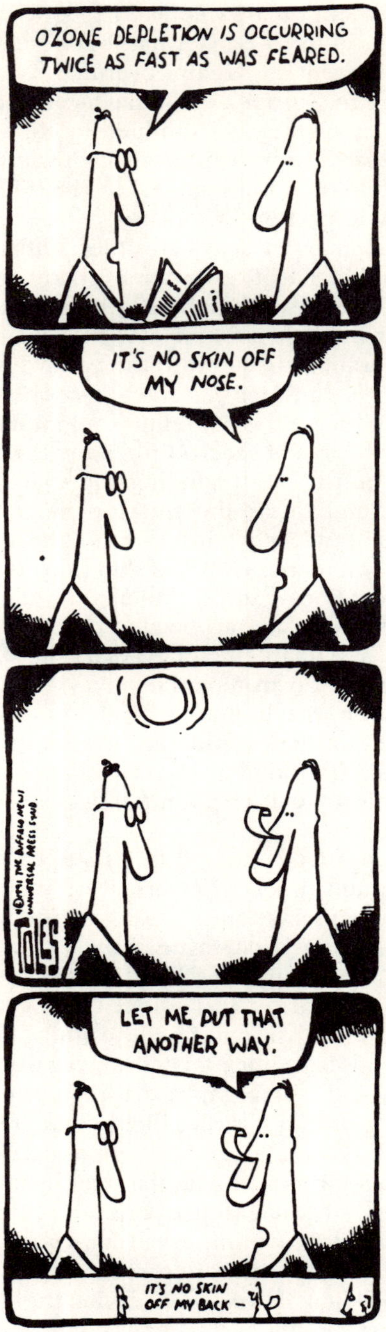

Figure 2.2 The impact of ozone depletion. (Toles copyright The Buffalo News. Reprinted with permission of Universal Press Syndicate. All rights reserved.)

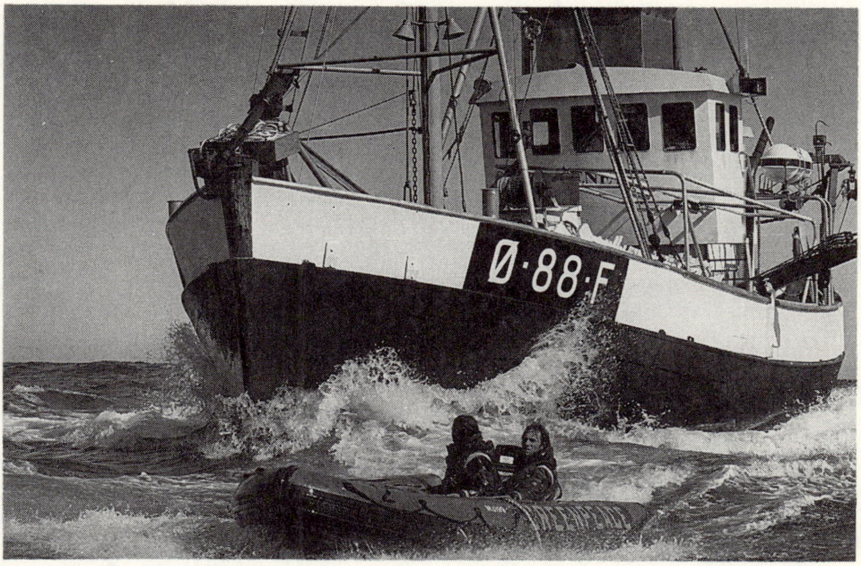

Plate 6 Active campaigning by Greenpeace has done much to increase environmental awareness. Here the Greenpeace inflatable from *MV Solo* confronts the Norwegian whaler *Senet* in the North Sea. (Courtesy of Greenpeace.)

by the traditional nature conservationists), Greenpeace and Friends of the Earth have now achieved respectability and understanding, their campaigns receiving huge public support. Membership has grown dramatically – Friends of the Earth (UK) had 2,000 members in 1971, rising to 100,000 in 1989 and 200,000 in 1994. Unafraid of questioning government decisions or of tackling big business head-on, Greenpeace and FoE address the fears that many people have about the changing environment, and take positive action to draw attention to those issues, frequently leading to changes in policies towards the environment. Their extremist image has been lost as environmentalism has become accepted and understood, and as environmentalists themselves have aged and matured. (See Plate 6)

Understanding of the issues has resulted in dramatic changes in behaviour, especially from the consumer in many developed countries. Green consumerism is now a reality in many developed countries, and 'environmentally friendly' products are a part of all our lives, whether they be aerosols without CFC propellants, drinks in recyclable containers, additive-free foods, or 'dolphin-friendly' tuna; biodegradable and recycled packaging materials are commonplace; labelling products (especially furniture) to indicate their country of origin and type of raw material is now as important to the manufacturer as it is the consumer. As Evans (1992) states:

> Green consumerism has achieved, in a couple of years, more than the Animal Liberation Front and the vegetarian movement have done in decades for farm

animals, and more for our environment as a whole than the nature conservationists have ever done.

Environmentalism – a global movement

The environmental movement is now world-wide, reflecting increasing interest and concern for the environment as global conditions continue to deteriorate. It is no longer the preserve of a small number of dedicated, far-sighted individuals, or of a growing number of conservation organizations. It is the concern of nations and their governments, of high ranking politicians. Even the United States Vice-President Al Gore has entered the debate with conviction, outlining a five-point plan for creating a sustainable environment (Gore, 1992).

The shift of environmental effort from national to international arena began with the International Office for the Preservation of Nature (IOPN) in 1938, revitalized in 1948 as the International Union for the Preservation of Nature (IUPN), which in turn became the International Union for the Conservation of Nature in 1956 (see Nicholson, 1970 and Fitter and Scott, 1978 for a historical account of IUCN and WWF). Based at Morges and then Gland in Switzerland, the IUCN has done much to encourage practical conservation, especially in Africa and south-east Asia, and in promoting the acceptance of international conventions relating to wildlife.

IUCN was revitalized with the launch of the World Wildlife Fund and 'its galaxy of energetic National Appeals' (Nicholson, 1986) and became increasingly effective through its expert teams which successfully tackled the expansion and upgrading of the National Park system, promoted Operation Tiger and enabled the Convention on International Trade in Endangered Species (CITES) to be accepted by the international community.

IUCN was central to organization of the UN Conference on the Human Environment at Stockholm in 1972. This meeting had several important outcomes, including the promotion of national environmental policies and the creation of environmental agencies and legislation, and at an international level the creation of the United Nations Environment Programme (UNEP) (see McCormick, 1989 and Strong, 1992). Several initiatives and conventions followed in the 1970s and 1980s – for example, the UNESCO 'Man and the Biosphere' programme (1971), the Ramsar Convention on wetlands of international importance (1971), CITES (1973), the Vienna Convention on the Protection of the Ozone Layer (1985) and its Montreal Protocol (1987). Several major reports were published, including the Brundtland Report (World Commission on Environment and Development, 1987), and the *World Conservation Strategy* (International Union for the Conservation of Nature, 1980).

The impetus created by these international calls for action culminated in United Nations Conference on Environment and Development (UNCED – more often referred to simply as the Earth Summit) in Rio de Janeiro in June 1992 – the first global acknowledgement that environmental quality and economic health are inextricably linked. More than 170 countries were represented at a meeting which searched for new strategies and new agreements to address the environmental challenges of the future. The preamble to Agenda 21 of the Conference outlined the predicament:

> Humanity stands at a defining moment in history. We are confronted with a perpetuation of disparities between and within nations, a worsening of poverty, hunger, and illiteracy, and the continuing deterioration of the ecosystems on which we depend for our well-being. However, integration of environment and development concerns and greater attention to them will lead to a fulfilment of basic needs, improved living standards for all, better protected and managed ecosystems and a safer, more prosperous future.

The prospect of combining global economic growth with environmental stewardship is an exciting one, and the number of countries represented, and the high level of commitment by them to the Conventions encouraging, particularly those relating to the control of 'greenhouse gases' (the Framework Convention on Climate Change) and species conservation (the Convention on Biological Diversity) (see Grubb *et al.*, 1993 and Rogers, 1993). Despite the signing of agreements at the highest level, many conservation organizations, especially Greenpeace, remained pessimistic about prospects for the future (see Rogers, *op. cit.*, p. 249). Perhaps the major achievement, as Reilly (1992) notes is that 'the Earth Summit marked the arrival of environmental concerns on the international stage as a major new consideration in foreign policy.'

Environmentalism – some definitions

Environmentalism is now all-embracing. The shift in thinking during the late 1960s and early 1970s about what 'environment' and the environmental movement really is, has been captured for many by Schumacher (1973):

> The environmental movement has, as an object of its ideological concern, the whole geosphere and biosphere including man himself. The continued long-term integrity of concern is its central value. The earlier preservationist manifestation of its concern, of course, persists in the support of wilderness areas, parks etc., for aesthetic and other traditional reasons.

Dictionary definitions are perhaps less helpful. *The New Shorter Oxford English Dictionary* (Brown, 1993) describes an environmentalist as 'a person who is concerned about or seeks to protect the environment, especially from pollution'. *The Dictionary of Human Geography* (Johnston *et al.*, 1986) gives the broad definition of environmentalism as 'The philosophies and practices which inform and flow from a concern with the environment.' *The Oxford Dictionary of New Words* (Tulloch, 1991) defines environmentalism as 'Concern with, or support for, the preservation of the environment; green politics or consumerism', noting that the term was first used in the USA in the early 1970s, at a time when the ecology movement was beginning to gain support. Originally associated with liberal politics and the 'hippie' generation, Tulloch notes the word originally had derogatory overtones, but has since achieved respectability as the environmental movement has developed.

These definitions, whilst being a useful compromise, fail to capture the reformist nature of the movement, 'the conviction that a better mode of existence is possible . . . opening up our minds and our organisations to new ideas about fairness, sharing, permanence and humility' (O'Riordan, 1981). Neither do they emphasize the search for compromise between opposing elements within environmentalism:

> individual freedom versus the common good; the protection of national sovereignty versus the need for global awareness; the rights of minorities versus majority rights; the protection of the rights of the present generation versus those of future generations (Pepper, 1984).

Various attempts have been made to categorize different elements of environmentalism – Cotgrove (1976), for example, distinguished between traditional and liberal environmentalists. Sandbach (1980) separated 'scientific' environmentalism with its adoption of ecological principles from 'anti-establishment' environmentalism which focuses on the alienation of man from the natural environment and the impact of technological society. The latter more spiritual approach is enshrined in the present-day Deep Ecology movement (see Devall and Sessions, 1985). Pepper (1984) examines the cline between the ecocentric deep ecologists and the technocentric 'cornucopians'. Merchant (1992) segregates Deep Ecology from Spiritual Ecology, Social Ecology (the green Marxists or red greens) and Radical Ecology, defining the latter as 'the cutting edge of social ecology. It pushes social and ecological systems towards new patterns of production, reproduction, and consciousness that will improve the quality of human life and the natural environment.' These various threads of environmental thought indicate the breadth and strength of the environmental movement of the 1990s, and refute earlier suggestions of a decline in interest in environmental issues (e.g. Bowman 1975).

So who or what is an environmentalist? Is it the more radical individuals engaged in the fight to prevent the logging of temperate rainforests in Northwest America – the self-styled 'eco-terrorists' attracted to the Earth First! Foundation (1980) and its declaration of ecological militancy? Their radical exploits are described by Nash (1989), who records the concern about their actions even amongst their fellow ecologists. Or is it the local naturalist content to take part in a Saturday outing to a nature reserve and sit in a damp hide watching birds? This latter individual probably would never consider him/herself to be an activist, similarly the eco-terrorist might be reluctant to be identified as a naturalist, yet they share much of the same philosophy. Age undoubtedly influences the place an individual takes on the 'environmentalist' cline – anyone who has lectured to a British natural history society is immediately aware of the above-average number of senior citizens in the audience; active environmentalists have a very different age profile, and attract young adults and families with children. It is interesting to note that in Britain recent active protests against the veal trade, live animal exports and new road building programmes appear to be changing these profiles. Pensioners voice their protest to police and the media in British ports; suburban housewives decry the destruction of woodland in the face of motorway extensions in Lancashire. It will be interesting to see if time continues to change the still notable polarity of young and old, activist and contented observer, environmentalist and naturalist.

Environmental issues and the need for continuing to promote awareness

Despite the efforts of the conservation movement, the future of the British environment is still unclear. As Evans (1992) notes:

> Britain can claim that no more than 7 percent of its citizens actively support the conservation bodies .... All the designated National Parks, nature reserves and country parks extend to no more than ten percent of this total (Land) area ... agriculture accounts for 77% of Britain, forestry 10% and urban land use about 10%. Wildlife habitat and species continue to disappear at their hands. ... eg Four of forty-six types of dragonfly were lost between 1950 and 1986 ... in that same short spell, over 100 plant species were lost.... Recent concern over the all-pervading threats of global warming, acid rain and the destruction of the ozone layer has created widespread comment but little meaningful action. There is political apathy and a general ignorance in official circles of the basic principles of nature conservation.

In Britain, although the support for nature conservation continues to increase and attract media attention, there is a need to direct greater attention to wider environmental concerns, the need to question the way we live. Such issues receive greatest debate when new environmental crises occur. Chernobyl, the wreck of the *Exxon Valdez*, the Gulf War oil disaster, immediately brought major issues into the spotlight.

The situation in Britain is reflected in most developed countries, and there have been significant changes of attitude to the environment. Individuals are increasingly energy conscious, family size is declining, there are more examples of corporate responsibility, environmental groups are more effective, environmental research is better co-ordinated and governments are becoming increasingly aware of their responsibilities. Individual awareness is high – a *New York Times*/CBS poll in June 1989 found that a remarkable 80 per cent of all respondents agreed with the statement that 'Protecting the environment is so important that requirements and standards cannot be too high, and continuing environmental improvements must be made regardless of cost' (quoted in Merchant, 1992).

The increasing activity of the environmental movement on the international stage is an indication of the recognition of the environment as a global concern. *The Global 2000 Report* (Council on Environmental Quality, 1981) outlined continuing environmental problems and examined some potential solutions; *Global Outlook 2000* (United Nations, 1990) provides a useful synopsis of current environmental problems, including greenhouse gases and global warming, ozone and ultraviolet radiation, acid rain, deforestation, desertification, water resources, genetic resources (biodiversity), hazardous wastes, and the environmental impact of new technologies and materials. The report concludes:

> The future of the world economy has not been decided, and no study can reveal it. There are limits to the possible but they are wide. At one extreme, there is the spectre of stagnation and environmental disaster. At the other extreme, there are opportunities for relieving poverty, improving the quality of life, safeguarding the environment for future generations, stabilizing the world economy, and moving towards a global community. Neither perspective can be dismissed as impossible. The world's actual course in the vast gulf between them will be decided by how individual people, countries and the international community meet these challenges.

Where do museums fit in?

The story of the rise of the conservation philosophy and its modern manifestation as environmentalism is a long and complex one. The two introductory chapters have sketched a brief account of a fascinating phenomenon which has influenced the way we perceive our world and our attitudes and behaviour to environment and wildlife; a philosophy which has culminated in many countries with laws to conserve the environment and the designation of specialist agencies to care for aspects of it. At first sight the role that museums have played in the environmental movement may seem minimal, and most historical accounts make only passing comments about museums. It would be presumptuous to regard museums (even National Museums) as major players in dictating environmental policy – clearly they are not. Yet throughout the past 200 years natural history museums have been part of the conservation scene. They have played an essential support role through the establishment and maintenance of collections and records which document biodiversity; taxonomic and biogeographic research on those collections have been a cornerstone of the biological conservation movement. Through the work of individual, dedicated curators, museums have helped to influence environmental policy decisions at local and even national level. Museums have frequently been at the hub of the growing network of conservation organizations, promoting co-operative research and practical conservation measures. Through their educational activities and exhibitions they have played a part in raising environmental awareness to new levels, especially amongst youngsters.

The following chapters will examine these roles with reference to the history of the conservation movement. However, in addition to examining the successful roles that museums have played in the past, it is important to consider how museums might react to the needs of the future. How, for example, can museum-based research be made more pertinent to environmental issues? And what types of museum displays should we be generating? – have museums so far failed to produce radical exhibitions which examine the root causes of environmental problems? Such questions need to be addressed in order to ensure a continued, progressive scientific and social role for museums in the twenty-first century.

References

Beckmann, P. (1973) *Eco-hysterics and the Technophobes*, Golem Press, Boulder, CO.

Bowman, J. S. (1975) 'The Ecology Movement', *Journal of Environmental Studies*, 8, 91–7.

Brown, L. (ed.) (1993) *The New Shorter Oxford English Dictionary*, Clarendon Press, Oxford.

Carson, R. (1962) *Silent Spring*, Houghton Mifflin, Boston.

Caulfield, H. P. (1989) 'The conservation and environmental movements; an historical analysis'. In Lester, J. P. (ed.) *Environmental Politics and Policy*, Duke University Press, Durham.

Cotgrove, S. (1976) 'Environmentalism and Utopia', *The Sociological Review*, 24, 23–42.

Council on Environmental Quality (1981) *The Global 2000 Report to the President: Entering the 21st Century, a Report*, Blue Angel Inc., Charlottesville, VA.

Devall, B. and Sessions, G. (1985) *Deep Ecology; Living as if the Earth really mattered*, Gibbs Smith, Salt Lake City.

Evans, D. (1992) *A History of Nature Conservation in Britain*, Routledge, London and New York.

Fitter, R. and Scott, P. (1978) *The Penitent Butchers*, Fauna Preservation Society, London.

Gilbert, B. (1990) 'Earth Day plus 20, and counting', *Smithsonian*, 21 (1), 47–55.

Gore, A. (1992) *Earth in the Balance; Ecology and the Human Spirit*, Plume, New York.

Grubb, M., Koch, M., Thomson, K., Munson, A. and Sullivan, F. (1993) *The Earth Summit Agreements: A Guide and Assessment*, Earthscan Publications and the Royal Institute of International Affairs, London.

International Union for the Conservation of Nature (1980) *World Conservation Strategy*, UCN, Gland.

Johnston, R. J., Gregory, D. and Smith, D. M. (1986) *The Dictionary of Human Geography*, Blackwell, Oxford.

Lowe, P. D. (1983) 'Values and Institutions in the History of British Nature Conservation'. In Warren, A. and Goldsmith, F. B. *Conservation in Perspective*, 329–352, John Wiley and Sons Ltd, Chichester: New York.

McCormick J. (1989) *The Global Environmental Movement: Reclaiming Paradise*, Belhaven Press, London.

Merchant, C. (1992) *Radical Ecology: The Search for a Liveable World*, Routledge, New York and London.

Nash, R. F. (1989) *The Rights of Nature: A History of Environmental Ethics*, The University of Wisconsin Press, Madison: WI.

Nicholson, M. (1970) *The Environmental Revolution*, Hodder and Stoughton, London.

Nicholson, M. (1986) *The New Environmental Age*, Cambridge University Press, Cambridge.

O'Riordan, T. (1981) *Environmentalism*, Pion, London.

Pepper, D. (1984) *The Roots of Modern Environmentalism*, Croom Helm, London, Dover NH.

Perring, F. H. and Farrell, L. (1977) *British Red Data Books. I – Vascular Plants*, 1st edn, Lincoln, Royal Society for Nature Conservation.

Pursell, C. (ed.) (1973) *From Conservation to Ecology: The Development of Environmental Concern*, Thomas Y. Crowell Co, New York.

Ratcliffe, D. A. (ed.) (1977) *A Nature Conservation Review: The Selection of Biological Sites of National Importance to Nature Conservation in Britain,* 2 vols, Cambridge University Press, Cambridge.

Reilly, W. K. (1992) The Road from Rio, *EPA Journal,* 18 (4), 11–13.

Rogers, A. (1993) *The Earth Summit: A Planetary Reckoning,* Global View Press, Los Angeles.

Rudig, W. and Lowe, P. (1986) 'The withered "greening" of British politics: A study of the Ecology Party', *Political Studies,* 43 (2), 262–284.

Sandbach, F. (1980) *Environment, Ideology and Policy,* Blackwell, Oxford.

Schumacher, E. F. (1973) *Small is Beautiful: Economics as if People Mattered,* Perennial Library, New York.

Strong, M. (1992) 'From Stockholm to Rio: A journey down a generation'. In *In Our Hands,* Earth Summit '92, UNCED Secretariat, Geneva.

Tulloch, S. (1991) *The Oxford Dictionary of New Words,* Oxford University Press, Oxford.

United Nations (1990) *Global Outlook 2000: An Economic, Social and Environmental Perspective,* United Nations Publications, New York.

Vesey-Fitzgerald, B. (1969) *The Vanishing Wild Life of Britain,* MacGibbon and Kee.

World Commission on Environment and Development (1987) *Our Common Future,* Oxford University Press, Oxford.

3. *Increasing environmental awareness – a mission for museums?*

> Museums exist to encourage and instil the joy of intellectual and aesthetic discovery.
>
> Lee Weber, Field Museum, Chicago

The challenge to museums

The growth of the environmental movement provides evidence of the establishment of the environmental ethic and an increasing realization that mankind has a huge responsibility – our own fate, as well as that of the millions of species we live with, is in our hands. Many environmentalists, whilst recognizing the need for international conventions and their enactment through national legislation, believe that a sustainable future will only be achieved by changing global (and hence individual) lifestyles. Jay Hair, of the USA's National Wildlife Federation is quoted by Gilbert (1990):

> Essentially the (environmental) problems have been created by how we live, and the solutions require that we change how we live. I think we can, while in the long term raising rather than lowering our standards of living. However, I don't think this can be mandated from above by government regulations and treaties. It has to come about because the people themselves *want* to make changes.

This identifies the crux of the problem – how to convince individuals that a high-quality environment, both physical and organic, is necessary for their survival, and as a consequence ensure that they respond appropriately. Individual perception of the environment is structured by culture and custom, desire and faith; Lowenthal (1974) identified that 'subjective, often unconscious, and culturally dominated forces play a major role in how we see the environment and how we act in it'. Wilson (1992) has argued that past, more 'primitive' societies were aware that the right action or responses to the world around them gave life and fulfilment, the wrong ones hunger, sickness or death. He notes that 'The imprint of that effort cannot have been erased in a few generations of urban existence', and points to a number of human cultural traits (including our passion for gardens, gardening and outdoor recreation), as examples of a phenomenon he refers to as *biophilia*, 'the connections that human beings subconsciously seek with the rest of life'. Gaining access to wilderness and wild places is perhaps the greatest manifestation of this human need, consequently it is not surprising that John Muir also identified this inbuilt, yet sometimes dormant, trait '... going to the woods is going home; for I suppose we came from the woods originally ... ' consequently ' ... there is a love of wild Nature in everybody, an ancient mother-love showing itself whether recognised or no, and however covered by cares and duties' (quoted in Nash, 1982). If museums are to foster an environmental ethic there is a continuing and growing need to reawaken and encourage the innate love of nature recognized by Muir in the mid-nineteenth century and Wilson in our own time.

It is imperative that people rediscover these ancient values if an attitude change is to be encouraged. Because the overall rate of environmental change has outpaced the ability of cultural evolution to modify the human default position, change is imperative – 'too many signs indicate that our ability to change the world is rapidly outstripping our ability to understand it' (Ornstein and Ehrlich, 1991). Knowledge is the major barrier to solving environmental problems – (Trudgill (1990) identifies the others as agreement, technology, economic, social and political) – identifying the evidence, interpreting it and then communicating it in an effective manner. In other words individuals will only be prepared to change their attitudes and behaviours if they are made aware of the problems – and understand how diverse and complex environmental issues are. They need to be made aware using techniques which make facts accessible, in non-threatening surroundings. People also need to be convinced by the arguments that are put to them, and know that they are founded on good data and reliable science, within the context of an institution with scientific credibility. Museums collect evidence of the environment and environmental change, have the skills to interpret it and make it accessible; hence the environmental debate poses unparalleled challenges and opportunities for them.

Considering the variation in museums around the world, it is inevitable that those challenges will be addressed in different ways and historical precedents, scientific responsibilities, geographical location, funding and staffing all influence the ways in which museums operate. Some indication of the range of museum activities, and the variation in those activities in natural history museums is shown, for example by perusing *The Manual of Natural History Curatorship* (Stansfield *et al.*, 1994). In essence, no two museums are alike, even in the same country. It is important therefore not to make generalizations about museums, and especially their conservation-oriented activities in different countries, where attitudes to the environment, and concerns for it, are equally diverse. The only common currency for all museums is their collections.

Museums pivot around their collections, and they are in many ways a strength in the sense of their scientific involvement with the environment, yet they can also be perceived as a weakness in their efforts to increase environmental awareness through interpretive activities. Their collections represent the past, and are best suited to tell stories of past people, their lives and their environment. Of course these messages are important, and can tell of past ecological problems, but the issues we must address are those of the present and the future. As Ornstein and Ehrlich (1991) noted 'While we can obviously still learn a great deal from history, it can be dangerous to focus too much of our attention on the past', and called for widespread public awareness and public debate about the present situation which would precede large-scale action. They are optimistic about achieving global change, and point to a number of major transformations which have already occurred including the shift to small families in American society, the transformation of wartime economies, innovative educational programmes, a more caring attitude to endangered species, and most importantly, the adaptability of the human mind. Perhaps the first item on the agenda for the future is to call everyone's attention to the reality of the situation, to enable individuals to make reasoned decisions about their behaviour. Museums must be at the forefront of this educative process, but in order to do so they need to shake off the shackles of conventional approaches to their essential functions of collection, curation and interpretation, to step beyond the boundaries imposed by the legacy of past collecting. As Griffin (1992) has observed 'Collections may make museums unique but they cannot dictate the business' and 'The fact is that some of us have got so hung up on collections and the narrow strictures of scholarship and professionalism that we have forgotten the outside world.' Museums, especially natural history museums, need a new direction in order to play a meaningful role in the environmental debate, a direction which enables them to confront present issues, to tackle the dilemmas head-on. The history and development of natural history museums have provided them with the basic tools to carry out this task – the collections, scientific ability, educational and interpretive expertise, access to the media and individual commitment – all that is required now is a change in philosophy, and a sense of urgency.

Museums before environmentalism

Environmentalism is a social phenomenon which is now reflected in the role of museums. David Allen (1976) has demonstrated that natural history in Britain was also a social phenomenon, the results of which include the formation of collections and the museums that house them. Natural history as a social pursuit spread world-wide in the nineteenth century – people wanted to read and talk about animals and plants, rocks and fossils, of their own and other countries, and understand them better. They also wanted to experience the natural world firsthand and natural history museums and collections (and zoological and botanical gardens) are a manifestation of that desire. Yet in a world still to be explored and exploited, issues relating to the conservation of species and habitats seemed hardly relevant, despite a growing recognition of change in both urban and rural environments. So when and how did museums begin to absorb the conservation ethic whose development we have explored?

The history of natural history museums has been well documented by many distinguished museologists (e.g. Alexander, 1979; Impey and MacGregor, 1985; Hudson, 1987; Porter, 1990). Detailed and entertaining histories of many individual museums, their collections and curators, have been written, providing insights into the origins of museums which vary in stature from those of international renown such as the Natural History Museum, London (Stearn, 1981), the Museum of Comparative Zoology, Harvard (Winsor, 1991), and the National Museum of Natural History, Washington (Yochelson, 1985) to smaller multidisciplinary museums with natural history collections of local or regional significance – for example, Leeds City Museum (Brears, 1989) or Ludlow, Shropshire (Lloyd, 1983). The literature on the general history of natural history (and its interactions with societies and museums) is large and diffuse. Numerous accounts of individual collectors and their collections have been published (e.g. Jackson, 1992; Ralph, 1993); the *Journal of the History of Collections* and *Archives of Natural History* (the journal of the Society for the History of Natural History) are particularly useful sources.

Cabinets of curiosity in the sixteenth century, travelling museums such as that opened to the public of London by Sir Ashton Lever in 1774, or the natural history collections established by learned societies such as Literary and Philosophical Societies in Britain and Lyceums in America in the early nineteenth century had little to do with environmental conservation, but everything to do with the upsurge of interest in the natural world.

Charles Willson Peale (1741–1827) capitalized on this interest to establish his Philadelphia Museum in 1786 (see Alderson, 1992; Kulik, 1992); recognizing the educational potential of his museum, his printed tickets bore the message 'The birds and beasts will teach thee!' – the museum received around 48,000 paying visitors in 1816. Private funding helped to establish many of America's best-known natural history museums, including the Peabody Museum (1866), the American Museum of Natural History (1869), the Field Museum, Chicago (1893) and the Carnegie Museum of Natural History in Pittsburgh (1896); James Smithson's bequest was, after some deliberation, used by Congress to found the US National Museum in 1879. The story of these great institutions in the nineteenth and early twentieth centuries is one of collecting (and rivalry) on a grand scale – conservation failed to materialize as museums, well financed by their trustees, plundered the American West and sent major collecting expeditions to all parts of the world. Porter (1990) mentions that:

> Thousands of animals for life groups at the Natural History Museum of Los Angeles County (established 1910) ... were gathered ... in East Africa .... In addition to big game trophies, the museum received entire families of chimpanzees and monkeys, birds, insects and plants.

Such indiscriminate collecting would now be unthinkable, but it is not difficult to understand why this practice was taking place – (even at a time when America was becoming increasingly aware of the significance of its own environmental resources) – competition and civic pride demanded that they be made.

Many of the specimens collected in Africa and elsewhere were collected specifically for major exhibits in American museums; many remained traditional systematic displays, but spectacular dioramas sought to bring exotic environments alive for the museum visitor. Alongside new approaches to presentation came educational initiatives. The Boston Museum of

Natural History introduced guidebooks, wallcharts and evening classes around 1870; the American Museum of Natural History developed a specimen loan scheme. As Kohlstadt (1992) notes 'the striking thing about museum developments in the 1870s and 1880s is the extent to which similar activities were initiated almost simultaneously in this country [USA] and abroad'. However, a later (1921) American initiative which broke new ground, and travelled only slowly to other countries, was the establishment by the American National Parks Service of its first site museum at Yosemite National Park. Interpretation had arrived, with new techniques and principles to explain the significance of animal and plant species *in situ*, as part of a living ecosystem.

Museums in America continued to flourish throughout the twentieth century; their not-for-profit status enabled then to establish a firm financial base through membership and endowments. New displays, excellent schools programmes, and diverse scientific interests have allowed them to flourish. As an example, the Field Museum, Chicago has attracted attendances of more than a million each year since 1926, and continues selective collecting activities in many countries; it has more than 10 million specimens in store, a huge scientific resource (Alexander, 1979). However, it is difficult to link the development of America's museums with a growing interest in the environment until the early 1920s, with the advent of the diorama; real change came in the 1950s, and particularly through the activities and writings of A. E. Parr (see Plate 7). What *is* evident is the growth of collections and increasing curatorial specialization, a determined taxonomic research effort (museums in America form the backbone of systematic research there), fresh and exciting exhibitions (though still mainly systematic), and innovative educational and public relations programmes.

As we have seen, the growth of British natural history museums is inextricably linked with the development of field natural history – British museums hold tens of thousands of individual collections, the majority made in the nineteenth century. Williams (1987) surveyed the state and status of British natural history collections and estimated that there were 296 institutions that housed biological collections, representing millions of specimens; regional collections surveys (e.g. Hancock and Pettitt, 1981; Davis and Brewer 1986) have identified individual named natural history collections held in provincial museums, many of considerable taxonomic significance.

The sheer size of collections formed by individual naturalists had a number of significant spin-offs, in particular the development of new means of preserving, preparing and storing specimens, but also in encouraging good record-keeping. However, the impact of collecting by Victorian naturalists on the British environment, and in the countries of the Empire, was considerable. Allen (1976), discussing the growing interest in keeping marine animals and ferns in a 'Wardian Case', (the forerunner of the aquarium), which occurred about 1850, notes:

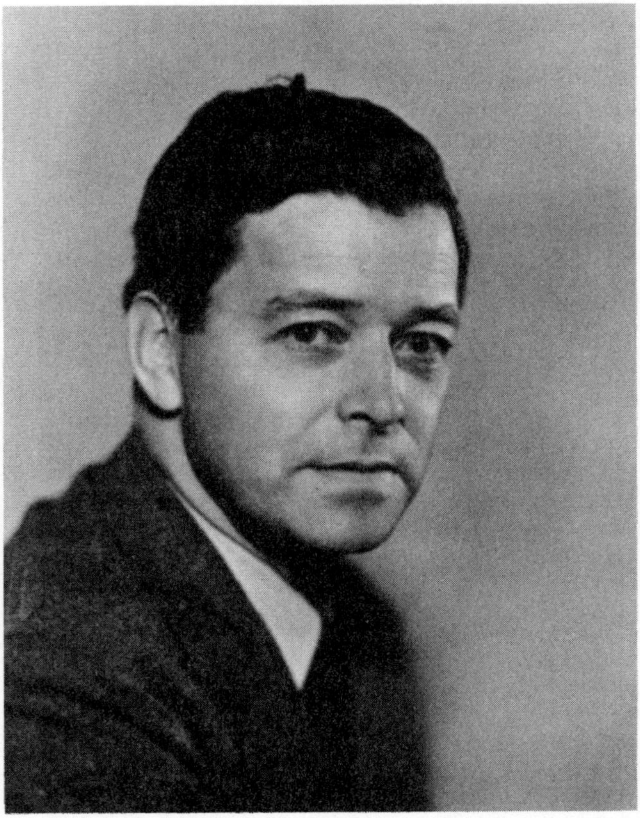

Plate 7 Albert Eide Parr, Director of the American Museum of Natural History from 1942–1959. A pioneer museologist he argued strongly for the environmental role of museums. (Courtesy of the Department of Library Services, America Museum of Natural History.)

> These ... were the people who rushed off in such masses in frantic search of sea anemones and ferns. Unschooled in the codes and avid for these untasted pleasures, they succeeded in perpetrating the most extensive damage .... Great stretches of the coast were largely stripped of their attractive inhabitants; whole areas were cleared of ferns, helped by professional touts who saw in this an easy means of turning the quick penny, loaded up great cart-loads from the wilder parts of Britain, and sent them off to the London markets.

It is probable that only a small proportion of the collections so made eventually found their way into a museum. There were a few voices of protest in the build-up to this collecting mania – Charles Waterton was dismayed at the loss of birds in Yorkshire in 1835; the Tyneside Naturalists' Field Club, founded in 1846, included in its rules a warning against over-collecting. (See Plates 8 and 9)

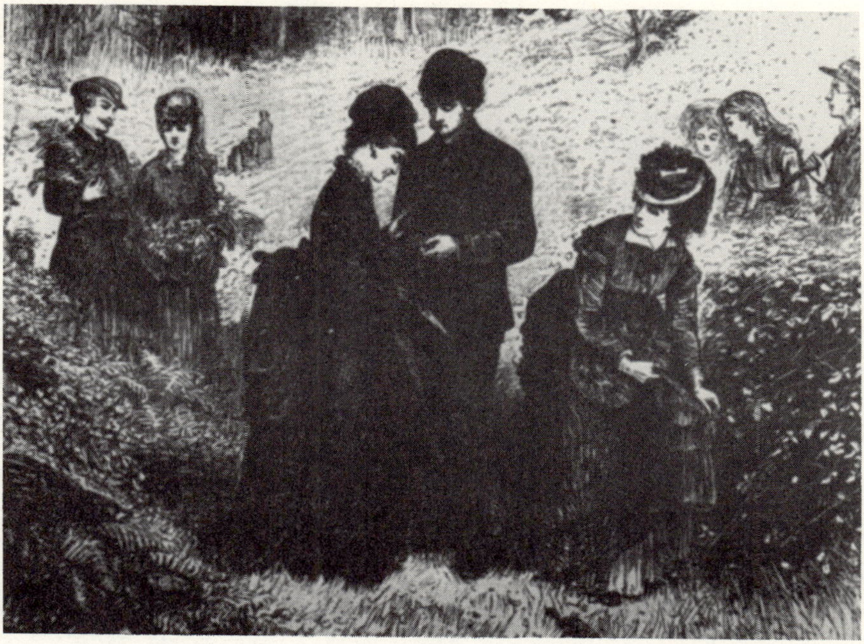

Plate 8 'Gathering ferns' from the *Illustrated London News*, 1871. (Reproduced courtesy of the Hulton Deutsch Picture Library.)

However, as with American museums, it is difficult to link early museum activities with the growth of the conservation ethic. What is undeniable is the continuing growth and enthusiasm for nature, and for collecting. The Field Clubs often generated their own museums, or its members were frequently associated with other societies (Literary and Philosophical Societies, Mechanics Institutes) which held collections. Members added generously to these museums, some of which still survive as private institutions, most of which were passed to local authorities to help found municipal museums; some became incorporated within university collections. Although the way in which Britain's systematic collections have been founded is somewhat piecemeal they remain a magnificent resource for today's biologists, rich in history, a detailed picture of past environments.

It is impossible to discover in the general literature on the history of museums more than the occasional hint that nineteenth-century museums had been affected or changed by environmental concerns, even less that they had tried to draw public attention to the issues. In exhibition galleries, despite the persuasive arguments put forward by distinguished individuals like Henry Flower, no distinction was made between research collections and those in the exhibition galleries – as Hudson (1987) notes, the nineteenth century ended ' ... with the great majority of [natural history] museums looking more like storehouses than exhibitions'. The lean years of the first half of the twentieth century, punctuated by two world wars, did

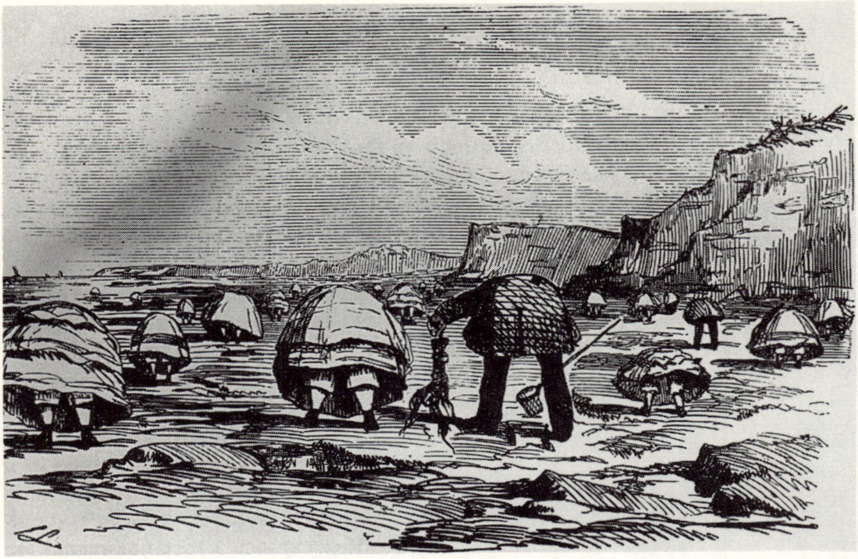

Plate 9 'Common objects at the seaside' by John Leech, *Punch*, 1857. (Reproduced courtesy of the Hulton Deutsch Picture Library.)

little to promote any change in attitude to the ways in which museums functioned.

Does the close examination of an individual museum reveal a different picture – a link with the conservation ethic? Let us consider one provincial natural history museum and chart its progress. The Hancock Museum, Newcastle upon Tyne, has been devoted to natural history and remained a private museum throughout its long history, and in both these ways is somewhat atypical of most provincial museums in Britain. However, because of its continuity and specialization it provides an ideal source to search for evidence of conservation-oriented activities. Like many larger provincial museums it has its origins in the collections of the city's Literary and Philosophical Society founded in 1795. The acquisition of the natural history collections of Marmaduke Tunstall (1743–1790) and George Allan (1736–1800) by the Lit. and Phil. in 1822 created deep divisions in its membership. As a consequence the Natural History Society of Northumberland, Durham and Newcastle upon Tyne split away in 1829, taking the natural history and ethnographic collections into its care and a small museum was built to house the growing collections in 1834. The interest in natural history was such that collections were being rapidly acquired; in addition the Society began to record the discoveries of its members in its *Transactions*, published from 1831. In 1835 the Society decided to open its museum to the public on one evening each month, the Committee minutes recording that '... the interest excited was so great, and the people came in crowds so large, as to fill the whole building, and totally defeat the end your Committee had in view.' (quoted in Goddard, 1929). In 1836 schools were admitted, and by 1838 annual attendance was in the region of 20,000. Natural history was in the ascendant.

The Tyneside Naturalists' Field Club, founded in 1846, was composed mainly of members of the Natural History Society, and Goddard (1929) makes specific reference to the Club's rules which promoted conservation:

> Rule XI – That the Club shall endeavour to discourage the practice of removing rare plants from those localities of which they may be characteristic, and of risking the extermination of birds by unwanton persecution.
>
> Rule XII – That on the other hand, members be requested to use their influence with landowners and others for the protection of the characteristic birds of the country, and to dispel the prejudices that are leading to their destruction.
>
> Rule XIII – That consequently the rarer botanical specimens collected at field meetings, be chiefly such as can be gathered without disturbing the roots of the plants; and, that notes on the habits of birds be accumulated instead of specimens, by which our closet collections would be enriched only at the expense of nature's great museum out of doors.

Goddard notes 'It is interesting and pleasing to note that north country naturalists even at this early date were endeavouring to protect wild life from senseless and useless ravages.' In these words we see the conservation ethic at work – in 1846 *and* 1929. The Tyneside Club published its own *Transactions* between 1848 and 1864 — an invaluable series of Catalogues which record the status of species in the local area; seven members of the Field Club were eventually elected Fellows of the Royal Society – an indication of the scientific profile of the organization. Many of the collections which were made during the course of this scientific work were incorporated into the Museum of the Natural History Society, which continued to attract the public in large numbers; the increasing popularity of the museum, and its growing collections led to the erection of a new museum in 1884. The driving force behind it was the ornithologist and taxidermist John Hancock (1814–1890), and the museum was eventually named for him and his brother Albany (1808–1873).

In 1896 the Society decided to award a Hancock Prize for a natural history essay – 'The object of the competition was not to form a passion for collecting but to stimulate observation and research' (Goddard, 1929). At this time the museum also began its loan scheme of specimens to local schools, and in 1897 made the museum's collections available to blind people; evening lectures for the public began in 1902. Formal school-teaching began in the museum in 1911 and henceforward – 'Each class visited the museum one afternoon a week for a course of six lessons on the Vertebrate Fauna of Great Britain. The lessons were planned by the Curator and were then rehearsed by him with the teachers who were to take the actual classes' (Goddard, 1929).

These educational initiatives indicate growing environmental awareness, but the museum's next move was a real sign that a conservation ethic was having an impact. Its involvement with the Farne Islands and its seabird colonies dates from 1923, when the Society hosted a major conference which ensured that funds were secured for their preservation. In July 1923 the museum first became involved in outreach, producing an exhibit on woodland fauna displayed at the Royal Agricultural Society's annual show in Newcastle. In 1924 the Natural History Society began its long lease of Gosforth Park Nature Reserve. Individual members of the museum's Committee began to be involved in conservation efforts, notably Abel Chapman (1851–1929), who helped to launch schemes which enabled the formation of the Kruger National Park in South Africa and protection of parts of the Spanish wetlands, the Coto Doñana.

By 1929 the Society and its museum were firmly involved in conservation. It was justifiably proud of its collections and its building, but it is the activities of its individual members which indicate that a new respect for wildlife was developing. The major use of the museum and its collections for environmental education and environmental research was yet to come, but already the conservation flag was being raised by the people associated with it. For example, T. Russell Goddard, author of the Society's *History* and Curator of the museum, was a member of the Farne Islands Committee and the International Committee for Bird Protection Wildfowl Enquiry.

From this time on individuals associated with the museum continued to play an increasingly important role in scientific research and conservation, and it became the venue for the origins of the Northumberland and Durham County Trust in 1961; a site museum was established at Kielder in Northumberland in 1963; the museum galleries were redesigned in the 1970s as old systematic displays made way for new interpretive approaches. Networking with conservation organizations was strengthened.

This example indicates that the adoption of the conservationist role by museums appears to have been a slow, subtle advance, running in parallel with the growth of the environmental movement, reflected in small changes in attitudes to collecting and research, exhibition and educational activities. Change *had* occurred, yet natural history museums in 1950 were still fulfilling essentially the same role as their predecessors one hundred years earlier, in the heyday of natural history collecting. They had developed and improved their networks with other conservation organizations, they had maintained and enlarged research collections and helped to document the environment, and through taxonomic displays given a glimpse of the components of the natural world to the public. The resources were in place, ready to capitalize on the phenomenon of environmentalism.

The environment as a mission for museums

James Oliver, addressing the symposium to celebrate one hundred years of the American Museum of Natural history in 1969, remarked that:

> The founders [of the AMNH] were authorised simply to establish and maintain a library and museum of natural history and to furnish popular instruction – a quiet, genteel goal. But the Museum, like a successful living organism, has continually evolved and adapted to its changing environment. Rooted in the past, with a primary task of preserving, studying and learning from the records of the

> ages, we are nonetheless oriented towards the future. We are keenly aware that in today's troubled world a dynamic museum must be an active force in the renovation of society, a determinant in the wise use of environmental resources, and a leader in the lessening of tensions and the promotion of understanding among men (Oliver, 1969).

This statement, made at the height of environmental activism in the USA which preceded Earth Day, reflects the dramatic change in attitude which had suddenly come into force within the world of museums. Questions were being asked about the ultimate outcome and use of research and exhibition programmes in natural history museums. The answer lay in the recognition of a new, underlying philosophy which gave museums a dynamic purpose, to foster the conservation ethic. The realization of new purpose which dawned in the 1960s has prompted a reassessment of museum activities and has led to marked changes in patterns of research and exhibition during the past thirty years. Museological literature has continued to stress the new role for natural history museums, and recently, possibly as a response to ever-increasing environmental problems and our awareness of them, the call for change has become more strident.

In 1950 a shaft of forward thinking cut through the gloom of the museum world. Albert Parr, always ready to challenge established positions, wrote:

> [museums] must find ... a new mission ... with a more direct bearing upon the urgent problems of contemporary existence. ... The side of nature which concerns society most of all is not undisturbed nature, but nature as the environment of man, and that is the field in which the educational efforts of the natural history museums could make their greatest contribution to human thought, welfare and progress today .... A knowledge of undisturbed nature is essential for an understanding of how it has changed and may change again.

This revolutionary message went largely unheeded. Engel (1962) examined the origins and role of natural history museums a decade later and suggested that they had 'served their term in natural history', and inferred that 'some people think of [natural history] specialists as odd specimens of the human species, fossilized in their childish curiosity for strange objects'. Natural history museums and their collections had declined in public appeal – even in the mid-1960s Parr (1966) could comment that:

> Many [natural history museums] have declined in community standing and in their status among cultural institutions. Their expositions, in a traditional vein, offer little relief from the pressures of world problems, and less help towards their solution .... Many are torn by inner conflicts between collections and exhibition, research and education ... the older institutions are struggling to overcome the handicaps of their inherited attitudes and traditions ...

Parr decided the goal of natural history museums was to help their public 'towards a better understanding of themselves, their natural opportunities and limitations', and was strongly in favour of stepping beyond the boundaries imposed by a museum's existing collection – a radical view at the time, but one often repeated since.

In the same year that James Oliver addressed the gathering at the American Museum of Natural History, Lothar Witteborg (1969) contemplated the future of museums against the background of the social revolution of the 1960s:

> a state of unbalance has been created leaving some of our social structures hanging on thin threads. Some age old social institutions are rapidly changing, values are changing, as are attitudes. Youth is in revolt. The museum as it stands today reflects the ultimate in conservative wisdom.

He concluded:

> We know that the establishing of a new philosophy and objectives for all natural history museums is mandatory if we are to survive and fulfil our social responsibilities as public institutions. We must also be able to move and shift with the social needs of society especially with regard to our public exhibition halls with their main emphasis on education and communication. As difficult a task as this may seem, our new philosophy must somehow reflect this aspect of flexibility and also communicate our desire to be a contemporary institution at all times. Natural history museums have a social responsibility to be relevant and contemporary.

He repeats Parr's call not to be constrained by collections, and calls for the breakdown of the traditional barriers between curatorial divisions and staff expertise in order, through exhibitions to convey environmental messages:

> Natural history museum exhibits have been so preoccupied with the wonders of idealized nature that they have tended to neglect the importance of nature as the priceless environment of man and the foundation of his existence. We must . . . base our thinking on the ideal that museums should influence man's rational attitude towards nature in the next five to thirty years.

Those 30 years are almost gone, but it is possible now to recognize in Witteborg's plea the impact of the environmental movement – in the space of a few years, against the social backdrop which had witnessed the birth of the new radical environmental movement, natural history museums had identified their new and vital purpose.

By 1973 it was declared that 'the conservation of nature is definitely the main concern of biologists all over the world' (Dorst, 1973), a sweeping and inaccurate statement, but one which indicates the depth of conservationist feeling and endeavour at that time. Calls continued to come for a rethink of the educational and research activities of museums. Parr (1973) was the first to recognize the role that museums might play in developing countries where the wise use of natural resources was necessary; he eloquently expressed the new philosophy which museums had now adopted, and which he had begun to promote 23 years earlier:

> There has also been another recent and radical change in the museum's functions in society. At the beginning there was, or ought to have been, an emphasis upon the diffusion of knowledge about the potentials of the natural environment as a basis for human life and progress. As mankind progressed and expanded its domain, a strong movement developed for the defence of nature

> against man's wants and desires, and even against his mere presence. Today the situation has changed again and we are faced with an urgent need to protect the environment, on behalf of man, for his present enjoyment and his survival in the future. In this struggle it falls to the natural sciences to try to defend the human condition and its supporting milieu against the damage inflicted by the exact sciences and the technologies they breed. The natural history museums are, or could and should be, among the strongest bastions of this defence, and therein lies a mission greater than any they have had before, since the beginning of their history.

Max Nicholson, in his address to the 10th General Conference of ICOM in 1975, reflected on the fact that like Darwinism, modern environmentalism exhibits 'an extricable blending of strictly scientific data and highly controversial value judgements' (Nicholson, 1975), which would continue to give environmental exhibitions vitality and variety. Nteta (1975) speaking at the same meeting, dedicated to 'The Museum and the Modern World', described the value of environmental displays in the National Museum to explain the impacts of overgrazing and water abstraction on the natural habitats of Botswana. At a time when most museum professionals sought to change the philosophy of their museums to address environmental needs, it is interesting to note that at this meeting de Silva (1975), facing very difficult and practical environmental problems in Sri Lanka, felt that:

> We should not reorientate existing museums to meet the requirements of present-day society. We should instead establish separate museums, specialised institutions, to meet this challenge .... Isolated museums cannot effectively deal with a problem affecting the entire present-day society or environment in general ... museums can help to find effective answers by forming part of a wider set-up involving all associated institutions.

By the 1980s the environmental role for natural history museums was firmly established. It surprised many museum professionals that it had taken so long for museums to discover a new purpose. For example, Lemieux (1981) expressed the view that museums had previously

> failed to apply to themselves the principles that they were discovering about the relationships between organisms and their community, and about the process of evolution that enables species to adapt to a changing environment. They ran the risk of extinction.

The new role for museums of natural history was encapsulated in the AMA's forward-looking review (American Association of Museums, 1984):

> Museums have another significant responsibility to educate people about the ability of our planet to sustain life. We live in an age in which plants and animals are becoming extinct at the most rapid rate in 65 million years. What this means is that up to one-sixth or more of the forms of life on earth may disappear forever during our life-times – a truly enormous change on a planetary scale, and one that may well seriously limit human potential in the future .... Museums can help bring about what futurists call a 'choiceful future', because they deal with both the possibilities and the dangers.

Sullivan (1993) has neatly summarized the new role for museums – all museums and not just those with a special interest in natural history – and has taken the environmental argument beyond research and exhibition, and into the ways that museums themselves function, a truly ecological approach:

> We must seek to develop ecologically literate visitors who understand the simple fact that they are fully implicated, connected, and interdependent with all the human and non-human beings that surround them. And we must teach/learn to care deeply about them. This respect for life and its preservation must be pervasive in all our policies, practices, exhibits, programs and behavior, whether in art, history or science settings. This ecological ethic of interconnectedness must infuse what we do, who we accept money from, what we come to symbolize.

Museums have responded to the perceived need for cultural change, rewriting their mission statements to emphasize their environmental role and setting new goals, many of them linked to major issues, especially loss of biodiversity. The mission of the National Museum of Natural History in Washington now (1995) states 'The National Museum of Natural History is dedicated to understanding, interpreting and conserving the integrity and diversity of the natural world and its peoples so that informed choices can be made about the interdependent future.' Its action statement reads 'We will support and encourage responsible decision-making about the use and care of the earth's resources so that the integrity, stability and beauty of this heritage can be sustained.' An active role in promoting conservation is evident in the goal ' ... to be a credible source of accurate, objective information to assist decision-makers in responsible use and conservation of the world's cultural and natural resources.' (National Museum of Natural History, 1993).Thus the decision has been made by this museum to engage in the environmental debate, to discuss global change and choices for the future, yet seeking to maintain academic credibility and objectivity. 'Knowledge for the next Millennium', the fund-raising campaign launched by the American Museum of Natural History, New York in 1993, stressed the interaction between its education/exhibition role and the needs of collections and research:

> While other institutions sharply reduce their commitment to natural history collections, the Museum's collections are yielding rich new lodes of information. As the newest techniques of molecular biology are being incorporated in systematics, Museum scientists have assumed an active role not only in understanding but also in preservation. And when the public needs, more than ever, to know what is at stake in issues such as biodiversity and global climate change, the Museum confronts its greatest educational opportunity – and most pressing responsibility. (American Museum of Natural History, 1993).

Annual Reports and other strategic documents from a diverse range of museums reveal an underlying conservation ethic. The Natural History Museum in London, arguably the world's premier taxonomic centre has a straightforward mission 'To further the understanding of the natural world using our unrivalled collection in world class exhibitions, education and research.' This is less overtly conservationist than many of the major museums in north America, where there is perhaps greater need to promote the museum to potential financial sponsors, to inform the world through glossy publications of the museum role. In 1993 virtually all Annual Reports obtained from natural history museums in the USA had an environmentally friendly mission. For example, the Cincinnati Museum of Natural History Mission Statement (1993) reads:

> The Cincinnati Museum of Natural History is dedicated to advancing the understanding of global change – past, present and future. In pursuit of this primary mission, the Museum also strives to accomplish several important goals: to advance the global quest for scientific understanding of our world as well as humanity's relations to it, to promote a greater public understanding of science and its benefits, to support the efforts of science educators, and to encourage preservation of nature's richness and diversity through land stewardship, innovative exhibits and educational programs.

Some museums in the United States have been established as part of an environmental institution, and its activities are consequently closely tied to its scientific objectives. An example is the Mississippi Museum of Natural Sciences, founded in 1971 as a branch of the State Department of Wildlife, Fisheries and Parks; its biological survey programmes and professional collections management help to support the work of the Department as a whole, with the mission 'To increase knowledge of the natural sciences particularly as they pertain to the conservation of Mississippi's wildlife resources, and to disseminate this knowledge among mankind.' (Shropshire and Shropshire, 1991).

Interdisciplinary museums have also adopted the environment as a cornerstone of policy. The Charleston Museum, the first permanent museum to be established in America, has significant natural history collections. It is also responsible for the Dill Wildlife Refuge, a semi-natural area including plantation sites and Confederate fortifications, which provides the museum with both research and educational potential stressed in its Long-Range Plan. (Charleston Museum, 1987). The Hancock Museum, Newcastle upon Tyne, which is primarily a museum of natural history, but also holds ethnographic and archaeological collections, ' . . . seeks to promote enjoyment, understanding and conservation of the natural world amongst people of north east England and visitors to Newcastle upon Tyne.'

The international perspective – messages from ICOM

The Natural History Committee of ICOM has debated at regular intervals the functions of natural history collections and the role of natural history museums in environmental conservation. Beijing played host to the ICOM International Conference on Natural Science Museums and the Environment in August 1993; the meeting held in Paris in May 1994 had 'Musées, Environnement et Education' as its theme. Biodiversity and the role of museums was selected for the meeting in Stavanger, Norway, in 1995. It is evident that environmental issues are centre stage.

A feature of ICOM meetings are the resolutions which are passed by individual Committees and ratified by the Conference. Shrouded in the mystique of icomspeak, they none the less provide a record of proceedings; they also reflect the concerns of their membership. Like all good bureaucratic measures, they resurface regularly in different guises, and little action is ever taken despite laudable objectives. As an example, Resolution 8 of the Natural History Committee of ICOM in Vienna, May 1979 read as follows:

> Recognising the vital and urgent importance of promoting broader public understanding of ecology, including human ecology and the conservation of natural resources,
> Convinced that natural history museums can contribute effectively to that understanding,
> Noting that while appropriately designed exhibition and extension programmes are needed in all parts of the world, the most serious problems of conservation and renewable resources are in the developing countries,
> Aware of the capability of the IUCN through its network of commissions and members to identify and define critical conservation issues throughout the world,
> Noting also that the committee encourages and promotes the philosophy of museums working together to solve global problems,
> Aware of the necessity that the local museum of a country is the best and most appropriate advisor on the interpretation of requirements of that country,
> Be it resolved that the International Committee of Natural History Museums of ICOM establish with IUCN a small working group to define methods and undertake pilot projects for co-operative museum based activities aimed at enhancing public understanding of ecology and conservation and to report its findings through the committee's newsletter and bring for discussion at the next meeting of the committee in Mexico City in 1980. (Anon, 1979.)

The 'small working group' was formed and named the Working Group on Environmental Education; it reported to the Mexico meeting, where a further resolution of support was passed for their work. The Natural History Committee meeting in Mexico addressed the other main role of natural history museums, taxonomic research. The case for increased resources for systematic work are well argued, but the resolutions can only be seen as optimistic:

> Recognizing the urgency of the present and projected loss of the world's heritage of natural habitats and species, described and undescribed,
> Recognizing the vital role that museums have in monitoring that loss,
> Realizing that sound environmental management is essential to the future welfare of all people,
> Recognizing that such environmental management, determined from ecological studies, can be biologically sound only if based on adequate taxonomic and systematic studies,
> Recognizing that a number of reports have recently been published in various countries drawing attention to serious gaps in our knowledge of the biotas of major geographical regions and of habitats being destroyed or adversely affected,
> Recognizing that the same reports note that the teaching and training of taxonomists is seriously insufficient for present and future requirements; and

> Recognizing the essential and fundamental role of Natural History Museums and departments in providing adequate facilities through the collections and libraries for the present and future training of curators, taxonomists and systematics,
> Be it resolved that the International Committee of Natural History Museums of ICOM:
>
> 1. Establish a working group to examine anew and to co-ordinate existing information on collections and future curatorial needs, especially those pertinent to the conservation of the world's natural resources,
> 2. Urges all countries to institute or expand training programs in their natural history museums, departments and research institutes; and,
> 3. Urges countries to provide greater employment opportunities for biological taxonomists and systematists, especially in museums. (Anon 1980.)

Laudable as these resolutions are, they were never achievable. Despite the well-argued case put forward by the Committee, the growing financial demands on governments, and on individual museums make any large-scale commitment to taxonomy impossible. Resolutions 2 and 3 were doomed to failure; Resolution 1 is a massive research project which as far as I am aware was never even begun. ICOM natural history meetings have continued until recently to produce convincing arguments for resolutions which will never attract the financial support necessary to make them viable. They do however demonstrate in a very dramatic way the impact of environmentalism – these arguments would never even have been made or understood in the 1950s. They are steeped in the philosophy of environmentalism, their naivety even retaining a hint of romanticism.

It is interesting to compare the resolutions of the early meetings of the Committee with the much more pragmatic approach of the 1990s. A policy statement issued by the Committee meeting in Paris 1994 reads as follows:

> The exploitation of the biosphere is becoming critical; it is our only 'living space'. The Committee, at its meeting in Paris, 27 May 1994 therefore affirmed, in a spirit of optimism, the following principles:
>
> 1. Natural history museums must effectively fulfil their vital and unique role in the study of biodiversity, conservation and environmental education.
> 2. These activities must be pursued in close collaboration with other ICOM Committees and where possible, with Universities and other organisations working in the environmental sciences.
> 3. Natural history museums must vigorously promote educational programmes and exhibition themes of high quality, but economically as possible, to create greater public awareness of environmental issues; they must also identify and attract that section of the public not hitherto touched by museum activities.
> 4. Finally, all these activities must be pursued without losing sight of those factors which, in the life of many cultures, express harmony with nature and not aggressive exploitation. (C. Hill, pers. comm.)

These succinct objectives are a far cry from the resolutions of the 1970s, and show increased awareness of the situation – the need to diversify the museum audience for example. However, they still need to be linked to achievable goals if real progress is to be made. Whether such goals are ever attainable within such a loose forum may be questioned. For example, there has for many years been a need to ensure taxonomic training for museum staff in developing countries, an issue which has surfaced regularly at ICOM meetings, without being resolved. Surely the approach which is needed is the twinning of museums in those countries with a major museum in Europe or North America; this would facilitate exchange of staff and expertise in systematics. Mares (1993), in his detailed comparison of museum resources in developed and developing countries has stressed the global nature of the museum community and the need for the development of such partnerships – 'International co-operation among museum professionals, especially taxonomists and educators, can lead to the solutions required to deal with the present-day challenges to our cultural and natural heritage.' Griffin (1992) has also made a plea for museums to develop 'future coalitions to achieve globally-important objectives'.

The international perspective – the Madrid conference

The International Symposium on the Preservation and Conservation of Natural History Collections held in Madrid in May 1992 brought together a wide range of expertise and opinion from the world of natural history museums. The main theme of the meeting was collections – their care and management, the increasing demands placed on museums by ever-growing collections, the uses of collections, and techniques of preservation – the full account is available in three excellent Symposium volumes. The meeting had been advertised as one which would address practical issues such as collection and preservation techniques and advances in collections assessment or collections management. This was the case, yet it was noticeable that a remarkable number of papers (e.g. Lovejoy 1993; May 1993; de la Penha 1993; Haas 1993; Emery 1993; Chalmers 1993; Talbot 1993; Mares 1993; and Griffin 1993) questioned the role of natural history museums, seeking a change in purpose, and outlining new strategies which oriented museums to the needs of environmental conservation. All recognized the dual museum roles of research and education, but also identified needs for co-operative research, exchange of technical and systematic expertise and greater focus on the value and longevity of natural history collections. The world-wide nature of environmental concerns and their interaction with natural history museums was stressed by Mares (1993), who proposed a global 'museum ethic'; the repatriation of collections and associated data is seen as an element in this philosophy. In looking to the future of museums he comments that:

> Museum curators should be leaders in conservation and research, environmental science, evolution, speciation, extinction, and many other topics that are vital to humanity . . . we need to build upon our great storehouses of data and apply this information to human needs.

If proof was required that the environmental movement had formed a symbiotic relationship with natural history museums, then it was available in Madrid.

A number of resolutions were agreed at the Madrid Symposium; they have subsequently been published (Anon, 1992). They focus on the uses of biological collections, particularly their value for the study of biodiversity, and these are discussed in detail later (Chapters 6 and 7). The resolution document is broken down into nine key issues:

1. recommendations to aid and monitor world-wide collection concerns;
2. facility and resource needs for conserving collections;
3. museums and educational needs in developing countries;
4. the need to escalate research and development of preservation and conservation techniques, and research in collections curation issues;
5. the need for trained systematists, collections managers and conservation specialists;
6. the need for databases;
7. education and public awareness;
8. postal systems and
9. endorsements – UNESCO, UNCED–Rio.

Each resolution is linked in some way to environmental conservation. The recommendations are detailed, ambitious and complex, and their implementation will be the task of the World Council on Collections Resources (WCCR) established at the Symposium and consisting of 'representatives from those world-wide organisations that are concerned about the preservation and conservation of natural history collections.'

An ecosystem model for museums

Natural history museums have reached a watershed. The collections made by the great museums which emerged in the nineteenth century presented a fragmented, exotic natural and cultural world, a view of nature progressing upwards from the simple to the complex, and culture similarly moving from 'primitive' society to the peak of civilization represented by western technological man. By visiting museums the urban public gained an impression of a world inaccessible to them before the advent of television and travel, and played a vital role in education and entertainment. However, in the twentieth century, and particularly from the 1960s onwards, our view of the world has moved away from the simplistic vertical model, shifting towards a paradigm of horizontal, global interdependence. Sullivan (1992) notes 'No longer viewed as the vertical masters of a natural order over which we were given divine dominion, Western peoples increasingly view themselves as participants in a horizontally inter-connected ecological system and an interdependent, pluralistic cultural society.' This new attitude reflects a realization that both natural and cultural systems are vulnerable and threatened. All museums, but especially those involved with the environment, need to respond to this change. There is a real need for active involvement, to shift the emphasis of our research, educational programmes and exhibitions into the environmental arena and to become active advocates of cross-cultural respect and tolerance. This 'ecosystem' model should be one to guide natural history museums into the twenty-first century, and ensure their relevance to the modern world.

If this is the model to follow, what practical goals and strategies might be required to achieve it? Much will depend on persuading governments and other funding organizations that museum-based scientific research is important – museums cannot afford to be marginalized – 'To many biologists, natural history museums have become insignificant islands in science, separated from the larger body of scientific endeavour' (Griffin, 1992). This means making our research relevant to today's needs, focusing on the taxonomy and ecology of animal and plant taxa which will supply the information we need to implement measures to protect biological resources. Allied to this research is the fundamental question of where (geographically and taxonomically) museums should focus their collecting efforts, and the need for better collections management. The variety of natural history museums in terms of scale, their geographical remit and scientific expertise is enormous, and each institution needs to set scientific goals which are both relevant and achievable. Much of the debate about the role of museums has focused on those that are considered the great institutions of natural history, but the place of small regional museums also needs to be understood and recognized – frequently their scientific expertise and networks have enabled local conservation measures of considerable significance. A goal to develop and enhance networks within each museum's scientific and conservation community must be seen as a priority.

New goals also need to be set for education and exhibition programmes in natural history museums if they are to achieve the changes in attitude which were discussed earlier. Griffin (1993) suggests the goal: 'Develop entertaining public programs which effectively communicate to the general public, information about the natural environment, the effect of human impact on it and strategies which are environmentally benign rather than destructive.' He also stresses the need for museums to take a more active role in promoting environmental awareness to the public, to participate actively in debate and become involved in decision-making on conservation matters.

We can no longer regard 'the environment' as a passing fancy; it has continued to be a major part of our lives since Earth Day, disproving the predicted issue-attention cycle for it (Dunlap, 1989). The diverse nature of environmental issues continue to make them newsworthy, they affect all our lives. There can be no doubt that environmental research and education will remain as a driving force for natural history museums in the decades ahead.

References

Alderson, W. T. (ed.) (1992) *Mermaids, Mummies and Mastodons; The Emergence of the American Museum*, American Association of Museums, Washington DC.

Alexander, E. P. (1979) *Museums in Motion*, American Association for State and Local History, Nashville.

Allen, D. (1976) *The Naturalist in Britain – A Social History*, Allen Lane, London.

American Association of Museums (1984) *Museums for a New Century*, AAM; Washington DC.

American Museum of Natural History (1993) *Knowledge for the Next Millennium – The Campaign for the American Museum of Natural History*, AMNH, New York, 36 pp.

Anon. (1979) 'Resolutions of the Conference of the Natural History Committee of ICOM', Vienna, 12–20 May, 1979, *ICOM Natural History Committee Newsletter*, No 2, (4), 4–5.

Anon. (1980) 'Resolutions adopted at the meeting of the International Committee of Natural History Museums in Mexico City on October 29, 1980' *ICOM Natural History Museums Newsletter* No 6.

Anon. (1992) 'Resolutions from the International Symposium and World Congress on the Preservation and Conservation of Natural History Collections', *ASC Newsletter*, 21 (4), 132–135.

Brears, P. (1989) *Of Curiosities and Rare Things: The Story of Leeds City Museum*, Friends of Leeds City Museum.

Chalmers, N. (1993) 'Achieving strategic change: natural history collections for the twenty-first century'. In Rose, C. L., Williams, S. L. and Gisbert, J. (eds) *Current Issues, Initiatives, and Future Directions for the Preservation and Conservation of Natural History Collections*, Vol 3, International Symposium and First World Congress on the Preservation and Conservation of Natural History Collections, 1992, Madrid, Spain, 143–147.

Charleston Museum (1987) *Long Range Plan*, Charleston, South Carolina, 14 pp.

Cincinnati Museum of Natural History (1993) *Annual Report, 1993*, Cincinnati Museum of Natural History, Cincinnati, Ohio.

Davis, P. and Brewer, C. (1986) *A Catalogue of Natural Science Collections in North East England, with Biographical Notes on the Collectors*, North of England Museums Service, Durham.

Dorst, J. (1973) 'The role of museums in furthering nature conservation'. In Engstrom, K. and Johnels, A. G. (eds) *Natural History Museums and the Community*, Universitetsforlaget, Oslo.

Dunlap, R. E. (1989) 'Public opinion and environmental policy'. In Lester, J. P. (ed.) *Environmental Politics and Policy – Theories and Evidence*, Duke University Press, Durham, 87–134.

Emery, A. R. (1993) 'Changing philosophies, roles and responsibilities'. In Rose, C. L., Williams, S. L. and Gisbert, J. (eds) *Current Issues, Initiatives, and Future Directions for the Preservation and Conservation of Natural History Collections*, Vol 3, International Symposium and First World Congress on the Preservation and Conservation of Natural History Collections, 1992, Madrid, Spain, 111–126.

Engel, H. (1962) 'Museums of Natural History', *Museum*, 15 (2), 124–127.

Gilbert, B. (1990) 'Earth Day plus 20, and counting', *Smithsonian*, 21 (1), 47–55.

Goddard, T. R. (1929) *History of the Natural History Society of Northumberland, Durham and Newcastle upon Tyne, 1829–1929*, Andrew Reid, Newcastle.

Griffin, D. (1992) 'Museums and the Environment: A challenge for us, not them'. In Saucier, C. (ed.) *Museums: Re-thinking the boundaries*, Proceedings of the 16th ICOM General Conference, Quebec, Canada, 1992, 48–58.

Griffin, D. (1993) 'Planning for the 21st century and preparing for the next 500 years'. In Rose, C. L., Williams, S. L. and Gisbert, J. (eds) *Current Issues, Initiatives, and Future Directions for the Preservation and Conservation of Natural History Collections*, Vol 3, International Symposium and First World Congress on the Preservation and Conservation of Natural History Collections, 1992, Madrid, Spain, 405–426.

Haas, J. (1993) 'Repatriation in a global context for natural history museums'. In Rose, C. L., Williams, S. L. and Gisbert, J. (eds) *Current Issues, Initiatives, and Future*

Directions for the Preservation and Conservation of Natural History Collections, Vol 3, International Symposium and First World Congress on the Preservation and Conservation of Natural History Collections, 1992, Madrid, Spain, 71–81.

Hancock, E. G. and Pettitt, C. W. (1981) *Register of Natural Science Collections in North West England,* The Manchester Museum, Manchester.

Hudson, K. (1987) *Museums of Influence,* Cambridge University Press, Cambridge.

Impey, O. and MacGregor, A. (eds) (1985) *The origins of museums; the cabinet of curiosities in Sixteenth and Seventeenth century Europe,* Clarendon Press, Oxford.

Jackson, C. E. (1992) *Prideaux John Selby: A gentleman naturalist,* The Spredden Press, Stocksfield, Northumberland.

Kohlstadt, S. G. (1992) 'Entrepreneurs and Intellectuals: Natural History in early American Museums'. In Alderson, W. T. (ed.) *Mermaids, Mummies and Mastodons; The Emergence of the American Museum,* American Association of Museums, Washington DC, 23–40.

Kulik, G. (1992) 'An American Visionary', *Museum News,* 71 (5), 56–57 and 69–70.

Lemieux, L. (1981) 'Museums of natural history and their social context'. In *The World's Heritage; The Museums' Responsibilities,* Proceedings of the 12th General Conference of ICOM, Mexico City, 1980, ICOM, Paris.

Lloyd, D. (1983) *The History Of Ludlow Museum, 1833–1983,* Ludlow Museum, Ludlow, Shropshire.

Lovejoy, T. E. (1993) 'The role of natural history museums in a changing world'. In Rose, C. L., Williams, S. L. and Gisbert, J. (eds) *Current Issues, Initiatives, and Future Directions for the Preservation and Conservation of Natural History Collections,* Vol 3, International Symposium and First World Congress on the Preservation and Conservation of Natural History Collections, 1992, Madrid, Spain, 27–34.

Lowenthal, D. (ed.) (1974) 'Introduction'. In *Environmental Perception and Behavior,* University of Chicago, Chicago.

Mares, M. A. (1993) 'Natural History Museums: Bridging the past and the future'. In Rose, C. L., Williams, S. L. and Gisbert, J. (eds) *Current Issues, Initiatives, and Future Directions for the Preservation and Conservation of Natural History Collections,* Vol 3, International Symposium and First World Congress on the Preservation and Conservation of Natural History Collections, 1992, Madrid, Spain, 367–404.

May, R. M. (1993) 'Global change: the need and concern for collecting and preserving'. In Rose, C. L., Williams, S. L. and Gisbert, J. (eds) *Current Issues, Initiatives, and Future Directions for the Preservation and Conservation of Natural History Collections,* Vol 3, International Symposium and First World Congress on the Preservation and Conservation of Natural History Collections, 1992, Madrid, Spain, 35–42.

Nash, R. (1982) *Wilderness and the American Mind,* Yale University Press, New Haven and London, 3rd Edition.

National Museum of Natural History (1993) *Mission Statement and Annual Report, 1992,* Smithsonian Institution, Washington DC.

Nteta, D. (1975) 'The environment'. In *The Museum and the Modern World,* Papers from the 10th General Conference of ICOM, Paris, ICOM.

Nicholson, E. M. (1975) 'The environment'. In *The Museum and the Modern World,* Papers from the 10th General Conference of ICOM, Paris, ICOM.

Oliver, J. A. (1969) 'Remarks by Dr. James A. Oliver at the Centennial Convocation of the American Museum of Natural History', *Museum News,* May, 28–30.

Ornstein, R. and Ehrlich, P. (1991) *New World, New Mind; Changing the Way We Think to Save our Future,* Paladin Grafton Books, London.

Parr, A. E. (1950) 'Museums of Nature and Man', *Museums Journal,* 50 (8), 165–171.

Parr, A. E. (1966) 'Yesterday and tomorrow in museums of natural history', *Studies in Museology*, Vol 2, 15–18.

Parr, A. E. (1973) 'The role of natural history museums in the life of man'. In Engstrom, K. and Johnels, A. G. (eds) *Natural History Museums and the Community*, Universitetsforlaget, Oslo.

Penha, G. M. de la (1993) 'Cultural change: the need and concern for collecting and preserving'. In Rose, C. L., Williams, S. L. and Gisbert, J. (eds) *Current Issues, Initiatives, and Future Directions for the Preservation and Conservation of Natural History Collections*, Vol 3, International Symposium and First World Congress on the Preservation and Conservation of Natural History Collections, 1992, Madrid, Spain, 43–50.

Porter, C. M. (1990) 'The Natural History Museum'. In Shapiro, M. S. (ed.) *The Museum – A Reference Guide*, Greenwood Press, New York, 1–29.

Ralph, R. (1993) *William MacGillivray*, The Natural History Museum/HMSO, London.

Shropshire, C. C. and Shropshire, R. T. (1991) 'Natural History Museums as a function of wildlife management'. In Cato, P. S. and Jones, C. (eds) *Natural History Museums, Directions for Growth*, Texas Tech University Press, Lubbock, 41–47.

Silva, P. H. D. H. de (1975) 'The environment'. In *The Museum and the Modern World*, Papers from the 10th General Conference of ICOM, Paris, ICOM.

Stansfield, G., Mathias, J. and Reid, G. (eds) (1994) *Manual of Natural History Curatorship*, HMSO, London.

Stearn, W. T. (1981) *The Natural History Museum at South Kensington*, Heinemann, London.

Sullivan, R. (1992) 'Trouble in Paradigms', *Museum News*, 71 (1), January/February, 41–44.

Sullivan, R. (1993) 'Lessons for the ruling class', *Museum News*, 72 (3), May/June, 54–55 and 70–72.

Talbot, F. H. (1993) 'Museums on the knife edge'. In Rose, C. L., Williams, S. L. and Gisbert, J. (eds) *Current Issues, Initiatives, and Future Directions for the Preservation and Conservation of Natural History Collections*, Vol 3, International Symposium and First World Congress on the Preservation and Conservation of Natural History Collections, 1992, Madrid, Spain, 147–154.

Trudgill, S. (1990) *Barriers to a Better Environment; What Stops us Solving Environmental Problems?*, Belhaven Press, London.

Williams, B. (1987) *Biological Collections U.K.*, The Museums Association, London.

Wilson, E. O. (1992) *The Diversity of Life*, Penguin, London.

Winsor, M. P. (1991) *Reading the Shape of Nature; Comparative Zoology at the Agassiz Museum*, University of Chicago Press, Chicago.

Witteborg, L. P. (1969) 'Natural history museums – a time for change', *Studies in Museology*, Vol 5, 1–6.

Yochelson, E. L. (1985) *The National Museum of Natural History – 75 Years in the Natural History Building*, Smithsonian Institution, Washington DC.

4. *Putting across the message*

In the end, we will conserve only what we love. We will love only what we understand; we will understand only what we are taught.

Baba Dioum, Senegalese conservationist

If you are thinking a year ahead, sow a seed. If you are thinking ten years ahead, plant a tree. If you are thinking one hundred years ahead, educate the people.

Chinese poet, 500 BC

Do not try to satisfy your vanity by teaching a great many things. Awaken people's curiosity. It is enough to open minds; do not overload them. Put there just a spark. If there is some good inflammable stuff, it will catch fire.

Anatole France

We are all in the environment; we eat it, we breathe it and sometimes we even make museums out of it.

Paul Martinovich

In 1808 a live grizzly surprised Charles Willson Peale in his kitchen, where he shot it dead; soon the curious flocked to his home to see the bear at close quarters. Ever a showman, Peale barred the public from further viewing, then skinned and mounted the animal in a natural pose; when installed on display in Peale's museum, the bear was more popular dead than alive. As a result, Peale began to create more 'habitat groups' using stuffed and mounted animals, and combined this idea with an earlier invention, that of painting backgrounds to give an illusion of depth – he had paved the way for the habitat group and the diorama. William Bullock's London museum, established in 1809 also used this approach, his main exhibit of large African mammals (his 'Pantherion') was set amongst accurate models of plants and used a 'panoramic effect of distance and appropriate scenery, affording a beautiful illustration of the luxuriance of a torrid clime' (Bullock, 1812; Lucas, 1921). Bullock, like Peale and Sir Ashton Lever before him, created spectacular and entertaining exhibits with the specific purpose of attracting and exciting visitors. A surviving Bullock museum example, 'combat' between a tiger and a boa constrictor (Hancock, 1980) provides us with a glimpse of these early efforts to re-create reality through taxidermy (see Plate 10). The reality was influenced by a need for action and drama,

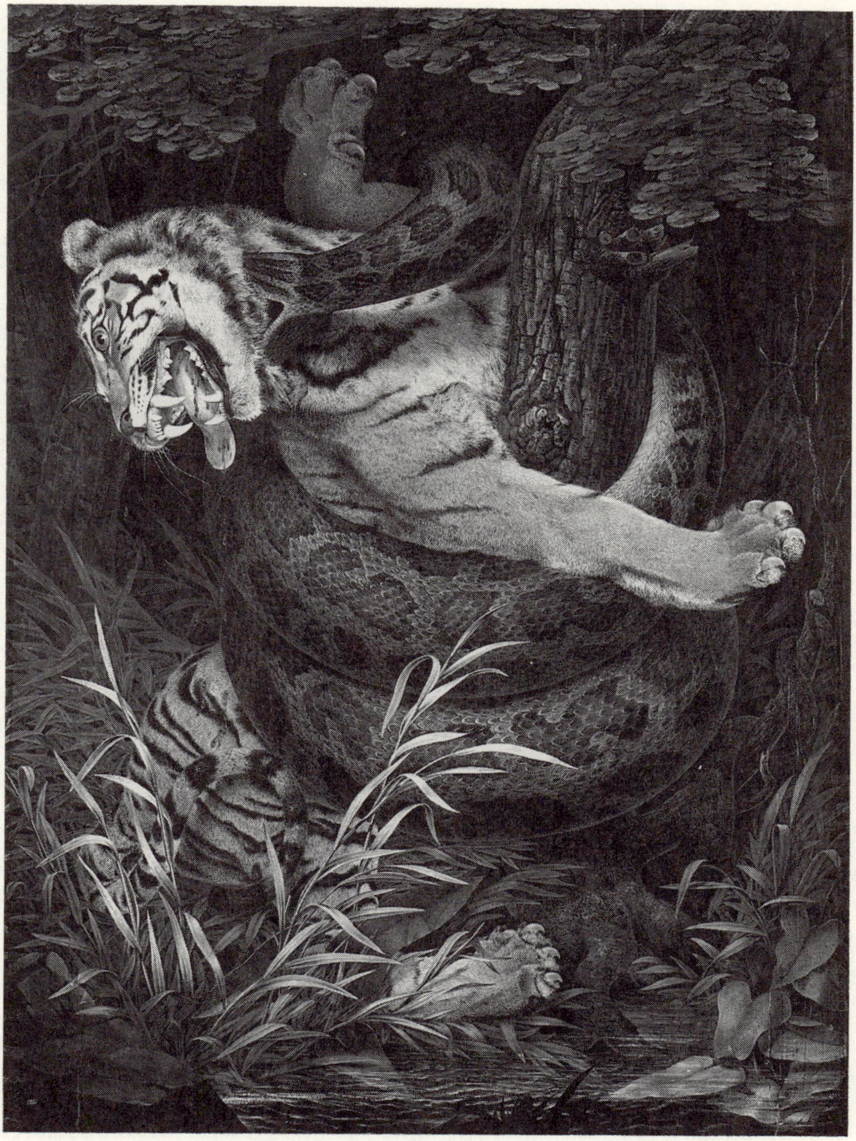

Plate 10 Alexandre Isidore Leroy de Barde (1814) 'Tigre Royal embroire par le serpent'. This gouache was based on an exhibit in William Bullock's London Museum, which survives today in Rawtenstall Museum, Lancashire. (Reproduced courtesy of the Department des Arts Graphiques, Musée du Louvre.)

and of course tended to reflect the widely held beliefs of the early nineteenth century that nature, especially in unknown countries, was frightening, and almost certainly dangerous. As a consequence the theatrical

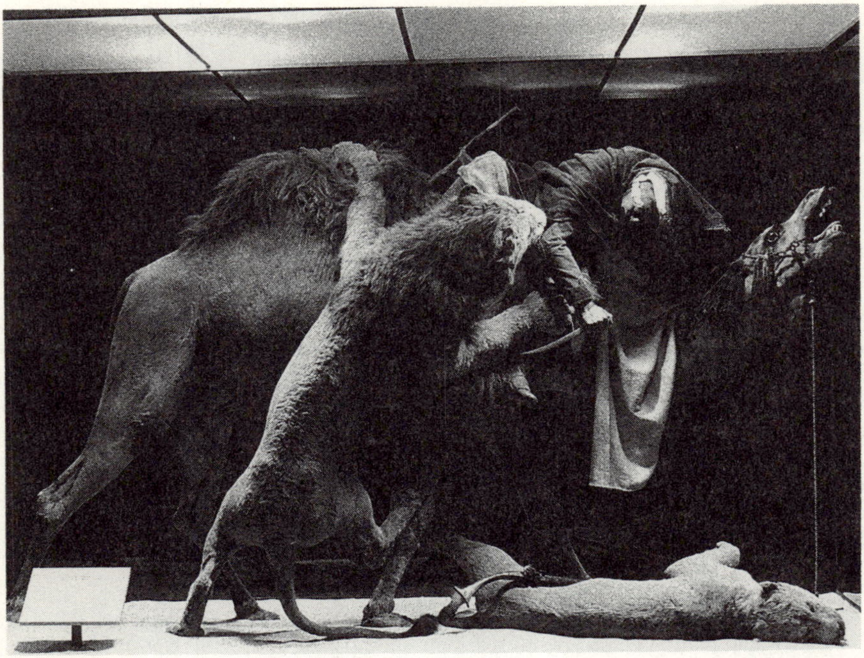

Plate 11 'Arab Courier attacked by Lions' – the dramatic work of taxidermist Jules Verreaux. (Courtesy of the Carnegie Museum of Natural History, Pittsburgh.)

approach to such animal groups continued. The creations of Jules Verreaux, French naturalist and taxidermist took these spectacular tableaux to the extreme; best known for his 'Arab Courier attacked by Lions' (see Plate 11) exhibited at the 1869 Paris Exposition (this dramatic work is now in the Carnegie Museum of Natural History, Pittsburgh), Verreaux reinforced the sensational appearance of his exhibits with lurid descriptive labels (Wonders, 1989).

The increasing interest in natural history shown by the Victorians, revealed by the activities of natural history societies and Field Clubs, also manifested itself in the demand for ornamental taxidermy – stuffed birds and mammals enclosed in bell-jars were popular, as were specimens crowded into a shallow case with a painted backdrop. Also familiar are the trophy heads mounted for sportsmen, travellers and adventurers, preparations which were the mainstay of many of the major commercial taxidermists, including London's Rowland Ward studios. These varied examples of the art of the taxidermist can still be seen in some museum displays (see Plate 12), but many museums and curators now regard such items as irrelevant to modern interpretation, even as mildly distasteful, with the consequence that most have been relegated to basement stores. Country houses and public houses still feature these sometimes grotesque and faded specimens – once the pride of Victorian drawing-rooms, they are now seen simply as relics of times past. Yet the actions of removing ornamental taxidermy from the public domain is to deny its history and significance, not only as an indicator of a growing fascination with the natural world during the Victorian era, but of the abilities of the taxidermists who created them.

Throughout the nineteenth century the skills and techniques used by taxidermists became evermore sophisticated, and the static postures of birds and animals seen in early work were replaced by dynamic creations which reflected behaviour in the wild. A good example is the work of the Newcastle taxidermist John Hancock (1808–1890), who, recognized as a leading ornithologist, used his accurate field observation (and skill with a gun) to produce for the Great Exhibition of 1851 set pieces which caused a sensation (see Plate 13). The surviving examples in the Hancock Museum, continue to excite visitors, and amaze today's preparators with their quality and vitality.

Perhaps surprisingly, many nineteenth-century museums did not regard habitat groups, or representations of nature, as fitting for their scientific or educational purposes. Comparative anatomy and systematics were serious business, so much so that one view in 1874 was that:

> 'Spread-eagle' styles of mounting, artificial rocks and flowers, etc., are entirely out of place in a collection of any scientific pretensions .... Besides, they take up too much room. Artistic groupings of an extensive collection is usually out of the question; and when this is unattainable, halfway efforts in that direction should be abandoned in favour of severe simplicity. Birds look best, on the whole, in uniform rows, assorted according to size, as far as a classification allows (quoted in Wonders, 1989).

Some museums can still be found which adopt such a policy, and arguably should never be changed, having become museum exhibits in their own right (see Plate 14).

Gradually, throughout the nineteenth century, the artistic nature and skills of the taxidermist began to be recognized; the Victorians were familiar with habitat groups at home, and as a consequence were receptive to them in a museum context. Wonders (1989) records how the acquisition of an orang-utan group prepared by William Hornaday was purchased by the American Museum of Natural History in 1883 at a cost of $2,000 – an action which Hornaday himself recognized as 'the beginning of an era in the progress of museum taxidermy'; such a radical change in policy by a major institution paved the way for similar changes in other museums – natural history displays had been 'popularized' through the habitat group, which gradually evolved into the more sophisticated and scientifically accurate diorama.

The word 'diorama' was coined by the pioneer photographer Louis Daguerre in 1822 (Peck, 1990), but the labour and skills necessary to create such displays in natural history museums mitigated against their widespread use in the nineteenth century. There was also a major philosophical change to be achieved before progress could be made, that of recognizing the educational role of museum displays – the far-sighted Henry Flower was one

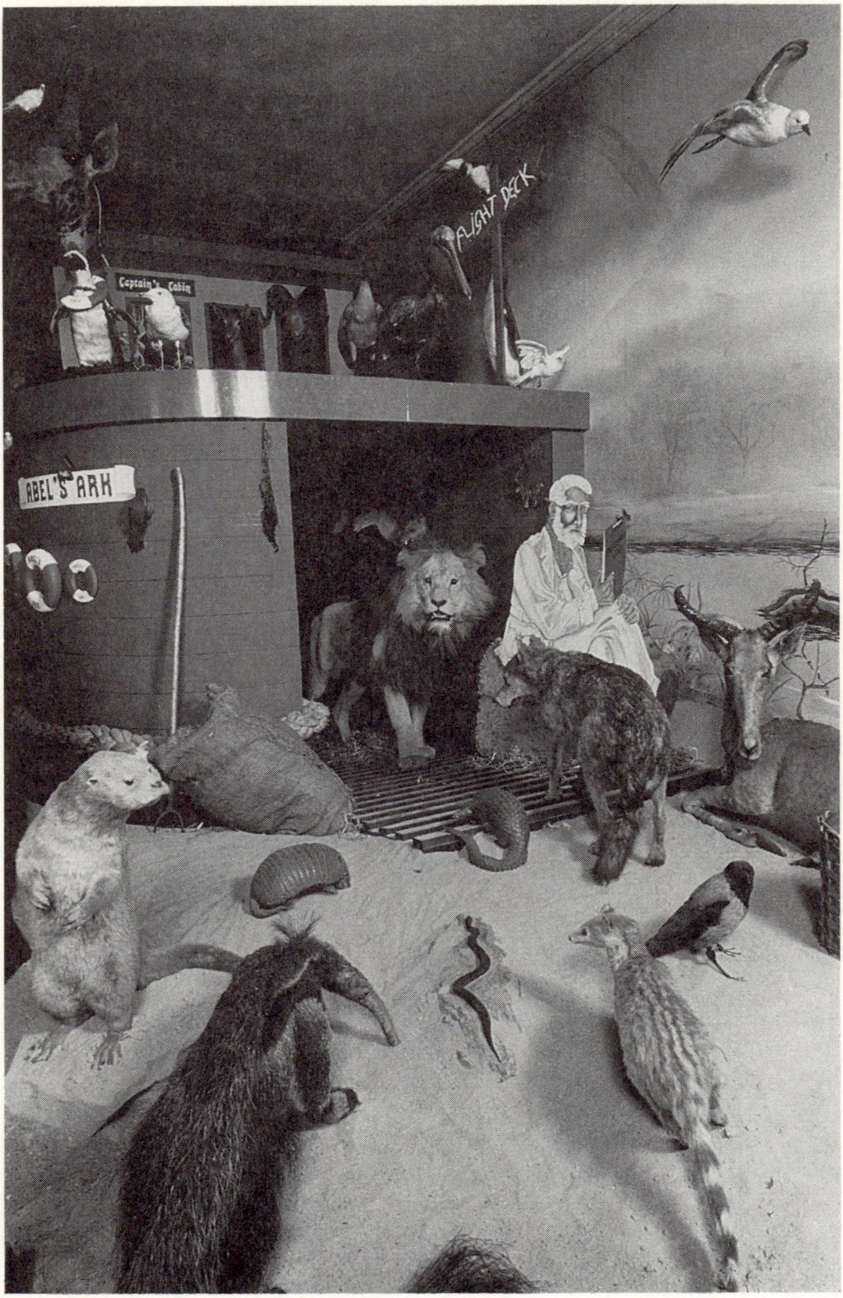

Plate 12 A unique use of trophy heads and mounted animals – 'Abels' Ark' at the Hancock Musuem, Newcastle upon Tyne.

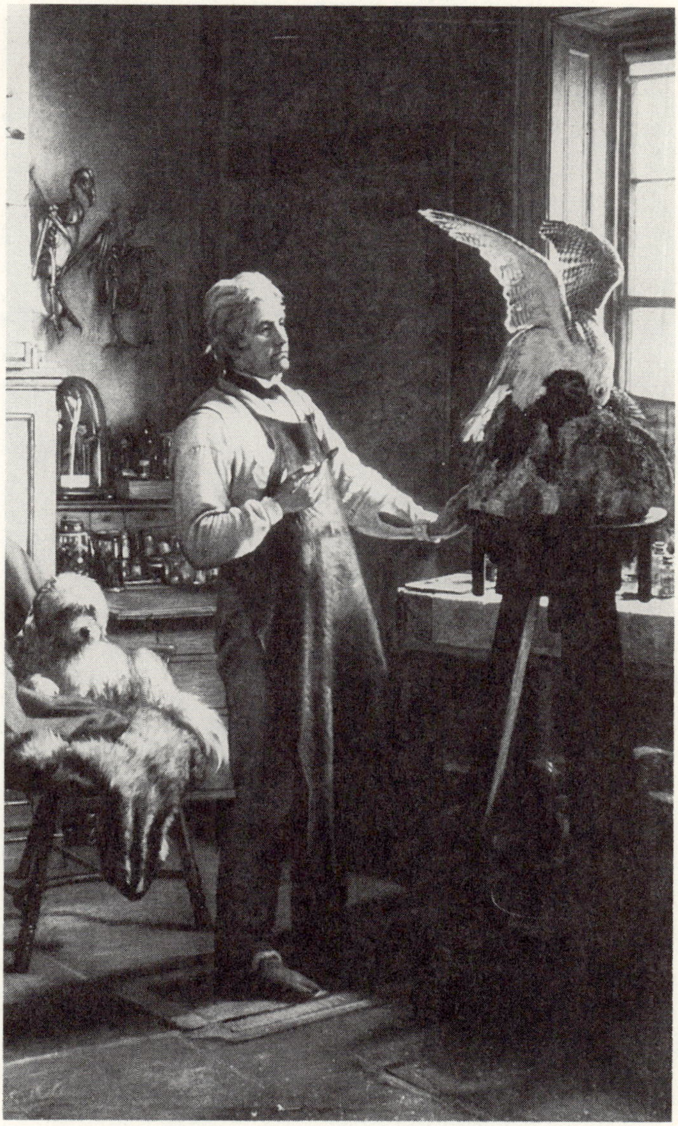

Plate 13 John Hancock (1808–1890) in his studio, oil-painting by H. H. Emerson, 1882. (Reproduced courtesy of the National History Society of Northumbria).

of the first curators to recognize the need for new *methods* of interpreting natural history in order to achieve educational goals (Flower, 1898), and that specimens needed to be obtained and prepared for that purpose, rather than for research goals. This was an advanced museological concept

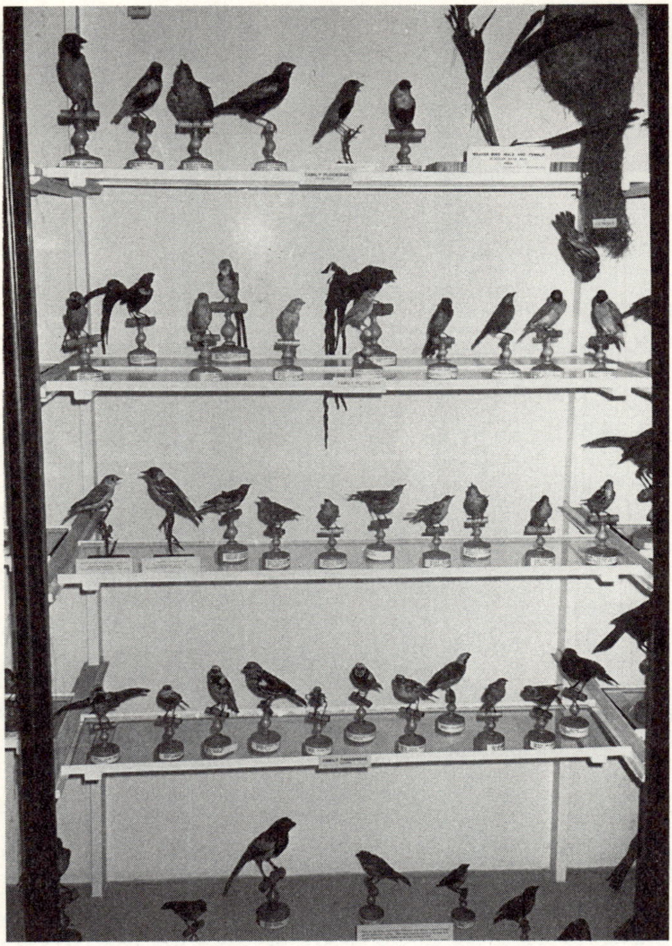

Plate 14 A prime example of a taxonomic display – the National Museum of Natural History, Dublin.

for the late nineteenth century. According to Hudson (1987) ' ... it was revolutionary in its day with the great majority of museums looking more like storerooms than exhibitions'; this advance meant the segregation of scientific and display collections (the study series and exhibition series in Flower's terminology), a distinction which continues to be recognized in many North American museums. Under Flower's direction the British Museum (Natural History) displays began to feature animals in 'artistic reproductions of natural environments'. George Browne Goode, Curator of the National Museum of Natural History, Washington, and a great admirer of Flower, pushed forward the boundaries of museum display by adopting the habitat group as the cornerstone of a new methodology – ' ... the most prominent form of exhibition technique during the period of great expansion of American natural history museums from 1900 to 1935' (Wonders, 1989). The term 'habitat groups' is attributed to Frank Chapman, curator of ornithology at the AMNH, and 'long before the concept of ecology became popular, habitat groups were recognised as holistic expressions of biological interdependence' (Wonders, 1989). By the late 1930s it was recognized that their purpose went '. . . beyond the reconstruction of a pretty picture . . . to be ecologic . . . that is . . . elucidate natural interrelationships between organism and organism, and between all and the physical environment' (Murphy, 1937).

It was in the early twentieth century that the American Museum of Natural History, New York, the Field Museum, Chicago and the Academy of Natural Sciences in Philadelphia realized the sophisticated habitat groups, (now more frequently referred to as dioramas) for which they are so well known (see Plate 15).The technique demanded extensive research in the field, and newly collected specimens – the National Museum expedition to Africa in 1909, led by Roosevelt, brought home 5,013 mammals, 4,453 birds and 2,322 reptiles and amphibians. By this time the segregation of collections in American museums – those collected for science, and those specifically obtained for exhibition purposes – was identifiable. Peck (1990) has argued that obtaining animals for dioramas in the early twentieth century had no major impact on wildlife populations, as:

> individual animals were hunted selectively and taken only as needed for display. Zoo collecting at the time often resulted in the deaths of many more animals. The individual animals that survived capture and the long stressful journeys to captivity often proved ill-suited to zoo life.

At a time when much of the African continent remained relatively unaffected by human populations, and game numbers were high, it is hard to imagine that museum collecting expeditions had any dramatic effect – though museum rivalry ensured collecting on a considerable scale. The end result was the creation in American museums of an enduring and accurate picture of the natural world at the time, one which now has increasing significance.

The diorama technique was also adopted in some European countries, notably in Sweden, where the Biological Museum in Uppsala was specifically designed for a series of habitat dioramas, and in Stockholm, where the concept of the large-scale open diorama was pioneered by Gustaff Kolthoff in 1880 (Wonders, 1990). Viewed from a two-tier central observation tower, several thousand mammal and bird specimens were set against a continuous background painting representing a panorama of the Nordic landscape. Other European museums followed a more conventional route of creating realistic dioramas behind glass, with the natural history museum in Berne, Switzerland, creating outstanding examples which take the art to its limit (see Plate 16). The museologist A. E. Parr observed a distinct difference between the 'reality' of European dioramas and a 'lighter spirit of sentimental realism' which characterized the North American versions:

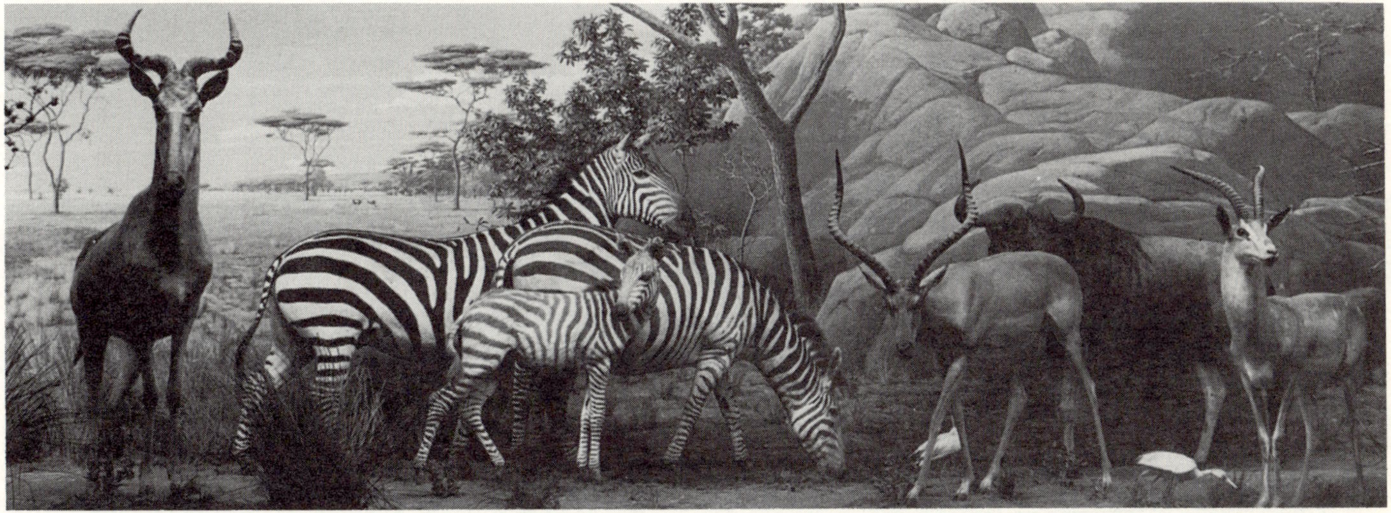

Plate 15 The diorama – an example from the Academy of Natural Sciences, Philadelphia. (Photograph by Owen McGoldrick reproduced courtesy of the Academy of Natural Sciences, Philadelphia.)

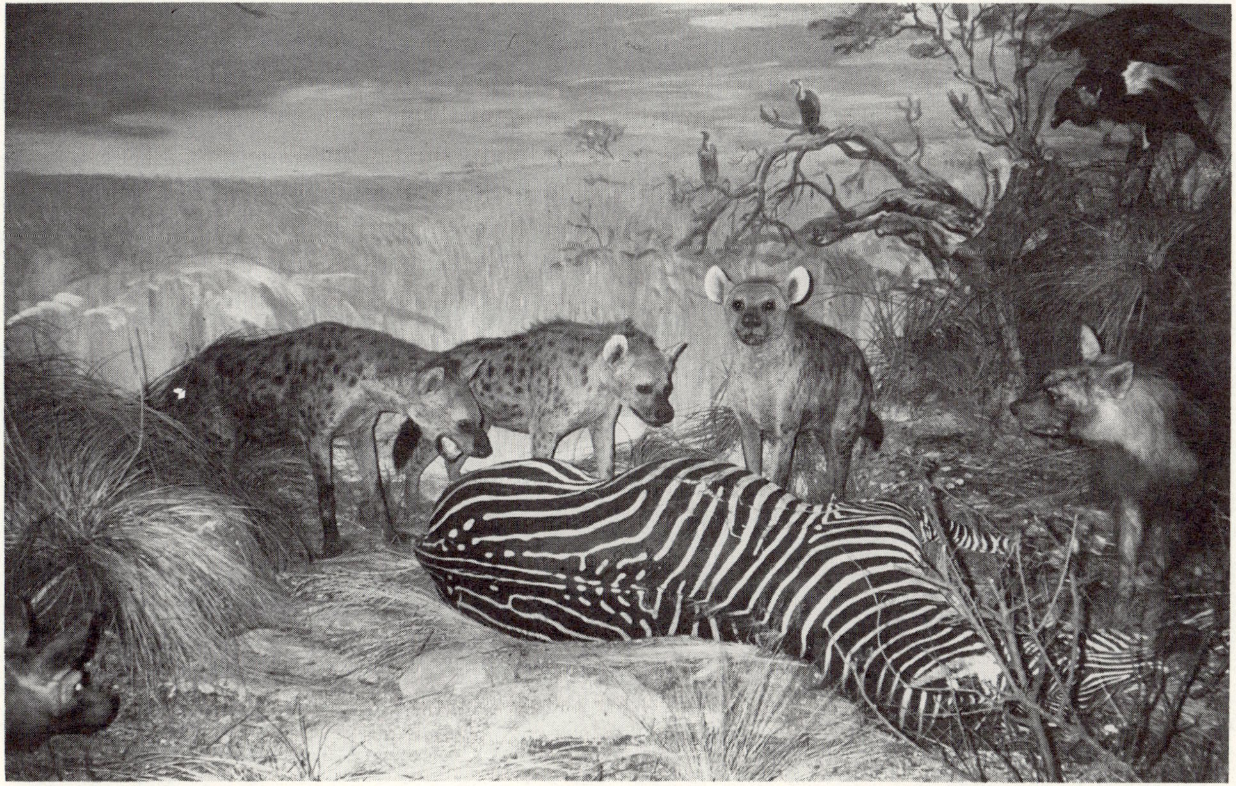

Plate 16 The diorama – some of the best examples in Europe are those in the Natural History Museum, Berne, Switzerland.

> He believed that when nature is interpreted so as to inspire the viewer's imagination, it possesses a special value that is destroyed by absolute verisimilitude. In North America, where a sentimental identification with the wilderness landscape and a tendency towards nature worship have strong historical traditions, it is not surprising that a romantic attitude toward nature prevails in the habitat dioramas (Wonders, 1990).

This link between the wilderness or nature movement and the development of the diorama is an important one – museums can be seen to be reacting positively to new concerns closely allied to the educational aims of a growing environmental concern. The taxonomic displays which had been the mainstay of the early nineteenth century were gradually displaced by the more dramatic habitat groups, and then by the realism of the diorama as the century came to its close and a new one began. By 1915 O. C. Farrington recognized that (in North America) there had been a revitalization of natural history museums, through the introduction of the art of the habitat group, the growing recognition of the value of 'nature study', and the 'realisation of how rapidly many of the forms of nature are disappearing' (Farrington, 1915).

The diorama innovation had a number of significant impacts. First, it breathed new life and purpose into the art of taxidermy, providing new challenges to a skill which dates back to the sixteenth century. No longer was it sufficient to mount a mammal or bird in a 'standard' pose, often totally unnatural or lifeless, attached to a wooden base or an artificial perch. The diorama demanded action and life. Perhaps more significantly, it placed the art of taxidermy at the forefront of museum education – animals could now be seen in a natural setting – museum visitors, many of whom had little experience of travel in their own country, had the opportunity to experience tropical rainforest, meet a polar bear in an Arctic landscape or witness the diversity of life on the African savannah. When the new dioramas of North American habitats were opened at the AMNH in 1942, they were able to offer ' . . . a valuable refuge and an opportunity for eyes weary of city streets to enjoy a grand tour of the North American continent. For a while at least the viewer can lose himself in a communion with Nature' (Anthony, 1942). It was this 'communion with nature' – the spiritual and emotional response to the diorama – that made it such a dramatic tool, not simply for display, but for nature conservation. As Coleman (1939) observed 'Habitat groups and the beginning of ecological study reflected the desire of the layman and scientist alike to accept nature whole. Appreciation of nature now became the theme. Conservation was taking the place of use as the ultimate motive'.

William Leigh, an artist responsible for the background paintings in the AMNH African Hall, remarked that the dioramas were 'a page of history that will survive, perhaps after much of this animal life has been wiped out – a record of something which never can be again – a document of inestimable value' (Leigh, 1938). These were prophetic words. Surviving dioramas are now in many cases a glimpse of the past, as species and habitats have indeed been lost. In one sense this makes dioramas an important historical source (and still an inspiration and perhaps even a warning), but in another sense it is a fabrication, a lie. If the intention is to re-create reality (as originally intended) should dioramas be brought up to date? Should the backdrop to a tropical rainforest exhibit feature the scars and blaze of deforestation? Should the foreground of a Rocky Mountain diorama contain human footprints, discarded packaging or rusting Coke cans? This would certainly be more honest, but would destroy a display of some historical significance. As a consequence, other means are being taken to make dioramas more pertinent to the late twentieth century – Berne Museum is rewriting all the labels associated with its dioramas to draw attention to the plight of threatened habitats and the declining range of the species exhibited (Lups, pers. comm.). Chicago's Field Museum dioramas are now 'set ... amid the latest in interactive, multimedia technology to convey the exhibit's central message – the interconnectedness of life ... the exhibit examines the impact of development and human activity, particularly human overpopulation that threatens the natural world' (Anon, 1993).

Beyond the diorama

It has been argued (McMichael, 1990) that 'social education' – an educational objective to develop or change visitors' attitudes on issues of social importance – must now be regarded as a central museum role, a view shared by many authors (e.g. Raven, 1989; Hall, 1991). With a Pacific perspective, McMichael uses examples of prejudice and misunderstanding which have been countered by museums – for example the new respect for the rich culture of Aboriginal Australians, and a recognition of their ecological success ' ... occupying a very wide range of environments in what is a generally harsh and difficult country to live in, without causing significant damage to the land or its flora and fauna.' Developing such 'social education' is central to the emerging role of natural history museums and their exhibitions – exhibitions which may have a basis in the past, but address current issues and make predictions (or offer choices) for the future of our environment.

To be effective in their displays, museums must reveal man's complicity in the fate of ecosystems; if the principal objective of an exhibition is to promote conservation, to go beyond the simple appreciation of beauty, oddity or magnificence that natural history provides, then it is imperative to begin to tell the human story which underpins the exhibit. Concern for the environment, and an appreciation of the rapid nature of technological advance means that in general people are more concerned with the present and the future, not the past – and by and large, museums have not adapted well to these changes, with many remaining deeply rooted in a nineteenth-century philosophy and approach. It was in an attempt to combat these entrenched attitudes that in 1961 A. E. Parr explored the concept of a museum of 'Man and Nature', which became a reality as the Manitoba Museum of Man and Nature in 1970 (Dubois, 1991). In a remarkable document Parr set out ideas and a philosophy which are as meaningful today as when they were written; Parr recognized that:

> Museums telling the story of man and his natural environment ... have a wonderful opportunity to create among all members of the community a better understanding of the world we live in, thereby helping us all to govern it more intelligently by the power of our vote under a democratic system of government.

Parr put collections secondary to the educational message for his proposed museum, a radical suggestion which like much of his thinking, was way ahead of its time – he had not only prescribed a new philosophy for natural history museums, but also produced a practical blueprint for action.

Whilst changes of such significance were taking place in North America, Swedish museums were also at the spearhead of the new movement, promoting travelling exhibitions such as 'Are we poisoning nature' in 1966 (Westerlund, 1991) in response to growing public concern and debate about the use of pesticides and fertilizers in agriculture. However, in Britain the environment was still not a central theme to new natural history exhibits, where even in the late 1960s, museums remained firmly within the strait-jacket of tradition. Although Stansfield (1972) was able to report on some progressive displays, notably in the provinces, he could only conclude that 'The fact remains ... that museums as a whole have been slow to take the initiative', with the responsibility for environmental education being firmly taken by the emerging County Trusts and Field Centres. In the 1970s the emphasis in local natural history museums was on the local environment, usually descriptive, habitat-based galleries which paid passing reference to biological conservation (e.g. Davis, 1978; Walker, 1978); galleries which reflect the local environment remain a feature in museums today. Particular subjects also attracted attention. Even as late as the 1980s – ' ... emphasis was still on single-subject galleries .... Recent single-subject galleries include the Bird Gallery at Manchester Museum in 1981 and the Fish Gallery at the National Museums of Scotland, Edinburgh' (Stansfield and Howard, 1994). In London, the Natural History Museum's new botanical gallery (1962) drew attention to the significance of National Nature Reserves, and Introducing Ecology (1978) examined the science of ecology. However, it was not until 1991, 30 years after Parr's pioneering ideas, that the Natural History Museum, London produced 'Ecology', its first real attempt to convey a conservation message in a permanent exhibition.

Why did it take so long for British museums to respond to environmentalism – to reflect a social trend in their exhibitions? There was change taking place, but it was a slow and relentless progression, rather than a dramatic shift of emphasis; this pattern is repeated globally in all natural history museums, even in North America, and appears to have been inhibited by three major factors, namely specimens, politics and the need for scientific accuracy. Museums have understandably demonstrated a reluctance to move away from 'specimen-centred' displays. For so long geared to the display of objects and specimens, museums have found it difficult to adjust to being driven by a storyline or the need to convey a particular ideology, to be an advocate for the natural environment. Secondly, the politics of environmentalism needed to be confronted – could museums which rely on government funding, or on politically astute benefactors really begin to tackle controversial environmental issues without suffering financial or political penalties? Finally, museums are conservative institutions, unwilling to 'rush into print' – and as a consequence have been prepared to await the scientific evidence before producing meaningful displays. Hudson (1987) draws attention to the way in which the Natural History Museum has reflected this careful approach, accusing it of remaining 'politically cautious, at a time when a number of biological topics have been the cause of great controversy . . . its eco, one might say, is not fully eco . . . It has remained . . . a *Times* rather than a *Guardian* museum'. With a wider perspective it is evident that by the late 1980s, when Hudson was making these observations, a sea change had already taken place. Against a continuing backdrop of public concern and debate these issues had been addressed – natural history museums had adopted environmental conservation as their major mission – the identify crisis perceived in the 1970s by Wittlin (1971) and more recently by Alberch (1994) had been overcome. Museums in Britain and elsewhere understood what they had to do and were creating meaningful exhibitions to promote conservation. The growing number of travelling and temporary exhibitions with a conservation theme staged in the 1970s included the Natural History Museum's 'Protecting Wildlife – 'The Conservation of Wild Creatures and Plants Act 1975', 'Britain's Threatened Wetlands' and 'Wildlife in Danger'; in Liverpool 'Man and the Environment' addressed pollution issues, whilst the impact of a single oil spill was addressed by the National Museum of Wales in the 'Christos Bitas' exhibition. Even the Victoria and Albert Museum had discovered the environment by the 1980s, with its exhibition 'The Discovery of the Lake District' (1984) looking not only at the aesthetic beauty of one of Britain's most scenic areas, but also at the increasing need for conservation policies to maintain the landscape (Victoria and Albert Museum, 1984).

The idea of using museums to inform and affect attitudes about the environment is now widely accepted, and it is easy to forget how this change has been achieved. One of the earliest exhibitions to address environmental issues was 'Deadline for Wildlife' mounted at the Carnegie Museum in 1953, which examined loss of species and habitat in Pennsylvania. Although by today's standards of presentation the exhibition looks amateurish (Anon, 1953), its messages were accurate and hard-hitting, drawing particular attention to the impact of the increase in human population – it was way ahead of its time. Another notable landmark in museology was the exhibition designed and constructed for the Centenary celebrations of the American Museum of Natural History in 1969. At this time the environment dominated the media, Earth Day was in preparation, and amidst some controversy 'Can Man Survive' opened to the public; using new media (six films, 800 slides and eight sound tapes), at a cost of $650,000 it was the most expensive single temporary show in the museum's history; it was also an exhibition which broke the mould of natural history displays. The focus was clearly on the damage which humankind was causing to the environment, and it reflected the growing interests in the wider environment, especially pollution. Specimens became secondary to the messages of environmental degradation it was designed to communicate; unlike most AMNH exhibits, it was not a comfortable environment to experience. Robbins' (1969) description is revealing:

> As the message develops, the exhibit environment alters accordingly: The floor levels change. Passages grow more restricted and claustrophobic. The air conditioning becomes less and less effective. One may read the warnings and assessments of our future, or examine the documentary photographs of our present. One may watch short film loops which illustrate the basics of the problem – air and water pollution for instance, or the deteriorating 'balance' of population and food supply. The exhibit that began as a quiet refuge from the New York City environment ends as an unsettling experience of the worst aspects of such environments. The backside of a city bus signals the final turn to the exit. The question 'Can Man Survive' is asked one last time, as one turns to see in a mirror who it is that holds the answer.

In North America Albert Parr had developed a new philosophy for museum exhibitions which had been realized in a new museum, and then taken a stage further in New York with 'Can Man Survive?' From this time onwards, the environment became a focus for exhibition activities in natural history museums; conservation messages became incorporated into new wildlife exhibitions, and some museums designed exhibits which addressed particular environmental concerns. Canadian museums have been at the forefront of change, some recent examples include 'Électrique' (an exhibit on the environmental impact of hydroelectric developments) and 'Autopsie d'un sac vert' (Recycling) at the Musée de la Civilisation, Quebec 1991 (Simard, 1991), 'Sable Island' at the Nova Scotia Museum, which examined the impact of man on a fragile ecosystem, 'Entre Terre et Eau' and 'Save our Species' at the Musée du Séminaire de Sherbrooke (Hall, 1991), and the travelling exhibition 'Save the Rainforest' (Winterhalder, 1991). The USA also broke new ground, with one of the first major rainforest exhibits at Milwaukee being opened in 1988 (Young, 1989; Korensic and Young 1991); Cincinnati Museum of Natural History continues the environmental debate with its most recent exhibition 'Our Weakening Web', a story of extinction which is made relevant to today by examining the biodiversity crisis and emphasizing the need for practical conservation measures (Gislason 1994). The rebirth of the 'Grande Galerie' of the Natural History Museum in Paris has seen the adoption of the impact of man on the environment as one of its three major themes (Adoutte, 1994), the others being biodiversity and evolution. In Britain the travelling exhibitions from the Yorkshire Museum have always had an underlying conservation message; arguably the best permanent environmental gallery is 'World in Our Hands' at the National Museum of Scotland, Edinburgh, opened in 1994 at a cost of £468,000 (see Plate 17). The aim of the project organizers was to present impartial information, present guidelines for a green lifestyle, and most important, to end on an optimistic note. Considerable imagination has been used to utilize specimens to good effect, whilst not compromising on environmental messages; it poses questions such as 'Why does wildlife matter?' and finally 'What can be done?' The Earth Centre, a proposal for a new museum

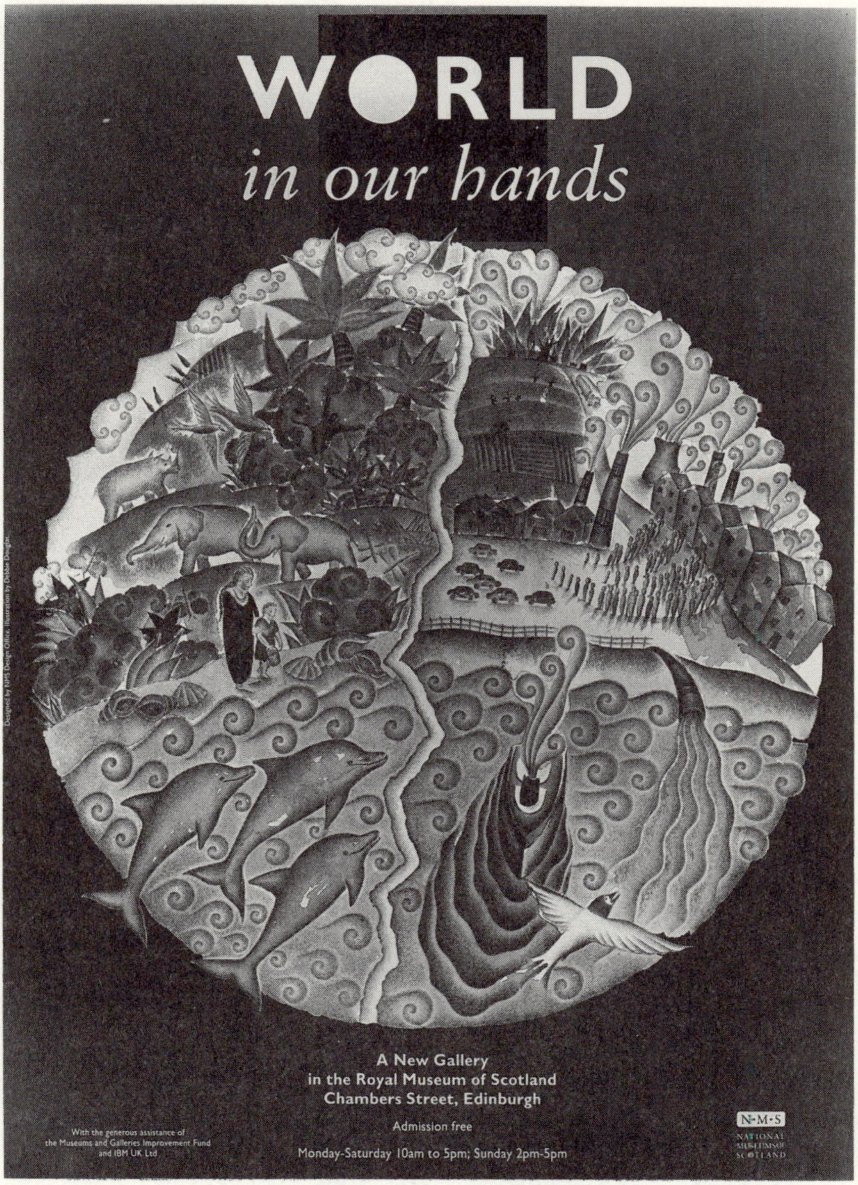

Plate 17 The promotional poster for 'The World in Our Hands', National Museum of Scotland, Edinburgh. (Copyright of the Trustees of the National Museums of Scotland, 1992.)

near Doncaster, is still in the feasibility stage; should it proceed, it will attempt to examine 'The Future of the World' (Blockley, 1993).

It is interesting to note that despite the advances in exhibition design and display, with greater emphasis on new media, especially information technology, and a growing interactive component, the diorama is still alive and well. As a technique for putting across environmental themes the diorama still has credibility, and continues to provide a dramatic element – the only major difference from the early 1900s is that the modern designer is not averse to using scientifically accurate reproductions of species, as in Milwaukee's rainforest exhibit (Kelly, 1991).

Dealing with controversy

The Museon, in the Hague, the successor to the Dutch Museum for Education, is devoted to telling ' ... the fascinating story of man and his world' and through a range of exhibitions fulfils that promise, using an interdisciplinary approach to discuss a variety of themes (du Weyden, 1989). Technology is used to good effect; for example in the rainforest exhibition (1989) where the loss of habitat is recorded in illuminated panels – each representing an area the size of a football field – and the gallery is filled not just with the taped sounds of birds and monkeys, but with the buzz of chainsaws and the crash of falling timber. At each crash, another football field lights up on the 'scoreboard'. The Ecos gallery in the Museon addresses key environmental issues in a hard-hitting fashion. In the museum guidebook it is described as ' ... a very unorthodox show, meant to make people aware of the environment in which human beings play a highly influential part'. The exhibition re-creates an environment reminiscent of any domestic situation; all the features of the dining-room, kitchen, bathroom are present; stairs lead to the bedrooms. A dual carousel slide-tape programme begins, a tightly scripted dialogue between a civil servant and a 'green'; as the programme progresses the room comes to life. At the mention of photosynthesis the rubber plant in the corner 'grows' six feet; the refrigerator door bursts open to reveal bubbling cylinders of brightly coloured chemicals – food is energy. When the discussion reaches the fate of that food, a cut-away dog descends from the ceiling, flashing lights marking the route of the alimentary canal; to make the point about the fate of waste products the toilet seat lifts eerily and water flushes automatically. Finally, the issue of population growth is discussed and emphasized as the root cause of many environmental ills, to the accompaniment of a rhythmically moving duvet in the bedroom. The technology is impressive, but the message is essentially a simple one – everything we do affects the environment – it is a clear and disturbing message. Major issues are tackled here in an uncompromising fashion, and a firm 'conservationist' stance is taken, no attempts are made to defend the polluting industrialist, the figure of authority in the slide presentation demonstrates a clear lack of understanding of biological and chemical cycles, and the inability of government to recognize the need for urgent action. This is radical display at its best – technically clever, humorous, liberal in its attitudes, yet hard-hitting –

something one might expect from the Dutch. But could a museum mount such a display in New York, Rome or Manchester?

All museums are faced with the dilemma of dealing with difficult or controversial issues, and some environmental concerns belong in this category – questions relating to nuclear energy, the creation of tidal barrages or the environmental impact of new roads are all issues which tend to polarize opinion. Should museums be prepared to take an environmental stand? Few people would fail to support the moral stance of an exhibit which damned badger-baiting for example, but would a similar response be gained by visitors to an exhibition which explored the negative environmental impacts of generating power from nuclear sources, especially if local jobs were dependent on that enterprise? At the height of environmentalism in the early 1970s, Hubendick (1972) was firmly of the opinion that museums should defend the environment at all costs '... Even if this means defending controversial viewpoints and a little fighting on the part of the museum official, it is certainly worthwhile. The standpoints defended by museums often represent some badly-needed counterweight against the short-sighted economical interest.'

Until museums begin to tackle these wider issues, and take a stand, our conservation messages will continue to centre on loss of habitat and species – whales and whaling, the effect of pesticides on birds of prey – they are important and still pertinent, but politically they are relatively 'safe' areas. Museums need to tackle new, current issues, going beyond questions of loss of biodiversity (which is the focus of many exhibitions as a result of a specimen-centred approach), to consider the other major environmental issues – sustainability, wise use of resources, global environmental change, aesthetic heritage and pollution for example. In general (with some exceptions, including 'The World in Our Hands' in Edinburgh, and 'The Delicate Balance' at Des Moines, Iowa (Johnson, 1992)) museums have failed to produce exhibitions which address fundamental global issues, the simple facts of overpopulation and the allocation of resources. Why do museums avoid these issues, or fail to make messages strong enough?

Rainforest exhibitions, in the main, have failed to deliver a powerful conservation message, emphasizing instead their beauty and value to mankind; exciting exhibitions such as The Yorkshire Museum's 'Monsters of the Deep', (Howard, 1991) having created an environment which puts the public into a receptive state of mind, noticeably fail to deliver a conservation message (see Plate 18). Visitors enjoy this exhibit, there is no doubt, and they must leave with an increased awareness of the marine environment – but does the theme music and clip from 'Jaws' reinforce the idea that sharks are evil creatures? Should the marine aquaria be 'eco-labelled' to avoid encouraging the proliferation of tropical marine aquaria with their dramatic effect on populations of marine organisms? Similarly, the Natural History Museum's Ecology gallery, whilst inspired in some areas, and heavily glazed in most, places its hard-hitting finale in the exit. Few people stop, few get the message. 'I counted 100 visitors through and only seven even paused before the screen' (Coles, 1991a). Why do museums make the most important messages an afterthought? It is because it is too harrowing to

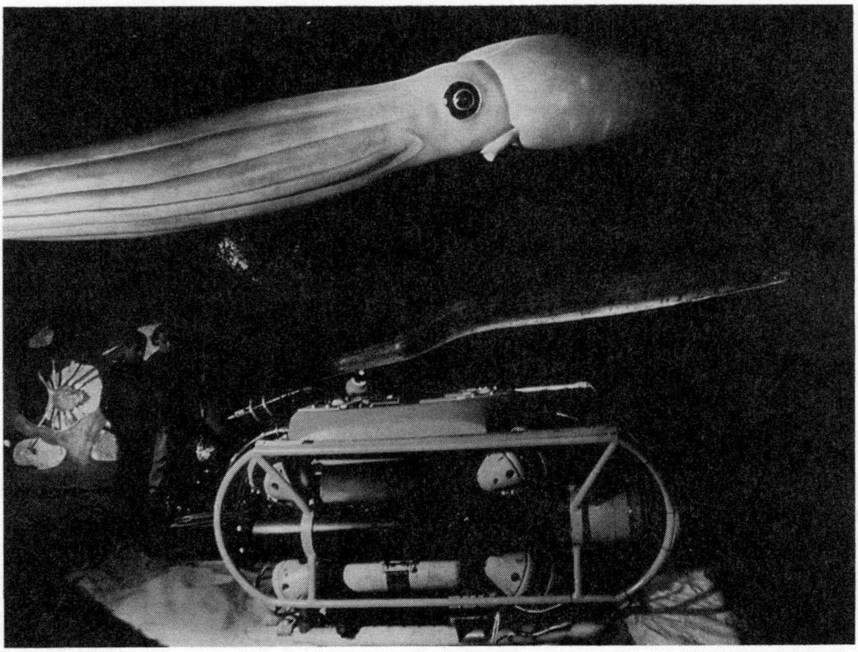

Plate 18 'Monsters of the Deep' – a Yorkshire Museum Travelling Exhibition. (Reproduced courtesy of The Yorkshire Museum.)

contemplate where humankind is going? Or is it simply that the need for increased visitor figures demands 'nice' exhibitions – after all, museums cannot afford to frighten people away can they? The Museon's Ecos demonstrates that it is not necessary to frighten people to relay a powerful message, and for this reason deserves more recognition for the significant advance it has made in our approach to discussing the environment.

If museums have noticeably failed to confront global issues, is it possible to find exhibitions which have successfully broached local environmental issues? Such matters are always in the news, whether it be bait-digging in a local nature reserve, a plan for opencast coal extraction or the erection of a wind farm in a site of high landscape value. Local museums may have an advantage, (being small and self-sufficient they can react quickly), but they also need to be politically aware – the good name of the authority may be at stake. Most environmental issues, even at a local level, are very complex, and it is important that the multifaceted arguments are presented clearly – to present only one side of a complex issue would be sadly misleading. A good example is the brave attempt made by the Pratt Museum in Homer, Alaska, to explain to the public the causes and consequences of the *Exxon Valdez* oil spill in the exhibition 'Darkened Waters – Profile of an Oil Spill' (Garfield, 1992). Emphasis is not on wildlife and ecology, but on local people – who give their view of the incident on tape recordings; links are made to the consumer society and the importance of oil to the US economy – the ultimate driving force behind a major environmental disaster. The exhibition does not lay blame at any one door, but leaves visitors to decide for themselves, having taken notice of the differing points of view of biologists, fishermen and oilmen. A similar approach was adopted when the California Academy of Sciences in San Francisco examined the human impact on fragile desert ecosystems by mining, agriculture, military use and recreation. 'The exhibit [Vanishing Desert] opened on Saturday. By Monday controversy raged. The American Motorcycle Association and others objected strongly and issued a list of demands' (Kulik, 1992). The vociferous biking community kept up the pressure for several months, demanding changes be made to the exhibition – the Academy refused 'Because the changes requested would have compromised scientific accuracy' and also due to the support received from the conservation lobby, who were well aware of the ecological damage that trail bikes can cause to fragile desert plants and burrowing animals. It was largely due to the careful research carried out prior to the exhibition that a balanced viewpoint had been given, and could be defended – by staff, directors and trustees, who had all been involved at an early stage in exhibition planning. Like the *Exxon Valdez* exhibition, Vanishing Desert had created controversy because it was close to home, unlike distant, vanishing rainforest, the plight of the desert was accessible and understandable.

In Britain one of the few examples of a local museum tackling a controversial environmental issue was Woodspring Museum's (1981) Severn Barrage exhibition (Coles, 1991b). A topic so politically controversial that it was ignored by all the larger museums in south-west England, the proposal to place a barrage across the estuary of the River Severn generated strong emotions. The local authority and Civic Society were against the barrage, local traders and those in the construction industry were for it. As it was perceived as a viable form of renewable energy, at the national level Friends of the Earth supported the barrage, the local branch of the same organization was against it. In this instance, what is the correct stance to present to the museum visitor? To support the barrage as a commitment to the wise use of a continuous source of power, or to condemn it as a project which would destroy a major ecological system? In effect Woodspring Museum could only sit on the fence (or the barrage), presenting all the arguments and leaving the visitor to make the decision – but at least they had the courage to tackle a major issue when all around them were ignoring it. The experience was regarded as worthwhile:

> Exhibitions of this nature benefit the museum by thrusting it into the public eye and showing it to be playing a contemporary role in green issues. They are not without pain however; the sheer number of organisations which had to be consulted to get hold of the facts was daunting; so too was sifting the facts from the rhetoric of the campaigners on either side (Coles, 1991b).

This echoes the experience of the Yellowknife (North West Territories) Museum when examining the controversial issues inherent in trapping and the fur trade in the exhibition 'Trapline, Lifeline' (Irving and Harper, 1988).

It is impossible to leave the subject of controversial exhibitions without referring to 'Seeds of Change; Five Hundred Years of Encounter and Exchange', the Smithsonian Institution's exhibition which celebrated the arrival of Columbus in the 'New World'. Accompanied by a lavish publication (Viola and Margolis, 1991) this exhibition created controversy on a considerable scale, primarily for the major omissions rather than its content. The five seeds or agents selected for the exhibition were sugar, corn, potato, the horse and disease – the latter, in addition to persecution, leading to the death of 90 per cent of the American Indian population. Not surprisingly, the feeling of the survivors' descendants was that little good resulted from the encounter, and many 'celebrated' the quincentennial by wearing black armbands. Should Columbus have been celebrated at all? The American Indian reaction to the exhibition was mixed; the black community also felt that slavery issues inherent to the introduction of sugar-cane had been carefully scripted. Perhaps most surprising, the story of tobacco had been omitted altogether – in its cultivation and impact on world health it has been one of the greatest forces for change. It was however, a brave exhibition – according to Sewell (1991):

> 'Seeds of Change' is about change – its sometimes purposeful but often random nature. In dealing with the actual results and subsequent meanings of 1492, the exhibition runs head into many controversial issues. The impact of the five seeds has been made not only on the land, but also on the people. Not everyone will agree on the extent of this impact, but the exhibition and its programs will allow a forum for discussion.

Active protest occurred at another Columbus exhibition – 'First Encounters' a travelling exhibition organized by the Florida Museum of Natural History. Demonstrations by students were followed by threats of legal action to close the exhibition – despite efforts on behalf of museum staff to address the negative side of the arrival of European culture (Milbraith and Milanich, 1991). Protests centred on the accusation that the blame for the decline in native populations had been shifted from Europeans to disease factors, despite the archaeological evidence for this. During its tour the exhibition has continued to arouse controversy, with some museums revelling in this, or deflecting criticism by producing additional information (reaction papers in Albuquerque for example), or holding special conferences so that views could be aired. It is refreshing to see that despite the difficulties with this exhibition Milbraith and Milanich (ibid.) state:

> Are we confined then to dull, middle of the road consensus in our presentation of exhibits – so that we can avoid controversy? Not in the least. In fact, we should welcome some controversy to make the public more sensitive to the issues and ideas presented.

Beyond the natural history museum

To be really effective, environmental exhibits need to put the focus on people. As a consequence, the best exhibitions in natural history museums have focused on the interrelationship between humankind and the natural world. Until recently, other museums, especially those devoted to history, and art galleries, have failed to capitalize on the upsurge in interest in the environment, or to begin to promote environmentalism through their displays. Yet these institutions could play a major educational role – as Martinovitch (1990) notes:

> Our failure to recognise the relevance of (say) a nineteenth century historic house to environmental conditions in the late twentieth century is symptomatic of many of our current problems; we fail to see the connections between events distant in time and space, and instead put human activities into separate, seemingly unconnected boxes.

He further elaborates the possibilities within an historic house – the web of global interconnections demonstrated through the types of woods used for furniture, the tropical agricultural products which graced the Victorian table, even the houseplants, for example. How were energy needs met? How did the local environment provide for the family's needs – and was their 'ecological footprint' less damaging than our own? How did Victorian 'recycling' schemes work?

Nilsson and Rosen (1988) have also examined the environmental role of 'cultural museums' in Sweden, noting that 'it is quite clear that cultural history cannot be properly understood without some knowledge of the preconditions and framework afforded to us by nature', and that 'Cultural history museums have not traditionally reflected the ecological aspects of our reality'. They correctly point out that the phenomenon of 'natural' disasters, including loss of ozone, desertification and flooding, are rooted in human activity; as a result historical museums, with their multidisciplinary and integrated approach, have a clear opportunity to educate visitors about human ecology and the exploitation of resources. Sweden's National Council for Cultural Affairs has supported this new role for museums, one which demands considerable research as well as a new orientation for displays. Under the project title 'The Spread of Ecological Knowledge' a major effort is being made to integrate an ecological approach within the framework of cultural museums and art galleries. This is 'ecosystem' thinking being put into practice, setting all museum activities (exhibitions, research, educational activities) within a larger framework bounded by the natural environment and humanity's influence on it. It should not be confused with the ecomuseum concept – all these proposed changes take place within the museum rather than being set in a wide geographical framework.

The potential role of art galleries was recognized by Pankhurst (1971), who felt that:

> Art Museums must point to and develop awareness of the broader ecological significance of all artistic endeavor. Without detracting from their focus on art as art, they can also stimulate day-to-day awareness of the aesthetic rewards or their deplorable absence, in each person's environment.

There is so much that art galleries could do in environmental education – from the relatively straightforward comparison of the landscape depicted in paintings of earlier centuries to that of today, or demonstrating the more environmentally sensitive nature of land management through the detailed observation of nineteenth-century rural scene – or using a smog-ridden or smoky industrial scene to depict the ways in which humankind had begun to affect the environment through the growth of industrial processes of the Victorian era – and also reflect on technological advances which have resulted in a return to cleaner air. A slight change to labels on selected paintings, perhaps accompanied by a small booklet, could take visitors on an environmental tour of an art gallery, revealing new meanings and truths hidden within the gallery. Such suggestions have been explored by Matilsky (1992):

> Exhibitions on the environment convey new insights and ultimately demonstrate the influential role that art and museums can play in contemporary society. Museums that broaden their mission to include these activities assume a powerful responsibility that potentially reverberates beyond the museum walls.

But – how many art galleries have taken up the challenge thrown down by environmentalism? Matilsky (ibid.) refers to 'Reverend Earth', an exhibition organized by the Centre for Contemporary Arts of Santa Fe, and 'Fragile Ecologies' a travelling exhibition from the Queens Museum of Art. Hall (1991) notes the Regina, Canada, Art Gallery exhibition 'Art for a Threatened Environment', the development of a new Canadian Wildlife and Wilderness Art Museum in Ottawa and the work of McMaster University Art Gallery, Hamilton in promoting environmental awareness.

Artists themselves have always found challenge and inspiration in wildlife and wilderness – artists have provided for us a magnificent record of past environments – and art can speak for the environment. For many artists in the 1960s and 1970s the environment became their art (Richard Long) or became an integral part of it (Smithson, Christo) – this relationship has been explored by Dery (1991), and by Matilsky (1992) who makes reference to 'ecological art' which emerged from the environmental art movement of the late 1960s and early 1970s – Patricia Johanson's *Leonhardt Lagoon* (1981–1986) in Dallas, Texas is cited as one of the most significant works to date. The lagoon is part of the artwork, which is seen as a combination of wetland habitat and organic, concrete sculpture.

Perhaps more accessible to most people is the work of wildlife artists, who continue to produce works which reflect their own interests in the natural world; often based on their own experiences these paintings and sculpture can have great strength and emotion. The celebrated wildlife artist George McLean is quick however to point out that '... my paintings are not about conservation. They are paintings for the sake of painting' (quoted in Barr, 1990). Despite this assertion, there is much truth in Barr's (ibid.) comment that the work of the best wildlife artists 'reaches our emotions and reminds us that we have a bond with these wild creatures. Wildlife artists can move us to decision and to action.' What of contemporary art which focuses on wilderness? Kernan (1990) has suggested that if we wish to establish what the next generation feels about the environment, we should look at contemporary works of art, and look to the artists ('... the antennae of the human race') for a lead. The inspiration of landscape is revealed in abstract work by (for example) Tracy Grayson, Neil Jenney, April Gornick or Joan Nelson, which unlike earlier American landscape painting avoids sentimentality and illusion. The New York art critic Suzanne Boettger remarks that 'They don't want to pretend you can really go back to that idyllic world. They're saying "How wonderful it is ... and you can't have it because you're ruining it" ... it's melancholic for the state of the world in general' (quoted in Kernan, 1990). Exhibitions of so-called 'new landscape art' dwell on the rapidly changing nature of the environment, and themselves are potent conveyors of environmental messages – though one suspects that the artists would probably deny any such meaning or objective in their art.

Involvement of all specialisms is a necessity, yet a situation which is common to many multipurpose museums is that of the fragmentation of disciplines, with 'art' often being particularly remote. Curators heroically defend their boundaries, and territoriality rules in stores and exhibitions alike. Whilst the former may be understandable, the latter is not – as Stansfield (1972) rightly argued, environmental messages are most easily propagated by breaking down traditional boundaries and by formulating new exhibition policies which guarantee co-operation. Thus far, examples are few, though the recent co-operation in Fresno, California between art gallery, museum and zoological garden to explore the theme of Rainforests (O'Donnell, 1995) is commendable. With such an approach the menu of opportunities from which to choose environmental messages is clearly substantial. It is unfortunate that as yet (with some exceptions) community, historical, art and even multidisciplinary museums have failed to take up the challenge.

The role of science centres in environmental education

In North America and the UK, raising environmental awareness has been adopted as a major role for science centres, and arguably issues such as pollution, recycling, the greenhouse effect and the loss of the ozone layer are best dealt with in such a learning environment. Such 'centres' can be museums which have adopted a particular interactive learning style but also curate major collections, or they can be purpose-built institutions without collections; alternatively, many museums have set up discovery centres in selected rooms within the original building. In essence it is the interactive nature of 'centres' which sets them apart – the opportunity for 'hands-on' experiences. As the methods pioneered in such areas have gradually been incorporated into museum exhibitions, there has been a gradual blurring of the boundary between museum and science centre, and collaboration has aided this change (Fowles, 1991).

Some centres seek to explore science in general, others, such as the Natural History Centre at Liverpool Museum (Greenwood *et al.*, 1989) and the California Academy of Sciences Discovery Room (Diamond *et al.*, 1988) confine themselves to the natural environment; the Chesapeake Bay Discovery Lab (White *et al.*, 1992) deals only with the marine environment and the Buxton Micrarium (Carter, 1987) concentrates on microscopic exploration of the natural world. These examples demonstrate the enormous variety of subject, size and mode of operation – what they have in common is that they are all exciting places to visit. Imagine for example a visit to the Union Terminal in Cincinnati, the last of the great municipal railway stations in the United States. Built in an extraordinary Art Deco style, it houses everything from an Omnimax theatre, lecture rooms, extensive library and a variety of restaurants as well as the attractions of a new natural history museum and Children's Discovery Centre. In this centre emphasis is on interactive displays, with demonstrators on hand to help and guide visitors through two major areas, one devoted to the human body, the other to man's environmental impact through time. Called 'Pathways to Change' it:

> explores the need for environmental ethics when these changes occur; it is the story of humans and our region through the past, present and future. Visitors are transported through time by means of light and sound time tunnels and an interactive video designed specially for the exhibit . . . . The final time tunnel brings visitors to 1990. A cutaway of a contemporary house explains heating and cooling, insulation and construction of modern dwellings, and focuses on shrinking resources and concepts of environmental conservation. A wrap-up area deals with balancing nature and human needs, and how technology can help us understand and predict environmental changes . . . . Visitors are given lists of ways in which they can help nature get back in balance. The pathway is complete, but the changes visitors can make are still in the future' (Anon, 1990).

Like Cincinnati, many of the purpose-built science centres have capitalized on the interest in environmental issues. New exhibits have been organized, or old ones modified, for example the New England Science Center has set up a GIS system as part of its Infospheres exhibition, which enables visitors to visualize changes in forest cover, rainfall or ozone levels; the Saskatchewan Science Center has opened a new display 'The Living Planet', and a number of Science Centres have collaborated to produce programmes which address the issue of global warming (Anon, 1993a). Virtually every recent issue of the *Association of Science-Technology Centers Newsletter* reports the success of a new environmental venture, ranging from new exhibitions such as 'Greenhouse Earth' (Anon, 1992) to research on the role of centres in the promotion of environmental literacy (Anon, 1993b). New technology, especially Internet links, are central to these developments in science museums (see e.g. Baumann, 1995). Many of the world's major natural history museums have recognized the value of discovery learning by installing permanent facilities – the Natural History Museum, London for example; a major feature of the revitalized Paris Museum of Natural History is 'une salle de découverte' (Guichard and Leclercq, 1994).

There can be little doubt that from the mid-1980s science museums and science centres have demonstrated their role in environmental education; by working together, science centres and natural history museums have the capability fully to inform visitors about environmental issues. However, it is essential that these 'scientific' messages are put within the correct cultural context if they are to be successful, and affect the behaviour of the museum visitor.

Can environmental exhibits change attitudes?

The zoologist Archie Carr, in *The Windward Road* (1979) wrote:

> I have always liked frogs. I liked them before I ever took up zoology as a profession; and nothing I have had to learn about them since has marred the attachment. I like the look of frogs, and their outlook, and especially the way they get together in wet places on warm nights and sing about sex.

This caring attitude towards amphibia may seem a little strange to most people, yet it is exactly this kind of concern that natural history museums need to encourage – not simply acquiring knowledge of wildlife but in addition, an emotional commitment. The most important goal in devising and executing museum exhibitions on environmental themes is to foster caring attitudes to wildlife and the environment. By exposing the public to the beautiful, wonderful and strange creatures which inhabit the earth, and explaining their morphology, behaviour or ecology, museums try to promote a better understanding, thereby encouraging respect. By focusing on the threats to ecosystems or particular species we can extend that understanding and respect into real concern and perhaps even positive action. This might be something as simple as informing friends of what they had seen, passing on their impressions and concerns, or perhaps joining an active local or national conservation organization.

However, museums must be cautious in their approach. Radical ecological movements have been accused of proliferating misinformation, which is often further accentuated by the media. Museums cannot tie their colours to particular causes, no matter how tempting it might be. There may not even be a sound scientific viewpoint – science is not always certain of the truth, and if science sits on the fence, then so must museums. This is not to say that we should leave our visitors wondering what to do next; having heightened their perceptions of the natural world we should encourage them to act. As Sullivan (1992) eloquently expresses:

> we must . . . avoid leaving visitors confused and depressed, impotent to effect any meaningful change. They must leave with a sense of hope, but also a sense of urgency to replace their complacency . . . we must get engaged in talking about options that people have and decisions they must make personally and as communities.

This might be our goal – but do museums really achieve such behavioural changes in our visitors? Do they leave the museum more aware of environmental issues, or do they simply remember one spectacular exhibit? Do they take away with them any environmental messages at all? Unfortunately few detailed evaluations of environmental exhibits are available to enable us to answer these questions. Those that have been performed rarely try to discover changes in visitor attitude to animals, plants or to the environment in general. It is something of an indictment that evaluation of exhibitions in the UK is rarely carried out except in the most cursory manner, success is often measured only by the number of visitors. It follows that the impact of individual displays or features in exhibitions is rarely assessed other than in a subjective fashion – for example we know that live animal exhibits or computer interactives act as honeypots – but how effective are they in conveying environmental messages? Many museum professionals make the assumption that their exhibition is simply so exciting that no visitor can fail to be persuaded that nature and environment are important, leaving as a more fulfilled and knowledgeable individual, ready to join the environmental crusade. But are these intuitive feelings really enough – and more significant, are they accurate?

Research on visitor behaviour in museums has been a growth industry in recent years (e.g. Csikszentmihalyi and Hermanson, 1995; Falk, 1991; Naqvy *et al.* 1991; Patterson and Bitgood, 1988), and certain guiding principles have been established that can help curators and designers produce more effective exhibits. For example, we know that family behaviour is orderly and predictable, that the public devotes a disproportionately large amount of time to the museum setting, that large, interactive or moving exhibits, or exhibits viewed early during a visit evoke a more concentrated behaviour, that there is a general fourfold pattern of orientation, intensive looking, exhibit cruising and finally, leave-taking. However, more research remains to be done to link these principles to the visitor's 'learning' performance.

It is important to attempt to define what sort of educational impact museum exhibitions are trying to provide – this 'may not be simply factual knowledge, but broad overviews, concepts, the stimulation of interests, new ways of thinking about the world' (Screven, 1976). Miles and Tout (1992) make a clear distinction between learning ('the addition of knowledge – of things, including abstract ideas, and of processes – to the memory') and understanding ('akin to what Bruner (1978) calls "grasping the structure of a subject ... in a way that permits many other things to be related to it meaningfully" '). They argue that this distinction is an important one, in that the empirical evidence to support museums as a learning experience is mixed, even contradictory (see, for example, Shettel, 1968; Screven, 1974; Borun, 1977), whilst there is a positive claim to be made for exhibitions as promoters of public understanding of scientific concepts. This subtle distinction between learning and understanding does not appear to have been made in the majority of evaluations made in natural history museums.

Some doubts have been expressed about the museum exhibition as a learning experience – for example, Falk (1988), whilst acknowledging that museum learning has been the focus of numerous research studies, notes that 'millions of visitors, hundreds of studies, and countless anecdotes later, little consensus exists on what museum visitors learn, or even by some accounts, whether they learn anything at all'. Falk's own research on what individuals recall from a museum visit seems to suggest that they remember the context of a visit, the time they spent in the museum, some features of the architecture of the building and its ambience, but recall of specific exhibits was limited to subjects they had been exposed to in other contexts – at school or elsewhere. In some respects this is disheartening. Are the exhibits that curators and designers work on with such dedication, and evaluate so carefully, so easily forgotten? On the other hand, if museums, through their exhibits, are reinforcing other learning experiences – and indeed promoting active learning by other means within the museum, – then all is not lost.

The Natural History Museum of Anniston, Alabama, has been used as a laboratory by the Psychology Institute of Jacksonville State University for some years and has benefited from the findings of the external evaluators, obtaining useful data relating to visitor profiles, exhibition design and patterns of visitor movement, and the holding power of individual exhibits (Pierce, 1989). The data gained has been used to good effect to change the stereotyped image of the museum ('adult-oriented', 'bland', 'formal', 'complicated') (Bitgood and Thompson, 1987) and alter visitors' perceptions. Ford and Bitgood (1992) have carried these early evaluative researches further by exploring the effectiveness of exhibits in this museum, especially in relation to visitors' beliefs about animals. The major finding was that the impact of a current visit was influenced by gender and previous history of visitation – the idea that learning is influenced by holding power of the exhibit, and previous knowledge or familiarity is also supported by Koran *et al.* (1989). Perhaps more revealing is the research on visitor attitudes to animals following visits to zoological gardens, (Bitgood, 1992) when a degree of 'positive' change appears to be influenced by factors which include age, gender, education, group composition, history of visitation, pre-knowledge, attitudes, skills, expectation and cognitive strategies. Focusing on a number of animal exhibits Bitgood posed questions before and after the visit relating to the perception of how attractive or dangerous a species was, and asked if it was worth conserving. For those visitors with little self-knowledge about animals, and who rarely read about them, Bitgood concluded that 'zoos may have a favourable impact on the audience that needs the most persuasion'. Interestingly, and perhaps surprisingly, the most knowledgeable visitors showed a sharp decrease in positive ratings. Bitgood suggests this is a short-term effect, and could be equated with the phenomenon of museum fatigue described by Robinson (1928). However, there is an important message here for all curators in all fields, that a knowledgeable visitor is likely to get bored, and will undoubtedly be the most critical.

The issues inherent in environmental exhibitions are frequently complex, and nowhere more so than in those relating to tropical rainforest, where issues such as biodiversity, deforestation, resource management and competing multicultural demands need to be addressed. 'Tropical Rainforest: A Disappearing Treasure' was organized by the Smithsonian Institution Travelling Exhibition Service in 1988; it was accompanied by a well-illustrated book *People of the Tropical Rainforest* (Denslow and Padoch, 1988). The exhibition did not dwell on the depressing facts and figures regarding loss of acreage or species, neither did it focus on the soils, plants and animals which make the rainforest special; it concentrated on people. As Robert Adams, Smithsonian Secretary comments in his Foreword:

> Central to the tropical forest issue are people – those who have lived there traditionally, those who see the uninhabited territory as the sole hope of land

> ownership they have never experienced, and those in the cities who make the decisions regarding its resources.

Neither book nor exhibition offer a solution to the 'rainforest problem', but they do help to explain the complex nature of the forest and people's use of it; the way in which tribal cultures and smallholders alike are threatened by the pressure of development and repeated incursions from outside; the consequences of government policies and economic realities on the future of the forest.

Doering (1990) carried out an evaluation of this travelling exhibition in order to discover if it effectively explained issues, changed perceptions and increased understanding. Selecting different types of people (visitors and non-visitors to the exhibition; those with no familiarity with rainforest issues through other media and others with prior knowledge), Doering sought to discover visitor perceptions about rainforest appearance, future prognosis, the impact of deforestation, causes of and solutions to deforestation, and the role of the individual in resolving the problem. The responses of exhibition visitors with no previous knowledge of rainforest issues was especially revealing; rainforests were seen as places of great beauty and importance, habitats which required protection and preservation – a testament to the powerful communicative strength of the exhibition. Media-aware visitors found their attitudes and opinions reinforced; these more knowledgeable visitors saw the complexity of the situation, its political dimension, and that rainforest loss affected them directly. Doering concluded that:

> The results of the Smithsonian study begin to suggest that exhibitions treating environmental issues can achieve educational goals by providing information that reinforces an individual's prior exposure or commitment to a topic. More importantly, such exhibitions appear to provide a valuable introduction for the visitor who happens to visit them accidentally and view them without any prior information.

Environmental education in museums

Exhibitions are usually only the starting-point for conveying messages to the public and to schoolchildren – the communication role of natural history museums extends into the realms of publication, lectures, guided tours, and activities for the general public and educational activities which are geared to the needs of formal education. Within the confines of prescribed curricula, museums have to address the needs of a particular age group (in Britain between the ages of 5 and sixteen), and react to those parts of the curriculum they are best equipped to support. In Britain the core subject of 'science' forms the basis for much of the formal education activities within the natural history museum, although opportunities can be found within subjects other than science, and as part of cross-curricular themes. In Britain the advent of the National Curriculum has had far-reaching implications for museums, but it has not (as first thought) put an end to school visits, but has required museum educators to encourage teachers to use museums and their collections in new, inventive ways (Goodhew, 1994). Environmental education is no longer a major feature of the National Curriculum in Britain, having been relegated to a cross-curricular theme which consists of three components: education about the environment (knowledge), for the environment (attitudes/values/actions) and in or through the environment (a resource). Suggested topics include loss of habitat and the exploitation of natural resources (National Curriculum Council, 1990). Museum visits continue to be popular with children, and the creativity of museum educators has ensured that natural history collections are fully exploited by teachers across a broad sweep of the curriculum.

It is outside the confines of the school curriculum that environmental education (perhaps more correctly environmental interpretation) has flourished. Inspired by exhibitions which promote the conservation ethic, museum educators have responded to the new challenges posed by the environmental debate, itself an umbrella for a variety of experimental learning patterns which include aspects of conservation, resource use and sustainability, even human rights and population education. For most natural history museums, environmental education is a core activity, one which has been endorsed by ICOM (Whiting, 1979), and for which the techniques have been well documented (e.g. Stansfield, 1994). The audience is not just children (though they are always well catered for): environmental programmes are for everyone. Some recent programmes from around the world give some idea of the range of activities – free workshops for science teachers at the Academy of Natural Sciences, Philadelphia on subjects as diverse as population education, the sun as source, people and rainforests; Business Seminar on Paper Recycling at the Baltimore Aquarium; Drama by Red Earth at the Natural History Museum, London exploring the relationship between humans and the environment; a lecture in Denver Museum of Natural History by Robert Ballard on marine conservation as part of a series of environmental lectures. The variety of media, and the number of events is quite remarkable, but the common aim is to increase environmental awareness throughout society.

If museums have a remit to promote environmental awareness by providing an educational experience which is 'for the heart as much as the head, as attitudes and values are at the core of our dealings with our surroundings and fellow men' (Opie, 1993), are they addressing the key issues? Is it sufficient to continue to open the public's eyes to the wonders of nature, or should we be adopting a more radical approach? Should we be looking not at 'the coral reef' or 'hedgehogs' but as issues which directly involve humankind? Opie (1993) has re-evaluated the concept of environmental education, stressing the significance of the following elements: a sense of place and time; personal exploration and inquiry; holistic thinking; global thinking for local action; personal values and citizenship. His list of objectives for environmental education, under the banners of life, resources, life skills and personal values relate to these key elements, and are aimed at creating an understanding about the environment in its broadest sense. To achieve these objectives Opie suggests some environmental themes for museums to explore through exhibitions and interpretive activities – human ecology (including birth control and family planning), famine and the displacement of populations, poverty and environment blight, consumerism, competition for resources, urbanization, soil erosion, sustainable development and the management of natural resources – it is argued that unless museums begin to tackle these difficult themes, they will remain 'the places where they store yesterday'.

Outreach – involving the community

Museums need to reach into their community in order to be most effective; for this reason museums have taken their environmental educational activities beyond the museum itself, through travelling exhibitions, outreach projects which involve particular (and sometimes forgotten) sectors of the community, by promoting volunteer programmes, involving local experts, and by developing apprenticeship programmes, co-operative ventures with other organizations, and action programmes which address particular environmental needs.

Travelling exhibitions are a long-established phenomenon; what is now evident is the increasing number of exhibitions which deal with environmental themes. Some recent examples include Henry Lowe Associates 'Great Rubbish Show' first shown at the Livesey Museum, London, described as an exploration of ' ... how we view and deal with rubbish of the past, present and future ... what we make of our waste today has become a world-wide political and social question.' (Lowe, n.d.). This exhibit has been successfully touring since 1991, only matched by the Natural History Museum's Travelling Discovery Centre, the popularity of which can be measured by the 12-month waiting list. This scaled-down version of the Discovery Centre in the basement of the NHM was initiated in 1990; the interactive learning approach has proved to be remarkably successful. Another approach by larger museums has been to provide user kits on a variety of environmental issues, (see Westerlund, 1991 for examples from Sweden), or a special lecture/activity service, for example the 'Eco-show on the Road', a venture promoted by the Academy of Natural Sciences, Philadelphia, ' ... travels to schools, community organizations, and private groups on weekdays, weekends and evenings. A Teacher/Naturalist, accompanied by a carload of live animals, slides and museum props, brings you an unforgettable program' (Academy of Natural Sciences, n.d.). The Eco-show provides a variety of different subject options (including rainforests and endangered species) and formats, from 'auditorium programmes' using slides and live animals, to individual school visits and even summer camps. Such activities carried out by museum professionals, whether within a prescribed format such as the 'Eco-Show' or through more informal talks and activities, has always been one of the key ways of promoting conservation. Virtually every museum promotional leaflet features some 'environmental' activity carried out beyond the confines of the museum – for example the Cincinnati Museum of Natural History brochure for 1993/4 featured weekends exploring the wildlife of Appalachia, explorations of cave systems and local rivers, and opportunities to travel further afield (Africa, Hawaii and China) with the 'Museum Auxiliary', as well as a host of museum-based activities with environmental themes.

One of the major ways in which North American museums have reached the community is through their activities with volunteers – a situation which is remarkably lacking in Europe. The scale of the enterprise is often extraordinary – for example volunteers at the Academy of Natural Sciences, Philadelphia broke all records in 1992, contributing over 25,500 hours – 531 people volunteered in total, with a large proportion of these attached to the Education Department (Thomson, 1992). Much of the success of the environmental educational activities in North American museums can be attributed to volunteers and their trainer/organizers – by putting real effort into providing for the training needs of volunteers, and by providing a certain status, they have tapped a rich resource; as a result the volunteers absorb environmental messages, which are in turn relayed to the museum visitor. 'Lab Rats' (Learning about Biodiversity, Research and Teaching Science) are the Cincinnati Museum of Natural History's main link with the public, volunteers trained to be exhibit interpreters, they are recognized as 'the official hosts for the museum' (Cincinnati Museum of Natural History, n.d.). Training is rigorous, continuous and demanding, and a Lab Rat is expected to commit a minimum of 185 hours throughout the year; the promotional literature emphasizes the benefits of becoming involved, including the development of communication skills, discovering career opportunities, building experience which can be added to a resume (CV), and interestingly, being able to 'Understand first-hand that YOU can make a difference.' It is noticeable that the majority of volunteers are male, white, middle class and well educated, and as a consequence the Women Into Natural Sciences (WINS) programme in the Academy of Natural Sciences, Philadelphia is particularly laudable. Aimed at disadvantaged young women enrolled in Philadelphia Public Schools, it provides 33 places annually for individuals to experience the working environment of the Academy, to learn about wildlife and environmental issues, and participate in the work of the educational and scientific teams. It is obviously difficult to assess the educational impact of such an experience – it is hard to imagine that taking new knowledge about wildlife and environment back into poor urban communities would reach a receptive audience in the light of more immediate and pressing social concerns. Yet an indication of success is the number of participants who return to the Academy long after their WINS year has ended, many of them returning as volunteers within the education section (Damstadter, pers. comm.).

Sharing resources and working in co-operation with bodies with similar aims (zoos, botanic gardens, urban nature reserves, statutory and voluntary conservation organizations) is also a significant form of outreach for museums, in order effectively to promote environmental education. Most museums in developed countries have extensive networks of contacts in (for example) other museums, national parks, nature reserves, and with the local scientific and naturalist community, which enable a two-way flow of information and personnel exchange which helps to achieve educational objectives. In Britain, the local museum is often a focus for such co-operative ventures, it is meeting place and information exchange. It is exactly this 'professional' networking which is increasingly recognized as a way forward in developing countries – for example in India (National Museum of Natural History, 1986) – where the few resources available must be shared in order to promote environmental awareness.

Indian museums have firmly accepted the challenges of environmentalism, promoting environmental education (Goswamy, 1995; Venugopal, 1995; Bhaumik, 1995), linking concepts of ecology to social organization and cultural integration (Som and Som, 1995), and even becoming involved in practical socio-environmental projects (Bagchi, 1995). The latter is a remarkable development in Puralia, West Bengal, an area which has been devastated by deforestation. The District Science Centre became the catalyst for major environmental and social improvement, providing resources and expertise to make wells, plant new crops and develop a new weaving economy using the leaves of palm, sisal and sal, as well as promoting health care through travelling exhibitions – on malaria for example. *Bagchi* (ibid.) records that 'The Science Centre stopped the development programme in the end of 1987 . . . in 1992 the Kheria Sabars sold articles made from wild date palm leaves worth more than Rs. 80,000 . . . Today [they] take their own decisions.' He ends his account of this remarkable museum experiment by suggesting that a change is required to the ICOM definition of the role of museums, that in addition to collection, preservation, documentation, research and education museums should be involved in 'transformation'. In the face of growing environmental and societal problems, the museum role in promoting change and implementing practical solutions is becoming increasingly significant; Ben Okri (1992), in *The Famished Road*, remarks that his spirit people, who from time to time must come and inhabit the Earth, fear the experience. He writes:

> There was not one amongst us who looked forward to being born. We disliked the rigours of existence . . . and the amazing indifference of the living in the midst of the simple beauties of the Universe. We feared the heartlessness of human beings, all of whom are born blind, few of whom ever learn to see.

The success of this project in India must give all museum professionals encouragement and hope for the future; museums, as part of the environmental movement, have demonstrated that not only can they reveal the wonders of the natural environment, but also encourage positive conservation action through community involvement.

References

Academy of Natural Sciences (n.d.) *Eco-Show on the Road, 1992–93*, Promotional leaflet, Academy of Natural Sciences, Philadelphia.

Adoutte A. (1994) 'L'evolution à la grande Galerie du Muséum national', *La Lettre de l'OCIM*, 33, 5–9.

Alberch, P. (1994) 'The identity crisis of natural history museums at the end of the twentieth century'. In Miles, R. and Zavala, L. (eds) *Towards the Museum of the Future; New European Perspectives*, Routledge, London and New York, 193–198.

Anon. (1953) *Deadline for Wildlife – The Story of Wildlife Conservation in Pennsylvania*, [Exhibition Catalogue], Carnegie Museum, Pittsburgh.

Anon. (1990) *The Museum Center at Union Terminal*, Supplement to *Cincinnati Magazine*, 24 (2).
Anon. (1992) 'Greenhouse Earth opens at Franklin Institute', *ASTC Newsletter*, 20 (2), 1–2.
Anon. (1993) 'Messages from the wilderness', *Museum News*, 72 (1), 19.
Anon. (1993a) 'Environment-related topics popular', *ASTC Newsletter*, 21 (1), 9.
Anon. (1993b) 'Science museums grapple with improving environmental literacy', *ASTC Newsletter*, 21 (3), 1–3.
Anthony, H. (1942) 'A Grand Tour of North America', *Natural History*, April, 190.
Bagchi, S. K. (1995) 'Museum and Community'. In *Museums for Integration in a Multicultural Society. Proceedings of the Annual Conference of the International Committee for Education and Cultural Action (CECA) of the International Council of Museums, Delhi–Calcutta, India, November 1993*, ICOM Asia–Pacific, Calcutta, 55–56.
Barr, D. (1990) 'Wildlife art: communicating an essential vision', *Rotunda*, 23 (3), 34–40.
Baumann S. (1995) 'CyberMuse: Science Museums on the Internet', *Museum News*, 74 (3), 46–48 and 69–71.
Bhaumik, P. K. (1995) 'Science Seminar – A Step towards a National Science Carnival'. In *Museums for Integration in a Multicultural Society. Proceedings of the Annual Conference of the International Committee for Education and Cultural Action (CECA) of the International Council of Museums, Delhi–Calcutta, India, November 1993*, ICOM Asia–Pacific, Calcutta, 82–83.
Bitgood, S. (1992) 'The impact of a zoo visit on attitudes: a preliminary report on interaction effects', *Visitor Behavior*, 7 (3), 7–10.
Bitgood, S. and Thompson, D. (1987) 'How do people perceive museums, parks and zoos?', *Visitor Behavior*, 2 (3), 9–10.
Blockley, M. (1993) 'Cultivate your green appeal and win new friends', *Museum Development*, July/August, 19–22.
Borun, M. (1977) *Measuring the Immeasurable: A Pilot Study of Museum Effectiveness*, Association of Science – Technology Centers, Washington DC.
Bruner, J. S. (1978) *The Process of Education*, Harvard University Press, Cambridge, MA.
Bullock, W. (1812) *A Companion to the London Museum and Pantherion, Containing a Brief Description of Upwards of Fifteen Thousand Curiosities, Antiquities and Productions of the Fine Arts*, William and Samuel Graves, London.
Carr, A. (1979) *The Windward Road: Adventures of a Naturalist on Remote Caribbean Shores*, University Press of Florida.
Carter, S. (1987) 'Buxton Micrarium'. In Pizzey, S. (ed.) *Interactive Science and Technology Centres*, Science Projects Publishing, London, 90–99.
Cincinnati Museum of Natural History (n.d.) *Lab Rats Program*, Promotional leaflet, Cincinnati, OH.
Coleman, L. V. (1939) *The Museum in America: A Critical study*, American Association of Museums, Washington DC, 3 vols.
Coles, A. (1991a) 'Through the looking glass', *Museums Journal*, 91 (8), 20.
Coles, A. (1991b) 'Taking Centre Stage', *Museums Journal*, 91 (8), 32–33.
Csikszentmihalyi, M. and Hermanson, K. (1995) 'What makes visitors want to learn?', *Museum News*, 74 (3), 34–37 and 59–61.
Davis, P. (1978) 'A new gallery on a limited budget: 'Local Wildlife' at Sunderland Museum', *Museums Journal*, 78, 70–72.
Denslow, J. S. and Padoch, C. (1988) *People of the Tropical Rainforest*, Smithsonian Institution, Washington DC.
Dery, L. (1991) 'Triptych – museums, artist, environment', *Muse*, VIII (4), 64–66.

Diamond, J., Smith, A. and Bond, A. (1988) 'California Academy of Sciences Discovery Room', *Curator,* 31 (3), 157–166.

Doering, Z. D. (1990) 'Environmental Impact', *Museum News,* 71 (2), 50–52.

Dubois, J. (1991) 'The Manitoba Museum of Man and Nature: A Museum with an Environmental Mission', *Muse,* VIII (4), 51–55.

Du Weyden, V. (1989) 'La Terre Notre Foyer', *Musées et Collections publiques de France,* 182–183, 48–52.

Falk, J. H. (1988) 'Museum Recollections', *Visitor Studies,* 1, 60–66.

Falk, J. H. (1991) 'Analysis of the behaviour of family visitors in Natural History Museums – The National Museum of Natural History', *Curator,* 34 (1), 44–50.

Farrington, O. C. (1915) 'The Rise of Natural History Museums', *Science,* 13 August, 206.

Flower, H. W. (1898) *Essays on Museums and Other Subjects Connected With Natural History,* Macmillan, London.

Ford, W. and Bitgood, S. (1992) *The Impact of a Museum Visit on Beliefs about Animals,* Presentation to the Annual Visitor Studies Conference, St Louis, MO.

Fowles, J. (1991) 'Museums and Science Centres; Collaboration for our environment', *Muse,* VIII (4), 21–23.

Garfield, D. (1992) 'Darkened Waters: Profile of a Spill', *Museum* News, March/April, 24–26.

Gislason, J. (1994) *Our Weakening Web – Information Pack,* Cincinnati Museum of Natural History.

Goodhew, E. (1994) 'Schools and natural history museums'. In Stansfield, G., Mathias, J. and Reid, G. (eds) *Manual of Natural History Curatorship,* HMSO, London, 255–265.

Goswamy, S. 'Environment Education: Museum Role'. In *Museums for Integration in a Multicultural Society. Proceedings of the Annual Conference of the International Committee for Education and Cultural Action (CECA) of the International Council of Museums, Delhi–Calcutta, India, November 1993,* ICOM Asia–Pacific, Calcutta, 74–75.

Greenwood, E. F., Phillips, P. W. and Wallace I. D. (1989) 'The Natural History Centre at the Liverpool Museum', *The International Journal of Museum Management and Curatorship,* 8, 215–225.

Guichard, F. and Leclercq, V. (1994) 'La genèse d'une salle de découverte, un espace enfants de la grande Galerie', *La Lettre de l'OCIM,* 33, 46–53.

Hall, N. (1991) 'Practising and preaching environmentalism', *Muse,* VIII (4), 14–16.

Hancock, E. G. (1980) 'One of those dreadful combats' – a surviving display from William Bullock's London Museum', *Museums Journal,* 79 (4), 172–175.

Howard, P. (1991) 'Monsters of the Deep', *Museums Journal,* 91 (8), 34.

Hubendick, B. (1972) 'Museums and Environment'. In ICOM, *The Museum in The Service of Man: Today and Tomorrow,* Papers from the 9th General Conference of ICOM, 39–48.

Hudson, K. (1987) *Museums of Influence,* Cambridge University Press, Cambridge.

Irving, S. and Harper, L. (1988) 'Not another fur trade exhibit? An inside look at *Trapline Lifeline*', *Muse,* VI (3), 38–40.

Johnson, W. H. (1992) 'The strange case of the disappearing moose', *Museum News,* March/April, 56–58.

Kelly, J. (1991) 'Gallery of Discovery', *Museum News,* March/April 49–52.

Kernan, M. (1990) 'A new generation of artists ponder nature's prospects', *Smithsonian,* 21 (1), 104–110.

Koran, J. J., Foster, J. S. and Koran, M. L. (1989) 'The relationship among interest, attention and learning in a natural history museum', *Visitor Studies,* 2, 239–244.

Korensic, M. S. and Young, A. M. (1991) 'The Rain Forest in Milwaukee: An Evaluation', *Curator,* 34 (2), 144–160.

Kulik, L. (1992) 'Disputed territory – California's vanishing desert', *Museum News,* March/April, 53–55.

Leigh, W. R. (1938) *Frontiers of Enchantment,* Simon and Schuster, London.

Lowe, H. (n.d.) *The Great Rubbish Show,* Promotional leaflet, Henry Lowe Associates, London.

Lucas, F. A. (1921) 'Glimpses of early museums', *Natural History,* 77.

Martinovich, P. (1990) 'Interpreting the environment: Turning the past into the future', *Museum Quarterly,* 18 (4), 5–40.

Matilsky, B. (1992) 'Art and Ecology; The Museum as Activist', *Museum News,* March/April, 47–49.

McMichael, D. (1990) 'Museums and altitude change in the Pacific', *Museum,* XLII (163), 31–4.

Milbraith, S. and Milanich, J. T. (1991) 'Columbian conflict', *Museum News,* September/October, 34–37.

Miles, R. and Tout, A. (1992) 'Exhibitions and the public understanding of science'. In Durant, J. (ed.) *Museums and the Public Understanding of Science,* Science Museum/COPUS, London.

Murphy, R. C. (1937) 'Natural history exhibits and modern education', *Scientific Monthly,* July, 78.

Naqvy, A. A., Venugopal, B., Falk, J. H. and Dierking, L. D. (1991) 'Analysis of the behaviour of family visitors in Natural History Museums; The New Delhi National Museum of Natural History, *Curator,* 34 (1), 51–57.

National Curriculum Council (1990) *Environmental Education,* Circular 7, National Curriculum Council, York.

National Museum of Natural History (1986) *Interdisciplinary Approaches to Environmental Education: A Report,* National Museum of Natural History, New Delhi.

Nilsson, B. and Rosen, B. (1988) 'Cultural History Museums and human ecology – a challenge to integration', *Museum,* XL (4) [160], 213–216.

O'Donnell, S. C. (1995) 'The Rainforest in Fresno', *Museum News,* 74 (3), 21–22 and 54.

Okri, B. (1992) *The Famished Road,* Vintage, London.

Opie, F. W. J. (1993) 'Environmental education: some world-wide trends and their implications for the South African Museum Service'. In De Jong, R. *Museums and the Environment,* Southern African Museums Association, Pretoria, 96–110.

Pankhurst, C. (1971) 'Art Museums'. In American Association of Museums. *Museums and the Environment: A Handbook for Education,* Arkville Press, New York.

Patterson, D. and Bitgood, S. (1988) 'Some evolving principles of visitor behaviour', *Visitor Studies,* 1, 40–50.

Peck, R. M. (1990) 'Diorama Drama', *Academy News,* 13 (2), 8–12.

Pierce, M. (1989) 'Four years of visitor evaluation at the Anniston Museum of Natural History', *Visitor Studies,* 2, 180–191.

Raven, P. (1989) 'Platform for Ideas', *Museum News,* 68 (6), 58–61.

Robbins, M. (1969) 'An exhibit asks the question of the century', *Museum News,* September, 11–13.

Robinson, E. (1928) *The Behavior of the Museum Visitor,* New Series No 5, American Association of Museums, Washington DC.

Screven, C. G. (1974) *The Measurement and Facilitation of Learning in the Museum Environment: An Experimental Analysis,* Smithsonian Press, Washington DC.

Screven, C. G. (1976) 'Evaluating the impact of museum exhibits', *ICOM Natural History Committee Newsletter,* (Ottawa, Canada), 37–43.

Sewell, M. C. (1991) 'When two cultures collide', *Museum News,* September/ October, 38–42.

Shettel, H. (1968) *Strategies for Determining Exhibit Effectiveness,* American Institutes for Research, Pittsburgh, PA.

Simard, C. (1991) 'At the Musée de la Civilisation: Environment in the Limelight', *Muse,* VIII (4), 48–50.

Som, S. and Som, S. (1995) 'The Co-ordinates of Integration'. In *Museums for Integration in a Multicultural Society. Proceedings of the Annual Conference of the International Committee for Education and Cultural Action (CECA) of the International Council of Museums, Delhi–Calcutta, India, November 1993,* ICOM Asia–Pacific, Calcutta, 47–49.

Stansfield, G. (1972) 'Environmental education', *Museums Journal,* 72 (3), 99–100.

Stansfield, G. (1994) 'Education and Interpretation in Natural History Museums'. In Stansfield, G., Mathias, J. and Reid, G. (eds) *Manual of Natural History Curatorship,* HMSO, London, 232–242.

Stansfield, G. and Howard, P. (1994) 'Natural History Museum Exhibition'. In Stansfield, G., Mathias, J. and Reid, G. (eds) *Manual of Natural History Curatorship,* HMSO, London, 243–254.

Sullivan, M. (1992) 'Trouble in Paradigms', *Museum News,* January/February, 41–44.

Thomson, K. S. (1992) *Annual Report, 1991–1992,* The Academy of Natural Sciences, Philadelphia.

Venugopal, B. (1995) 'Nurturing young naturalists: A case study of the summer programme for teenagers of the National Museum of Natural History, New Delhi, India'. In *Museums for Integration in a Multicultural Society. Proceedings of the Annual Conference of the International Committee for Education and Cultural Action (CECA) of the International Council of Museums, Delhi–Calcutta, India, November 1993,* ICOM Asia–Pacific, Calcutta, 76–77.

Victoria and Albert Museum (1984) *The Discovery of the Lake District,* Victoria and Albert Museum, London.

Viola, H. J. and Margolis, C. (1991) *Seeds of Change: A Quincentennial Commemoration,* Smithsonian Institution Press, Washington and London.

Walker, B. (1978) 'Natural history, geology and man in a steel producing town', *Museums Journal,* 78, 57–58.

Westerlund, S. (1991) 'Article on travelling environmental exhibitions in Sweden [no title]', *Museums Journal,* 91 (8), 35.

White, M., Mast, C. and Barco, S. (1992) 'Chesapeake Bay Discovery Lab', *Curator,* 35 (4), 285–297.

Whiting, J. (1979) 'The role of museums in environmental education', *ICOM Natural History Committee Newsletter,* No 5. 7.

Winterhalder, K. (1991) 'Save the Rainforest and Explore the Rainforest', *Muse,* VIII (4), 67–68.

Wittlin, A. S. (1971) *Museums: In Search of a Usable Future,* MIT Press, Cambridge, MA.

Wonders, K. (1989) 'Exhibiting Fauna – From Spectacle to Habitat Group', *Curator,* 32 (2), 131–156.

Wonders, K. (1990) 'The illusory art of background painting in habitat dioramas', *Curator,* 33 (2), 90–118.

Young, A. M. (1989) 'The Rain Forest in Milwaukee', *Curator,* 32 (3), 229–244.

5. *From interpretive centre to ecomuseum – museums beyond the walls*

> The whole environment is one, large museum if we recognise it as such, and if we apply museological principles.
>
> Udo Kusel, National Cultural History Museum, Pretoria, 1993

> Through interpretation, understanding; through understanding, appreciation; through appreciation, protection.
>
> Freeman Tilden, 1957, *Interpreting Our Heritage*

> Ecomuseums and ecomuseology .... New incarnations of the museum? Modish neologisms? Alibis for our ability to transform an out-dated institution? None of these judgements is absolutely true, nor absolutely false either.
>
> Hugues de Varine, 1985

In Britain, one of the earliest manifestations of environmentalism was the upsurge of interest in countryside matters – agriculture, forestry, landscape and wildlife interests became the subject of scientific investigation and public concern. The provision of better access to the countryside and more recreational facilities became part of the debate as increasing numbers of people travelled to places of scenic beauty:

> In some respects, in some places, at some times, the town invasion of the countryside seems total. This ever-increasing use of the countryside for the satisfaction of urban needs for recreation and leisure occupations is viewed by many countrymen as part of an all-embracing take-over of the countryside by the townsman (Bracey, 1970).

In the face of these growing expectations, many landowners and managers, especially voluntary conservation bodies and government organizations responsible for managing areas of natural and cultural interest, began to realize that informal education and information needed to become central to their operations. This was in part to provide a service, informing visitors about wildlife and landscapes, people and rural economics, but also to influence patterns of behaviour and movement of visitors on their land.

The latter is now a key element in the management of heritage sites, and has received considerable academic attention (e.g. Harrison (ed.) 1994). To many people the concept of 'managing' semi-natural or natural landscapes may seem strange, yet in effect most of the 'wild' landscapes of Britain are the product of particular kinds of management. Some of the most valued wildlife habitats – heather moorland or species-rich haymeadows for example, only exist because they are subject to a specific, controlled management regime. As natural habitats, even in protected areas, face increasing pressures it is essential that they be managed in order to ensure that ecosystems are maintained, but also so that they are not adversely affected by recreational pressures. As part of this management, visitor behaviour is influenced in part through better publications and more efficient public relations (Gilbert, 1964), but also by providing an infrastructure in the countryside, features which most people now take for granted, the interpretive provision which includes nature trails, guided walks, activity weekends, nature centres or trailside museums designed to 'educate' the visitor about the rural environment. The belief is that 'they will not only enjoy what they see more, but will do less damage to it if they are given information about what they are seeing, when they are seeing it' (Beazley, 1969).

'The Countryside in 1970' conference, held in November 1963, encapsulated the concerns for the countryside and for conservation in the heady days of the 1960s, and education was recognized by delegates as being of special significance – pleas were made for science to go 'outside the classroom and away from dead specimens', for an enlarged Conservation Corps and for more field centres, and the observation made that 'we are not thinking only in terms of [promoting] general visual education and ... awareness .... I think we have to arouse feelings of discontent' (Longland, 1964). This hint at dissatisfaction with the situation at the time permeated the conference proceedings, and although the rival factions (conservationists such as the SPNR, National Trust and RSPB, and landowners, farmers, foresters, ramblers, field sports' enthusiasts and many others) defended corners with vigour, the demand for change and a realization of the need to inform and even arouse the public, was considerable. Government response was to pass the Countryside Act 1968, which strengthened the powers of the Countryside Commission (created in 1965 from the National Parks Commission), enabling more opportunities for informal and formal recreation through the provision of grant aid. A specific proposal was the creation of Country Parks – informal recreation areas in green spaces close to urban areas. From the late 1960s the provision of recreational facilities (footpaths, trails, car-parks, viewpoints, access areas, visitor centres) had a significant visual impact on the countryside, and also created a whole new profession of countryside recreation managers with knowledge and skills in areas as diverse as woodland management, footpath construction, landscape policy and environmental law. It also created a need for information officers and professional interpreters with abilities closely allied to those of the museum profession – people who could inform visitors about their natural and cultural heritage.

Interpreting the environment

Explaining the complex processes of nature to day-visitors or tourists on-site required new techniques and it is noticeable that 'interpretation' – a new educational methodology to explain the significance of sites and monuments – was adopted in much of the developed world during the mid-1960s. This is probably a reflection of the fact that the demand for countryside recreation seen in Britain was mirrored in other countries; America for example also experienced a boom in 'nature tourism' in the 1960s. Travel with the experience of wilderness or wildlife as a central purpose has continued to grow ever since, and is sometimes referred to as ecotourism (defined by the Worldwide Fund for Nature as 'tourism to protect natural areas, as a means of economic gain through natural resources preservation') or biotourism.

The philosophy and methods of interpretation owe much to Freeman Tilden who, in *Interpreting Our Heritage* (Tilden, 1957) argued that the purpose of interpretation was to 'reveal meanings and relationships through the use of original objects, by firsthand experience, and by illustrative media, rather than simply to communicate factual information' and introduced six essential 'principles of interpretation' (see Figure 5.1), which have stood the test of time. Tilden also realized that interpretation was something new, a different kind of educational experience, writing 'We are clearly engaged in a new kind of group education based upon a systematic kind of preservation and use of national cultural resources. The scope of this activity has no counterpart in older nations or older times.' In addition to education, interpretation also had a moral purpose, it was an appeal for an appreciation of the beauty and wonder of the natural world – an attempt to jog the memory of city-dwellers, a reawakening of deep-seated knowledge. It would do this through a new methodology which involved many diverse activities including the use of trails, guides, actors, games, signs, story-telling, slide shows, and film.

These techniques, and many others, had been developed in National Parks in the United States and Canada as a means of increasing visitor awareness, understanding and appreciation of the natural environment. As early as 1919, guides at Lake Tahoe, California were using interpretive techniques to assist visitors (Aldridge, 1975), following a tradition which had already been established in Europe (Everhart, 1972). By the 1920s many Parks had initiated education programmes; Parks Canada established its Education and Interpretation Section in 1959, and in the same year park rangers in the Banff National Park, Alberta, opened a trail with signs located at significant points on the route which enabled the identification of trees, habitats and geological features – possibly the first nature trail in North America (Wilson, 1992). Interestingly, Banff was also the site for the first park museum on the American continent; established in 1895 it has been restored as a period-piece (Fortier, 1988).

In the 1960s interpretation was adopted in many other countries as a viable technique, including Australia (O'Brien, 1988), the Far East (Marsh,

The Six Principles of Interpretation

- Any interpretation that does not somehow relate what is being displayed or described to something within the personality or experience of the visitor will be sterile.
- Information, as such, is not interpretation. Interpretation is revelation based upon information. But they are entirely different things. However, all interpretation includes information.
- Interpretation is an art, which combines many arts, whether the materials presented are scientific, historical or architectural. Any art is in some degree teachable.
- The chief aim of interpretation is not instruction, but provocation.
- Interpretation should aim to present a whole rather than a part, and must address itself to the whole man rather than any phase.
- Interpretation addressed to children (say, up to the age of twelve), should not be a dilution of the presentation to adults, but should follow a fundamentally different approach. To be at its best it will require a separate program.

Figure 5.1 Freeman Tilden's six principles of interpretation. (From Tilden, F. 1957 *Interpreting our Heritage*.)

1988) and the Pacific rim (Jacobsen and Harris, 1988). As well as new educational techniques, interpretation demanded a certain amount of new 'hardware' – information centres, signs and notice-boards, trails and hides. Increased recreational demand called for picnic sites, camp sites and toilet blocks; new thought had to be given to the siting and design of such facilities in natural settings (Beazley, 1969 and 1971). Structures and techniques developed rapidly, becoming increasingly sophisticated and geographically widespread. Interpretation of the natural and cultural environment quickly became institutionalized and professional; Tilden's principles were revised and 'interpretive plans' became integral to the management of national parks, nature reserves and archaeological sites (Aldridge, 1975); new techniques and new methodologies were developed, and have been summarized by Pennyfather (1975). The range of interpretive provision is considerable, and frequently surprising. New situations often require innovative methods; for example the mainly nocturnal mammals of Australia became the focus for 'spotlight walks' in natural areas – wombats detected by their eye shine 'can be followed as they waddle or gallop to their large underground burrows'; underwater interpretation required the development of the snorkelling trails which are so popular on the Barrier Reef, and the use of glass-bottomed boats (pioneered there in 1937) (O'Brien, 1988). Interpretation has also made the most of technological advances, for example in the use of video equipment to relay images from inside a woodpecker nest to the general public (Davies, 1987), or more spectacularly, to convey images of fish and invertebrates from deep ocean trenches to sea-life centre visitors (Kool and Trayner, 1990).

The use of new technology in interpretation is just one way in which, according to Vander Stoep (1991) the whole face and character of interpretation have changed dramatically in recent years in response to the needs of organizations and the expectations of visitors. Games, interactive video and other gimmicks, living history, craft demonstrations, drama and story-telling are increasingly used, consistent with the high-tech/high-touch trend in society (Naisbitt, 1982). Changes in funding have led to a decline in the number of interpretive programmes, which has in turn caused the contracting out of visitor services and more co-operative ventures. There are also fewer interpreters, as authorities move towards holistic interdisciplinary training and modes of operation; yet the demands on those interpreters has increased and expanded to encompass major environmental issues such as acid rain, biodiversity loss and pollution; outreach activities have also become widely accepted.

The visitor centre

As the desire to visit the countryside has grown, National Parks have felt the pressure, with the popular and accessible Peak and Lake Districts each recording in excess of 20 million visitors per year in 1987; in the same year, on any summer Sunday 17 million people visited the countryside, and the demand has not yet levelled out (Bromley, 1994). Reasons for using the countryside for informal recreation are physical, socio-economic and psychological – getting back to nature, rehabilitation, self-discovery and fulfilment are all part of the experience (Torkildsen, 1986). Factors which make it possible include increased car ownership, income, and annual holiday entitlements. In Britain, this demand is met by more than 50 groups and organizations which provide opportunities for countryside recreation, including all the major conservation organizations such as the National Trust, English Nature, the Wildfowl and Wetlands Trust and the County Conservation Trusts. Many major conservation sites have a visitor centre at the heart of their operation; in Britain's National Parks there are now 65 centres, visited by more than 3.5 million visitors each year (Blunden and Curry,

1990). Though written in 1947, and despite the innovations in interpretive media, the concept of the visitor centre proposed by the National Parks (Hobhouse) Committee, still holds true today:

> we have in mind a combination of the facilities provided by the best local museums and libraries, the Field Study Centres ... and the new County Colleges. They would provide material, in the form of books, maps, models and museum exhibits, for enquiry by the visiting public into the topography, history, natural history, rural economy and general culture of the Park and its neighbourhood, together with facilities for reading, field study, and similar pursuits, for periodical talks and exhibition of cinema films, and for the sale or distribution of National Park literature and maps. In short, each Centre would be a focus of interest and a source of infinite educational value for those visitors ... who seek a deeper understanding and appreciation of the significance of unspoilt country. (Himsworth, 1969.)

Today's centres fulfil a variety of purposes, including site orientation, education and interpretation, the location for toilets, cafeteria and shop; they may fulfil operational needs, providing offices and stores. They can also provide a useful function for disused or redundant buildings (Barrow, 1988). The centre is usually a means of providing an introduction to a site, but is occasionally more significant than the site itself – Beazley (1969) notes, for example, that:

> the visitor to the Grand Canyon ... may learn a lot about its natural history by visiting the Center but nothing in the world matches the experience of the canyon itself. At Kittyhawk on the other hand there was nothing to see except the flat ground over which 'man first flew'. In such circumstances the Visitor Center is everything; the Wright Brothers Memorial and the Center are synonymous.

There can be little doubt that interpretation is best carried out on-site, and there are many excellent examples of visitor centres interpreting a wide range of natural and cultural sites, including battlefields (Culloden, Gettysburg), archaeological sites (Housesteads, Stonehenge), and industrial buildings (Abbeydale, Sheffield) in addition to areas of wildlife and scenic value, such as 'Europe's first visitor centre', Landmark, near Inverness (Aldridge, 1971), which aims to interpret 'Man in the Highlands'. To select only one good example of a visitor centre from such a range is invidious, but Wilson (1992) cited the Frank Slide Interpretive Centre in the Crowsnest Pass of Alberta for its valuable 'contemporary nature education'; a Rocky Mountain valley, it is rich in the interdependence of natural and cultural history:

> It is the site of a major railway crossing of the Continental Divide, several abandoned coal mining towns, and one of the world's largest rock slides. The centre has been built at the edge of the slide rubble. Its construction directs the visitor's gaze out over the valley as exhibits tell the stories of railroad construction, coal and oil extraction, and the collapse of the mining economy just before the First World War. Outside, paths radiate out to platforms where exhibits tell about railroad alignments and the arrangement of the town sites – all within the context of the valley's physiography and geology.

It is evident that there is a range of interpretive facilities in the countryside, and that trying to classify them as visitor centres, trail-side museums, or interpretive centres is an impossible task. However there is an important distinction to be made between a pure visitor centre (which seeks to provide for the more domestic needs of visitors) and an interpretive centre; the latter must have as its main concern an explanation of the significance of a landscape, natural site or monument, and foster an understanding and awareness of how the area has changed – why it looks like it does today. An interpretive centre should also act as a link into a site; it provides orientation and an overview, but should encourage visitors to widen their experience through other activities such as self-guided trails. In all these ways the best examples of interpretive centres parallel museums, the only major difference is the absence of permanent collections and the need for object-centred curatorial activities.

An indication of the common purpose of centres and museums is evident in the recent guidelines for country parks developed by the Countryside Commission (1995), which includes a checklist for effective operation of visitor centres, and includes reference to staff, circulation, provision for disabled visitors, and the need for displays to be up to date, informative, interactive and appealing to youngsters, advice which could equally be applied to museums. Jullien (1989) has suggested that there is a cline of 'museums beyond the walls' and cites six categories of museums which range from the urban museum with only a prospect of urban open spaces to those museums which interpret particular habitats, scientific sites or landscape features. As we move along the cline there is increasing emphasis on place, and on the natural environment, but less consideration for collections. In effect the museum metamorphoses gradually into an interpretive centre.

Distinctions become even more blurred when taking environmental education centres into consideration. Often located in sites of considerable wildlife and landscape interest, they interpret the significance of an area, but to a particular audience (school groups) and in a more formal manner. A good example is the Schuylkill Center for Environmental Conservation, Philadelphia, established in 1965 'to serve as a specialised education facility primarily concerned with natural systems and their relation to mankind'. Located in some 500 acres of woodland, meadow and wetland, it has an interpretive gallery, holds representative natural history collections, a major environmental library, and operates a 'loans box service' on topics such as 'Extinct and endangered species', 'freshwater ecosystems and wetlands' and 'Air pollution'; its 'Ecovan', a 32-foot mobile classroom is designed to conduct educational and interpretive programs at sites beyond the Center (Schuylkill Center, 1993). Here we have a combination of museum, educational facility, interpretive centre – and much else besides.

Interpretation and the management of natural areas

In his speech to the 'Countryside in 1970' conference, J. L. Longland of the Development Commission remarked that the conference emphasized 'that very curious paradox about the English, that they combine a perfectly genuine passion for the countryside with an almost unequalled skill in destroying it'. Longland was referring to destructive changes caused by agriculture and forestry, but the increasing number of tourists and day-trippers enjoying the countryside is now also known to be a significant form of disturbance. Active physical pursuits, recreational hunting and fishing, even simply enjoying scenery or observing wildlife can have dramatic ecological impacts (Edington and Edington, 1986). The effectiveness of National Parks and other protected areas in promoting recreational use has had detrimental effects on the very resource which is being conserved; recreation and conservation are uneasy partners. Yet, in the UK there is a general consensus that exclusion of the public from areas of nature conservation interest, or other fragile areas or structures, should be avoided wherever possible, in the interests of education and developing public support. Management of natural areas and cultural sites therefore requires management decisions – initially assessing which areas are of special significance and especially vulnerable, and then adopting spatial and temporal zonation of land as a key technique; within these zones different interpretive strategies (or none at all) can be employed, relaying different messages. Interpretation must be seen as an integral part of a holistic approach to resource planning which can help to avoid perceived conflicts between resource use and recreation. Interpretive planning (see Figure 5.2) therefore lies at the heart of the management of natural and cultural sites, and setting clear objectives, and monitoring and evaluating interpretive provision is essential. Evaluation is now seen as an important part of the interpretive process (Marsh, 1988a) and a number of useful evaluation guides for interpreters are available (e.g. Hodgson and Fritschen, 1984, Marsh, 1986, Roggenbuck and Propst, 1981); observations by practitioners and others on the evaluation of interpretive plans and programmes are summarized in Machlis (1986).

Interactions – museums, visitor centres and site interpretation

How have museums responded to these developments in the countryside? – and what has been the impact on the way in which they now operate? Stansfield (1967) noted that:

> It is disappointing to find no mention of museums within the three hundred page report of this [Countryside in 1970] Conference, and we might deduce that, to date, museums had failed to make any impact in this field of activity.

Despite this omission, the truth is that museums did respond, they became involved in a variety of ways; some set up new outstations and site museums themselves, many provided specialist help and advice to the emerging information and interpretive services, and all learned, adopted and modified the techniques of countryside interpretation. Museums had made a

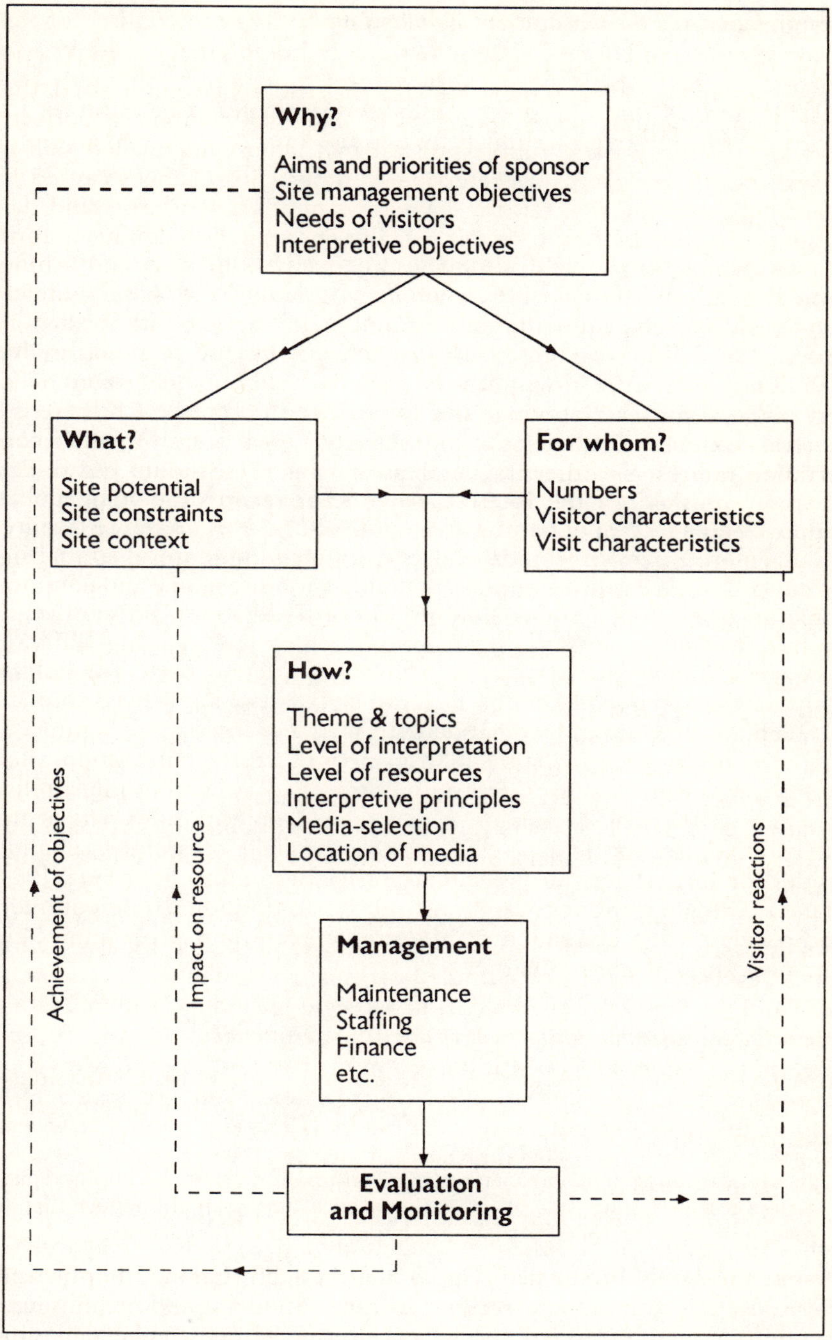

Figure 5.2 The interpretive planning process.

commitment to the environment by stepping beyond the confines of the building; by the end of the 1960s there was a realization that a museum was much more than a place where objects were curated and exhibited. The museum and its staff had a responsibility to research and interpret its environment – museums without walls became a reality.

Field museums

The establishment of 'field museums' was a first indication that environmentalism had influenced the museum profession in the UK. In 1965 a Natural History Museum and Nature Trail was established by the Ulster Museum at Tollymore Forest Park in Northern Ireland to describe the wildlife of the Mourne Mountains (Deane, 1965); at about the same date, Newcastle's Hancock Museum joined forces with the Forestry Commission to establish the Border Forest Park Museum at Lewisburn in Northumberland, illustrating the social and natural history of the Pennine upland and its man-made forests (Tynan, 1969). Curry (1969) records the co-operative venture between Plymouth Museum and the Devon Trust for Nature Conservation to establish the Warleigh Wood Field Museum on a 35-acre site in Devon; the former Tamerton Foliot Railway Station acted as the museum building. In all these ventures the expertise of museum curators, their expert knowledge of natural history and their own area, coupled with a familiarity with display techniques and access to specimens was essential to success. Although these early ventures showed much promise, few museums in the UK had the resources to follow their lead; some did, and continue to do so (Scunthorpe Museum's Normanby Park and Smithhills Hall, Bolton (Greenwood, 1976) for example), but few museums operate on-site interpretation away from base. With the advent of the Countryside Act, and the greater availability of financial help, the early steps taken by museums were superseded by other organizations; many (including voluntary bodies such as County Wildlife Trusts) although not necessarily having greater resources, were unfettered by regulations and were eligible for grant aid; others (especially government-funded organizations such as the National Parks and the Forestry Commission) had a regional and recreational remit which encouraged action. As a result the mantle of responsibility for developing countryside-based information centres and interpretive facilities passed from museums to other organizations, and proceeded at a considerable pace – at the 1969 Countryside Centres Conference it was reported that the Forestry Commission had opened seven centres that year; some 30 information centres functioned in National Parks (Stansfield, 1970). The demise of certain museum-run centres, and the rise of environmental interpretation elsewhere has been summarized by Stansfield (1983). However, it should be stressed that despite these changes, at the practical level, museums – or more correctly individual curators – were still involved in countryside interpretation, providing information and specimens, writing leaflets, and taking guided walks, a contribution which is rarely recognized in the literature, but one which is of considerable importance.

Lewis (1993) records that field museums had a much earlier genesis in the United States. As in Canada, the National Park Service had the remit and the funding to enable the founding of museums at some of the world's most spectacular locations, including the Yosemite Valley Museum, the Yapavai Observation Station Museum on the rim of the Grand Canyon, and the Bear Mountain Museum in the Palisades Interstate Park near New York. Lewis regards the arboretum established in Yosemite in 1904 as the first 'museum', though a centre for field studies was established in the same park around 1915, and featured stuffed and mounted specimens. Much of the enthusiasm for the establishment of these 'field museums' came from museum professionals – the Yosemite Museum was the brainchild of Joseph Grinnell, Director of the Museum of Vertebrate Zoology at the University of Berkeley, California. According to Stansfield (1967):

> These museums set the pattern for a whole series of museums within the United States National Park Service covering natural history, geology, archaeology and military history. Similar museums, dealing only with natural history, are maintained by the National Audubon Society (nature centres) and by the Natural Sciences for Youth Foundation as well as many private enterprises.

Embracing a new philosophy – connections and interrelationships

Lowenthal (1988) observed that:

> On the whole, we admire nature as previous to history yet also in line with the present; wilderness and other natural sites offer a heritage at once age-old and up-to-date; diurnal and seasonal rhythms are comfortingly familiar. By contrast, the fascinations of human history lie in its uniqueness, contingent surprises, unexpected confrontations, and the transience of people and events forever past ... nature seems essentially other than us; we may feel at one with its life-supporting fabric, but we seldom put ourselves in nature's place or project ourselves into non-human lives. By contrast the cultural heritage promotes empathetic communion.

Due to this contrast, interpretation of the natural environment faces a real challenge. There is a need to combine themes and stories of cultural, social and natural history, to make the natural world accessible by making it more relevant to the everyday life of people by seeking to explain interrelationships and the significance of nature to man. Good natural site interpretation does this, as for example Aldridge's (1981) interpretation of the components of a desert ecosystem (see Figure 5.3).

These ideas of linking themes, and making messages relevant to people have gradually been adopted by museums. The traditional natural history museum simply looked at the diversity of life, nothing more; new museums – or new displays in old museums – have embraced the primary purpose of interpretation to develop and explain interconnections, the parallel history of the natural and social environment. New understandings are also developed by tackling more complex biological ideas – food webs, photosynthesis, soils, speciation and genetics, community succession, or by addressing specific environmental concerns, or the scientific background to socio-medical issues such as AIDS. It is not just new techniques which

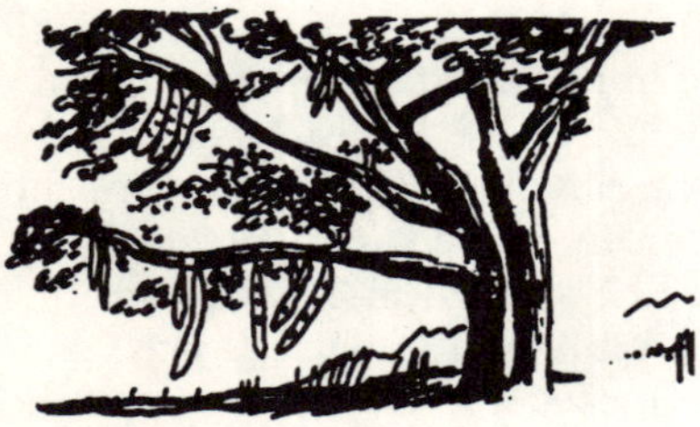

1. Here we are at the first stop mesquite shrub over 3 metres tall and gives valuable shade to reptiles

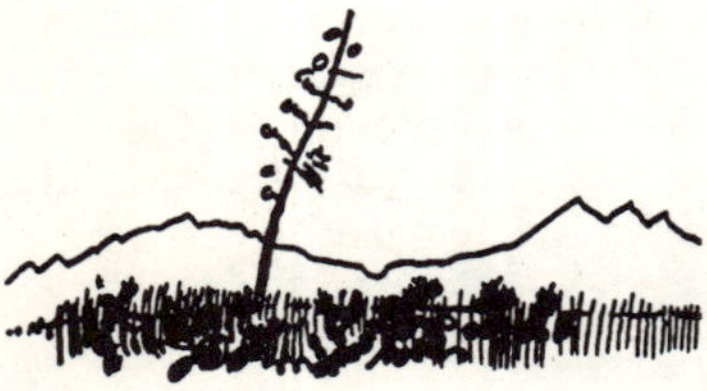

2. Ah! An agave in flower, I guess you know this is a member of the Amaryllis family?

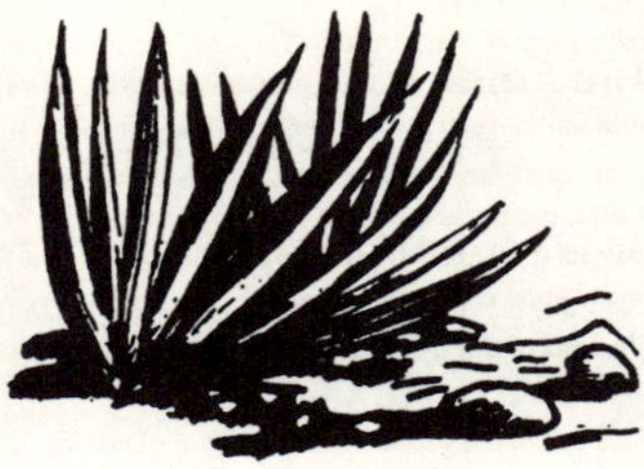

3. Would you believe that the yucca is a lily?

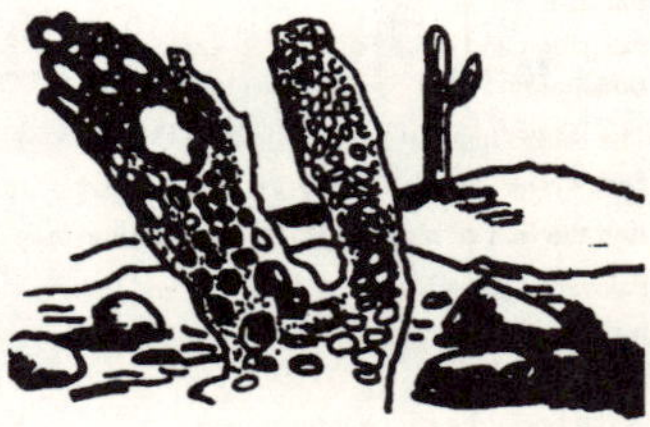

4. and that inside many cacti the skeleton is solid wood?

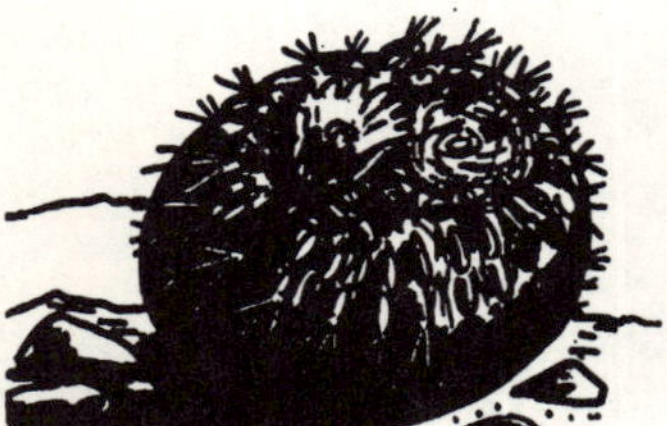

5. Here we have a barrel cactus in flower . .

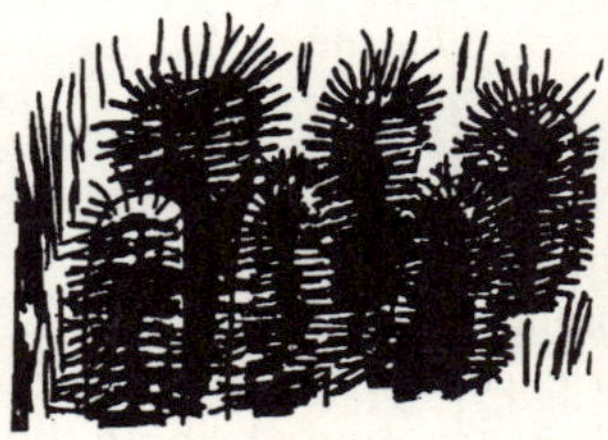

6. See how the hedgehog cactus is adapted to the desert environment

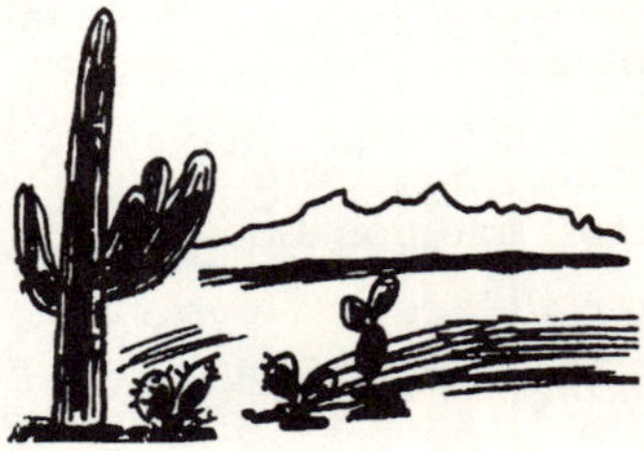

7. The saguaro can live for 200 years, it also adapts – thick skin, leaves rolled into spines

If we were to interpret this place only by means of that nature trail then we should miss its significance, for TONTO can tell a story of the Salado Indians – a tribe remarkable for the way they adapted to the desert environment. So let us take the trail once more, for this same trail can tell us more:

1. Mesquite beans were used to make alcohol and meal cakes, the sap was used in repairing pottery and making sweets (prehistoric American chewing gum?). The wood made good tools and weapons.
2. The tips of the agave leaves made needles for textile work and
3. The tips of the yucca made awls for shoemaking. Its buds, flowers, fruits and seeds were eaten and its fibres made into floor coverings, rope and shoes. Soap was made from its roots.
4. The skeletons of old saguaros made good roof timbers and firewood.
5. Because the barrel cactus faces south it was used in navigation and its red spines were often made into ornaments
6. The Salado Indians ate the fruit of the hedgehog cactus early in the year
7. And the fruit of saguaro later in the year and they made
8. Paloverde seeds into flour for bread and used the
9. Prickly pear fruit as a jelly or preserve

This is the beginning of site interpretation at Tonto, as told by the United States National Park Service but you need to be there to comprehend it fully.

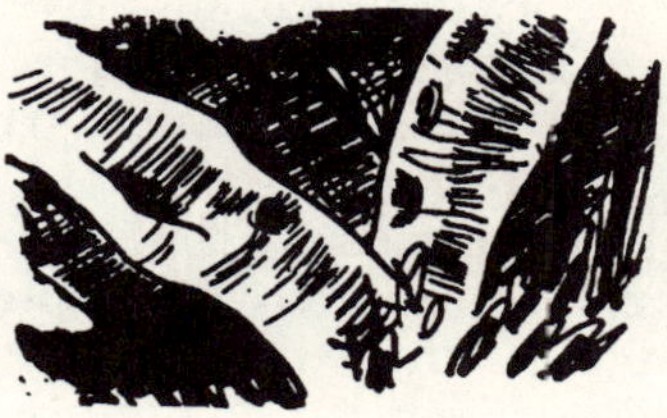

8. This yellow-green shrub (paloverde) sheds its leaves in the hot season and the bark makes chlorophyll

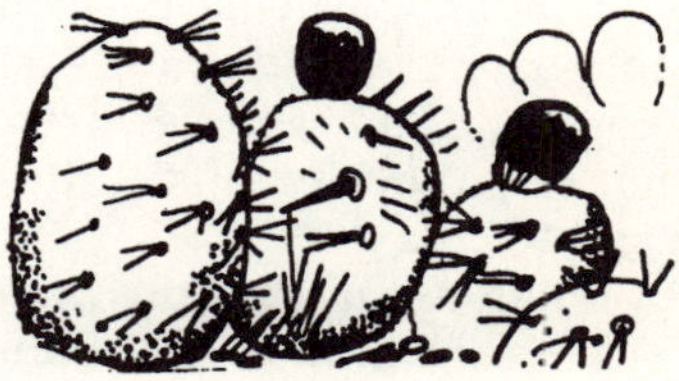

9. This is the prickly pear which survives extremes better than other American cacti and consequently has a wide range

Figure 5.3 Don Aldridge's approach to the interpretation of a desert ecosystem. (Based on the cartoons in Aldridge, D. 1981 *The Monster Book of Environmental Interpretation*; reproduced with the author's permission.)

interpretation has given to museums, but a new philosophy too, based on interrelationships and relevance.

An excellent recent example of this changing philosophy comes from South Africa. In describing the development of a new museum at Soutpan, and research being carried out by the Cultural History Museum of Pretoria to establish on-site interpretation centres, Kusel (1993) notes that in a time of unparalleled change in South Africa, 'we have to follow a more holistic approach in order to cover the complete conservation field and start seeing the integrated interaction between various disciplines and how they form a whole. We have to get rid of the old ideas and think afresh on our museums.' The overriding emphasis is on the need to develop connections and cross traditional boundaries.

Embracing new techniques

It is interesting to note Geoffrey Stansfield's (1970) comment that at the 1969 Countryside Centres Conference, 'It was ... clear that the philosophy of interpretation as described in the American literature is not clearly understood, and its application even less so.' The UK was evidently at the bottom of a steep learning curve, yet the speed with which the opportunities were taken, and new techniques absorbed and adopted during the next five years was quite remarkable. What is particularly important is the way in which museums observed the techniques being adopted at interpretive centres, and quickly adapted them to their own needs, and we should not forget that the variety of interpretive media which museums now employ so readily have their origins outside the walls – this feedback, and the exchange of ideas which has continued to occur, is of considerable significance. Features of interpretation which were pioneered in places such as Landmark, including multiple slide-projection programmes, open displays and interactive exhibits which encourage the use of all the senses were quickly adopted by museums. Interpretive techniques used first to good effect at cultural sites, particularly 'living history' events using actors, are now commonplace in museums. Evaluation of interpretation, first used within the context of the wider environment, has now been adopted by museums under the guise of exhibit evaluation, and is a growing field of museological research.

Although natural history interpretation in museums is carried out 'off-site', the surroundings still stimulate the senses; the frequent availability of re-created environments and the use of real objects mean that interpretive techniques can be used to good effect to encourage appreciation and understanding of the natural world. Interpretation in museums is not used as a management device, but at its best might change attitudes and influence behaviour in the long term.

In the natural environment, the resultant comprehensive new facilities (visitor centres, theatres, trails) and activities guaranteed that protected areas would, in the words of a Parks Canada brochure, 'be explained and interpreted as living museums of nature, where people can observe and appreciate the beauty that surrounds them'. The use of the term 'living museum' for the natural environment does much to explain the adoption of interpretive techniques within the traditional museum context. However, the role of the interpreter, whether in a National Park or confined to a museum, remains the same, 'to provide direct opportunities for recreational, inspirational experiences with our heritage, and nothing should substitute for the real thing' (Peart, 1988).

The ecomuseum concept

The countryside interpretation movement encouraged a new philosophy which examined interactions between the natural and cultural environments; ecomuseums took this idea a stage further and placed that interaction within a community context. The ecomuseum paradigm was developed in the 1960s by two Frenchmen, Georges Henri Rivière (1897–1985) and Hugues de Varine, with the latter the driving force behind the idea that museums 'become democratic institutions with local communities taking a far more active role in their planning and functioning' (Corsane and Holleman, 1993). Rivière, with a background in folk museums, was anxious to interpret human history, and the objects and artefacts associated with that history, in an environmental context (Hudson, 1992; Rivière, 1973). Museum projects which evolved in response to the development of the French Regional Natural Parks from 1967 (Hubert, 1985) such as The Camargue Museum (Duclos, 1980) and the Ecomuseum of Marqueze, Sarbres, within the Landes de Gascogne (Moniot, 1973) did much to put their ideas into practice. The term ecomusée (ecomuseum) was devised by de Varine for use by Robert Poujade, the French environment minister in addressing the 1971 ICOM Conference (Wasserman, 1989), though:

> it is widely argued that ... Rivière ... laid the foundations of the ecomuseum movement a quarter of a century earlier in planning the Bretagne Museum in Rennes. Poujade, de Varine and Rivière all wanted to provoke the then very traditional and centralised French museums to look beyond their front doors and crusade for both the environment and the renewal of the museum as an institution – to create 'a new museology' (Boylan, 1992).

Rivière (1985) defined an ecomuseum as:

> an instrument conceived, fashioned, and operated jointly by a public authority, and its local population. The public authority's involvement is through the experts, facilities and resources it provides; the local population's involvement depends on its aspirations, knowledge and individual approach. It is a mirror in which the local population views itself to discover its own image, in which it seeks an explanation of the territory to which it is attached and of the populations which have preceded it, seen either as circumscribed in time or in terms of the continuity of generations. It is a mirror that the local population holds up to its visitors so that it may be better understood and so that its industry, customs and identity may command respect. It is an expression of man and nature. It situates man in his natural environment. It portrays nature in its wilderness, but also as adapted by traditional and industrial society in their own image. It is an expression of time, when the explanations it offers reach back before the appearance of man, ascend the course of the prehistoric and historical times in which he lived and arrive finally at man's present. It also offers a vista of the future, while having no pretensions to decision-making, its functions being rather to inform and critically analyse. It is an interpretation of space – of special places in which to stop and stroll. It is a laboratory, insofar as it contributes to the study of the past and the present of the population concerned and of its total environment and promotes the training of specialists in these fields, in co-operation with outside research bodies. It is a conservation centre, insofar as it helps to preserve and develop the natural and cultural heritage of the population. It is a school, insofar as it involves the population in its work of study and protection and encourages it to have a clearer grasp of its own future. This laboratory, conservation centre and school are based on common principles. The culture in the name of which they exist is to be understood in its broadest sense, and they are concerned to foster awareness of its dignity and artistic manifestations, from whatever stratum of the population they derive. Its diversity is limitless, so greatly do its elements vary from one specimen to another. This triad then is not self-enclosed : it receives and it gives.

This definition includes some demanding and sweeping objectives, and has significant implications for the way in which the ecomuseum would work –

traditional boundaries and power bases had to be dismantled, and multidisciplinary and interdisciplinary teamwork substituted as a norm. On the ground, the open-air museum movement pioneered in Skänsen, Sweden (Kavanagh, 1990) had considerable influence, in particular the techniques of relocating buildings, re-creating period settings and the use of costumed interpreters to demonstrate craft skills and interact with visitors. However, all open-air museums are not ecomuseums, the difference being one of emphasis on the natural environment and the significance of place. That place is not confined to the museum site:

> incorporating a much larger territory in a fairly open-ended manner. Here the territory is not simply defined in geographical or administrative terms, but rather as any whole unit where the inhabitants share a common way of life, culture, occupation or traditional custom. Run from a central headquarters the ecomuseum has a series of 'antennae' within the territory of the museum. These antennae form a network through which the activities of information-gathering, research, display and education can be performed (Corsane and Holleman, 1993).

The concept of what constitutes a collection, or an exhibition, is also redefined by the ecomuseum. According to Van Mensch (1993) 'it has become increasingly difficult to discern where the museum stops and the real world starts'. Everything within its territory becomes a part of the collection and part of an exhibition, and not just landscapes and underlying geology, animals, plants, people, buildings and objects, but also more intangible items which help to build the network of interrelationships between them, including customs, traditions, local skills, behaviour patterns and social structures. For these reasons the ecomuseum cannot adopt a conventional acquisition policy (de Varine, 1973), as objects are preferably left on-site, where they can be interpreted in context, or if taken into the museum, following research and documentation they are then returned to owners or sites (Querrien, 1985). To adopt such an approach requires the double-input system that de Varine expounded for the ecomuseum, the combined efforts of curatorial staff and the local community which would lead to greater understanding of place and enable rational choices to be made for the future. In terms of increasing environmental awareness through education, encouraging people to examine their lifestyles and values and their consequences, and thereby influencing and adopting sustainable strategies, a successful ecomuseum should have much to offer.

The philosophy of the ecomuseum has proved to be an attractive one; the ideas were first tested with the development of the Museum of Man and Industry centred on the communities of Le Creuseot and Montceau-les-Mines, north of Lyon, which opened to the public in 1974, and acted as a model for the ecomuseum movement, and encouraged other countries in Europe to embrace the idea (Evrard, 1980). Today ecomuseums are found as far afield as Sweden (Engstrom, 1985), Portugal (Nabais, 1985), Mali (Konare, 1985), Brazil (Almeida-Moro, 1985), and Canada (Rivard, 1985; Stevenson, 1982; Choquette, 1988; Binette, 1991); as Rivière suggested, the 'diversity is limitless', and it is often difficult at first sight to see why some museums have adopted the title, or even the common goals they share. It is doubtful if any ecomuseum has achieved the standards and objectives set by the founders of the movement, even in France, where as Hudson (1992) notes 'it is safe to say that none of the 28 institutions which are officially recognised as ecomuseums ... today carries out the complete ecomuseum task as it was originally conceived.' Even the bold experiment of Le Creuseot now only survives as a poor imitation, and the ecomuseum of Haute-Alsace (Grodwohl, 1987), once perceived as a model example, has suffered cutbacks – according to Hudson (1992) a victim of its own success. Even the term 'ecomusée' seems to have fallen out of favour in France, being displaced by the 'community museum' for new initiatives (Boylan, 1992).

Interestingly, some UK museums which Rivière saw as outstanding models of the ecomuseum concept, such as Ironbridge or Leicestershire Museums, with museums scattered across a number of sites, pay scant regard to the idea. Recent developments in the North Pennines to interpret the lead-mining history of Durham, Northumberland and Cumbria at major sites at Allenheads, Nenthead and Killhope (see Plate 19), have also avoided an ecomuseum label. The North Pennines was Britain's biggest and most successful lead-mining area, it is a landscape and a people dominated by mining. Thirty per cent of Britain's lead was mined here, and the two major companies employed thousands of men and boys; the remains of the industry – abandoned buildings, smelt mill chimneys and spoil heaps are everywhere to be seen. It is these sites that have acted as the key interpretive facilities where buildings and equipment have been carefully restored, but the driving force behind the venture is the local community, which sees the museum developments as an integral part of a plan to regenerate the local economy. This has been particularly evident at Allenheads in Northumberland, and Nenthead in Cumbria, where the local community has been at the centre of fund-raising, and carried out much of the conservation and interpretive work. There can be no better example of an ecomuseum in Britain, having all the necessary ingredients, yet the North Pennines Heritage Trust has no plans to adopt the title ecomuseum. According to Lawes *et al.* (1992) the dismissal of ecomuseums by most English-speaking countries is because 'the ideas are inappropriate, or the language is off-putting to [their] empirical and pragmatic cast of mind'. It could of course be something quite simple, the English mistrust of anything French.

A case study – ecomuseums in the Cévennes, France

The Cévennes National Park (see Plate 20) covers some 91,500 hectares in Lozère and Gard; beyond this boundary is a 'zone périphérique' which extends the area to more than 237,000 hectares, and into the Ardèche. The landscape of the Park includes the upland limestone plateau of the Causse Méjean, the granite massif of Mont Lozère, the schists and granites of the Bougès mountains, the valleys of the Gardon and its tributaries, and the mountains clothed in forests of pine, spruce and beech which rise to Mont

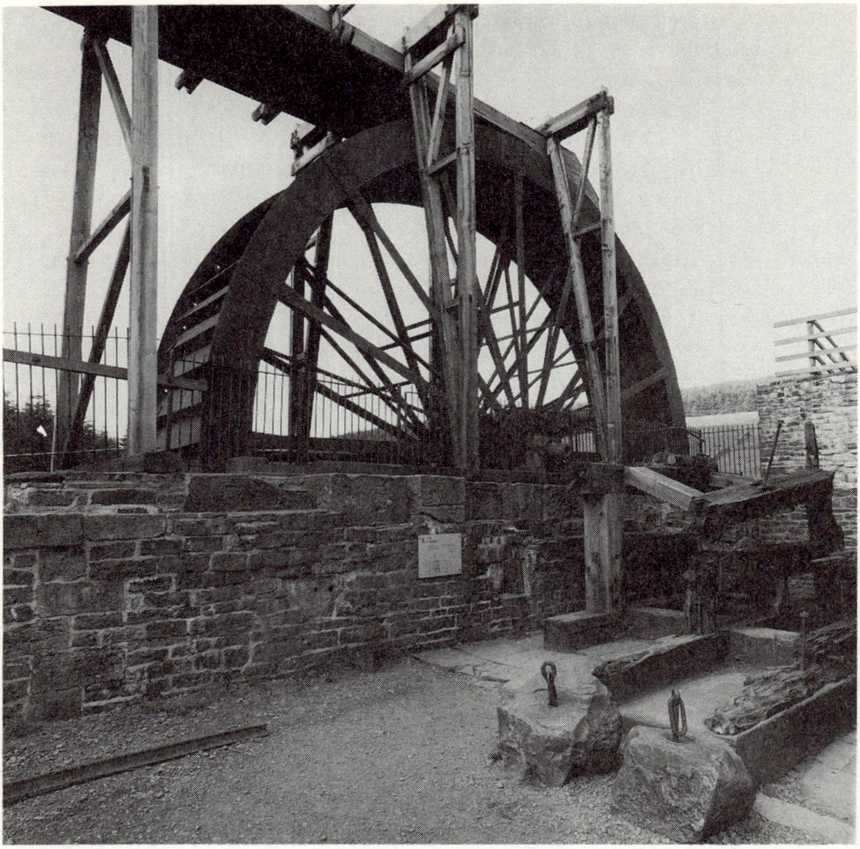

Plate 19 The overshot water wheel at Killhope Wheel Lead-mining Museum, Weardale, County Durham. (Reproduced courtesy of Durham County Council.)

Aigoual, some 1,560 metres above sea level. It is an area of wild scenic beauty characterized by mountains clothed in groves of sweet chestnut (once a staple crop), and dotted with red-roofed farm buildings. Tourism, rearing sheep, cattle, and goats, and forestry are the major occupations there today. The silk industry, for so long a dominant part of the economy of the Cévennes, is all but extinct, although it has left its mark on the landscape, with mulberry a common sight, as are disused 'magnaneries' (buildings where silkworms were reared) and spinning mills. The natural history of the Park, as expected from the variety of habitats and non-intensive agricultural practice which is a feature of the area, is extraordinarily diverse, with over 1,200 flowering plants (including ten endemics), a mammal fauna which includes wild boar and beaver, and a spectacular insect fauna. Because of its wildlife and landscape the Cévennes was designated as a biosphere reserve in 1985.

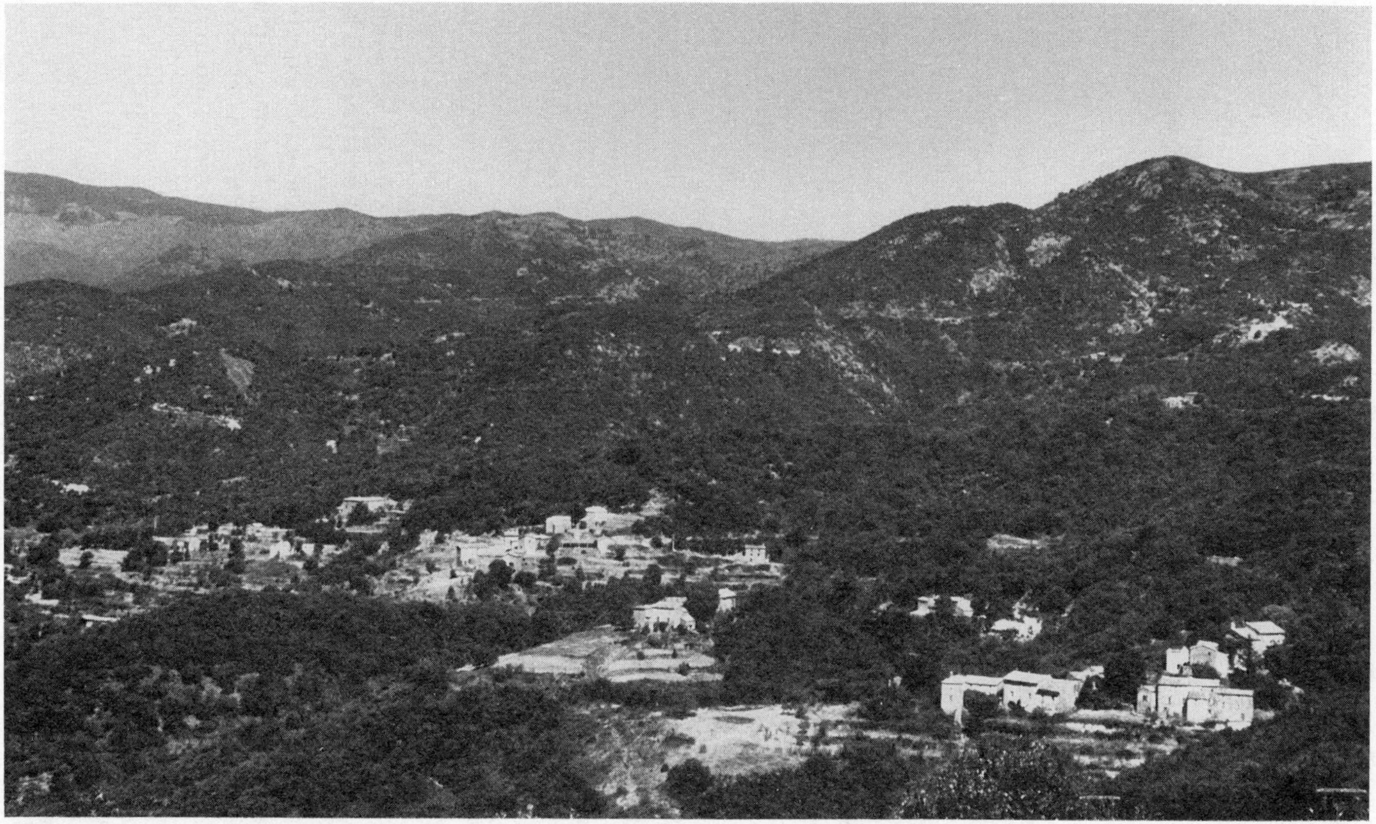

Plate 20 The landscape of the Cévennes National Park, France.

Established under the UNESCO 'Man and the Biosphere' programme, biosphere reserves are built on the idea of an ecosystem, a self-regulating community of plants, animals (including man) and their physical environment. Because of their multiple use, they are seen as a significant resource for education, and even as a means of reinterpreting the traditional idea of a protected area or park. Perhaps it is not surprising then that the area has adopted ecomuseums as a means of communicating ideas to the public. There are three ecomuseums in the region, based respectively at Mont Lozère, in the Cévennes and on the Causse Méjean.

Mont Lozère is the highest mountain in the Cévennes at 1,699 metres, a cold and windswept area which has been dramatically altered by man over thousands of years. The ecomuseum attempts to tell the story of this harsh environment, through its headquarters building, the Maison de Mont Lozère, in the town of Pont-de-Montvert, where there is a permanent exhibition, and at the nearby Ferme de Troubat which has restored buildings, including a watermill, bread ovens, threshing areas and barns. Visitors are encouraged to visit other architectural elements in the area, including fountains, bell-towers and entire hamlets, to get an impression of the way of life of its former inhabitants. Exploration of local landscape and wildlife is made possible by self-guided routes at Mas de la Barque and Mas Camargues, which pass through upland pasture, beechwoods, coniferous plantations and sub-alpine grassland. The various elements which comprise the ecomuseum are described in leaflets; it is in the trail guides that the best explanations of man and environment are found.

The Ecomusée de la Cévenne is a geographically widespread consortium of museums, trails and cultural sites which was founded in June 1989 with the objective of being 'Une organisation de moyens, un lien entre les musées existants et l'espace, une mise en commun des connaissances et des idées, une présentation commune facilitant la démarche du visiteur' (Parc national des Cévennes, n.d.). The 11 organizations involved concentrate primarily on the themes of silk and sweet chestnut, with attractive museums of local history at Saint-Hippolyte-du-Fort, Saint-Jean-du-Gard and Le Vigan. The dinosaur footprint site at St Laurant de Trèves and the two major trails, provide a biological and geological input to an ecomuseum dominated by social history. Like Lozère, the emphasis is on the past, with little comment about the nature of the region today and the threats imposed by tourism, forestry, water abstraction and mining operations. The sites and museums which form the ecomusée of the Causse Méjean are equally geared to the past, with an archaeological exhibition at La Borie, and a preserved 'Caussenarde' farm at Hyelzas.

Girard Collin, curator of the Ecomusée of Mont Lozère, remarked that:

> We must be able to leave the consumer society behind and go to places where the natural and human heritage looms large – offering us models of equilibrium and lessons on how to live. And then perhaps it may be possible for us to imagine another kind of time, a slower time ... when, having learned how to live with our cultural and natural environment, taking it into account in our everyday life, and in our projects of technological development, it will not be necessary to legislate the setting aside of parks and protected areas. These measures only exist because

> the heritage of the past – our heritage – is neglected or poorly loved (quoted in Wilson, 1992, p 241–2).

One could argue exactly the opposite – that the Cévennes National Park, and other protected areas exist precisely because people care about such areas, that they are recognized as natural and cultural resources. It is evident that much is being done in the Cévennes through the ecomuseum movement to preserve and interpret past environments and cultures, to generate understanding and concern, but as yet they fail to communicate ideas of present and future. Collin's suggestion that we can learn from these remote and beautiful areas, reflecting on our lives and attitudes is undoubtedly true, but to encourage a change of behaviour we must also be given a glimpse of the future, and the consequences of maintaining current lifestyles. The idea that one day we might regard our total environment with respect, and that protected areas will be unnecessary, still seems somewhat optimistic.

Ecomuseums, ecology and the environment

Biologists have reservations about the term 'ecomuseum', perhaps because it seems to be an attempt to hijack their area of interest. The implication of the prefix 'eco' is its link with ecology, which is itself derived from the Greek *oikos* – a house or living space. However, the science of ecology, at first associated with the study of animal and plant populations, and especially questions of population dynamics and energetics, has itself moved on to embrace wider issues. The usage of the term itself has changed; biologists still use the term 'applied ecology' to describe the scientific methodology used to solve practical problems associated with the natural environment, but beyond the confines of science the word 'ecology' has been used somewhat indiscriminately, particularly as environmentalism has gathered pace. So we see phrases such as 'the ecology of the area' or 'altering the whole ecology' – in other words ecology is seen as embracing the concept of the whole environment. Few people today would make any distinction between ecology and environment – in the public eye they are perceived as the same thing. Fortuitously, it is a broad definition of ecology which is demanded by the ecomuseum philosophy, which emphasizes the need to place museum exhibits within a larger environmental framework. It is also a realization that the environment is an intricate web of interconnections between physical and biological systems and human activity – 'Protagonists of the ecomuseum ideal believed that in all the activities performed by this type of museum the ecological and environmental aspects should be taken into account and emphasised' (Corsane and Holleman, 1993). The word 'ecomuseum' is therefore a realistic one; however, the question of whether ecomuseums address ecological (or environmental) issues is another matter. Are ecomuseums really 'an expression of man and nature ... [portraying] nature in its wilderness ... a conservation centre ... [offering] a vista of the future'? How ecological are ecomuseums?

The simple answer is that they rarely are, and many ecomuseums seem to have adopted the title more as a marketing strategy than as a philosophy; many ecomuseums in France are no more than open-air museums renamed. For example, the Armorique Regional Nature Park in Finistère has a network of museums, many of which adopt an ecomuseum title (Le Doaré, n.d.); the Monts D'Arrée Ecomuseum consists of the Moulins de Kerouat (a restored water mill and tannery) and Maison Cornac, a sixteenth century farmhouse; the Ouessant Ecomuseum is a restored house at Noiu Hella, preserving traditional architecture and furniture. Little reference is made at any of these sites to the environment, yet the Mineral Museum at Presque'île de Crozon, which has not adopted the word ecomusée, refers to mineral resources, geology and landscape, provides a starting-point for geological tours, and has a garden which re-creates natural plant habitats of the area. The Game and Hunting Museum at Scrignac examines the relationship between wildlife and hunting, past and present. There seems little logic for the adoption of an ecomuseum label in this region of France – the so-called ecomuseums fail to fulfil the demands of the environmentalist, or indeed the 'community' ideal of de Varine.

Even when ecomuseums display the cultural heritage with reference to important natural resources, (including wildlife, woodland, coal, lead or other minerals, and water) the emphasis is always on the past – the ecomuseums of the Cévennes are a good example. They say little about today's environment and nothing at all about its future and as a consequence fail to achieve one of the major cornerstones of the ecomuseum philosophy, that of providing a framework and guidance for future use of the locality's resources. Current environmental issues are rarely debated; indeed all too frequently the 'environmental' component of an ecomuseum is no more than a traditional case of stuffed and mounted birds.

There are some encouraging exceptions however, including the museum at Jokkmokk in Swedish Lapland, which tackles the difficult issues of maintaining the traditional Lapp economy based on the reindeer, whilst exploitation of the natural environment, its forests, mineral resources and water develops at an alarming rate. Tourism, new roads and new technology have also had an impact on the Lapps, and the museum is acting as a considerable focus of activity to document the old techniques and traditions, but also to educate and inform about the current situation, and helping local people to 'set about solving for themselves the problems at issue' (Engstrom, 1985). Proposals for a Kootenay ecomuseum in British Columbia, Canada, also put the natural environment centre stage; here the ecomuseum has been redefined – 'In its truest sense ... an Ecomuseum is the interpretation through museological devices of the relationships between culture and nature incorporating the dimension of time' (Choquette, 1988). An area of great scenic beauty and considerable biodiversity, it was first populated by a culturally distinct group of hunter-gatherers known as the Kootenay People. The ecomuseum focuses on palaeoecological and archaeological research in the region, and describes how, in this environment, people were able to live self-sufficiently – a model of sustainability. It is significant that in Canada the ecomuseum concept is now being seen as an aid to strategic planning – in essence bringing the ecomuseum concept as an additional element into the process of interpretive planning (Wood, 1991).

The world map of ecomuseums is firmly based in the northern hemisphere, but it is encouraging to see the philosophy being adopted as far afield as the Caribbean and South America. In 1980 the Montoso Ecomuseum of the Rainforest in Puerto Rico was established as an 'outdoor museum park that interprets the changing patterns of plant, animal and human life in the Puerto Rican Rain Forest from prehistoric times to the modern era' (Flores, 1980). The local community was involved from the start in this project, which makes environmental education its key objective – in a country with little museological expertise and few museums, it was a bold and imaginative venture. Whereas in most ecomuseums, the region is seen in terms of its biophysical resources, in Montoso the significance of place takes on a mystical meaning, because there:

> Man first cleared a magic dance circle in the forest – The Sacred Grove – and entered into a permanent and continuing dialogue with nature. And in the Sacred Grove there was buried an oracle and he spoke for the Spirit of the Earth. So for us in Puerto Rico, surrounded by a turbulent sea of shifting sands and beliefs, listening to the Spirit of the Earth is one of the essential ideas to be explored by our ecomuseum. (Flores, 1980)

Flores describes how through the exploration of the interaction between forest and humankind the museum provokes the visitor to think about choices for the future, where 'the soft path leads to Climax Forest. The hard path leads to the end of the species and the death of the planet'. In Brazil, the Itaipu Ecomuseum is situated in an area altered dramatically by dam building for a major hydroelectric scheme and tourist pressure in the Iguassu Natural Park; as a region it still has islands of high biodiversity, and populations of indigenous peoples – the museum was established to preserve these, to aid environmental restoration, and to promote a feeling of community amongst people from a variety of ethnic and cultural backgrounds (Camargo-Moro, 1993). These two examples appear to be much closer to the original concept of ecomuseums, with community involvement and high environmental goals.

Holistic museology

Existing ecomuseums, through their variety of modes of operation and differing objectives, reflect the difficulty of defining them – each ecomuseum appears to have its own opinions about what its role should be, and few fulfil the dreams of de Varine and Rivière. It would be an easy matter to dismiss the idea out of hand. However, despite a discernible decline in popularity, the movement has resulted in an important re-evaluation of the museum role, a catalyst enabling museums 'to bypass conservatism and to make better sense of objects surviving from the past' (Hudson, 1992). As a result, from the ecomuseum philosophy, and from new ideas for 'integral museums' and 'neighbourhood museums', a 'new museology' has emerged (Mayrand, 1985) with the community as a central tenet. It is only a small mental leap from new museology to something which encompasses community and environment, the 'holistic museology' proposed by Corsane and Holleman (1993); they argue that within the framework of an evolving South Africa, there is a need for museums and curators to move away from an interpretive approach which is Eurocentric to one which is multicultural, and the holistic paradigm enables the history museum to celebrate cultural differences set against an environmental backdrop. Here museology is setting itself social and environmental goals, and looking to the future, rather than the past. To the new museology, which has promoted social and political awareness, holistic museology adds an extra dimension, environmental awareness.

A holistic approach could be used in any museum and by all disciplines, and in extensive geographical areas with a recognizable cultural and natural identity, involve more than one type of organization. In such a situation museums could work in co-operation with National Park authorities and voluntary conservation organizations, as has been described for the Cévennes and suggested for Dartmoor (Endacott, 1992). Combining progressive museum activities with other interpretive facilities in the countryside or urban areas provides the opportunity to emphasize the threads which link natural resources with cultural identity, nature with humankind.

References

Aldridge, D. (1971) 'Landmark Visitor Centre, Carrbridge, Inverness', *Museums Journal*, June, 3–6.

Aldridge, D. (1975) *Guide to Countryside Interpretation: Principles of Countryside Interpretation and Interpretive Planning*, HMSO, Edinburgh.

Aldridge, D. (1981) *The Monster Book of Environmental Interpretation*, Geo Abstracts/Council of Europe.

Almeida-Moro, F. (1985) 'Sao Cristova: a district ecomuseum, *Museum*, 37 (4), 237–241.

Barrow, G. (1988) *Visitor Centres: An Introduction in Environmental Interpretation*, Centre for Environmental Interpretation, Manchester.

Beazley, E. (1969) *Designed for Recreation: A Practical Handbook for All Concerned with Providing Leisure Facilities in the Countryside*, Faber and Faber, London.

Beazley, E. (1971) *The Countryside on View*, Constable, London.

Binette, R. (1991) 'L'Ecomusée de la Maison du Fier Monde', *Muse*, VIII (4), 8–9.

Blunden, J. and Curry, N. (1990) *A People's Charter?: Forty years of the National Parks and Access to the Countryside Act 1949*, HMSO, London.

Boylan, P. (1992) 'Ecomuseums and the new museology; definitions', *Museums Journal*, 92 (4), 29.

Bracey, H. E. (1970) *People and the Countryside*, Routledge and Kegan Paul, London.

Bromley, P. (1994) *Countryside Recreation: a Handbook for Managers*, E. & F. N. Spon, London.

Camargo-Moro, F de (1993) 'From the ecomuseum to the integrated site museum', *ICOM News*, 46 (2), 11–13.

Choquette, W. (1988) 'Developing a Kootenay Ecomuseum'. In Lunn, J. (ed.) *Proceedings of the First World Congress on Heritage Presentation and Interpretation, Banff, Canada, 1985,* Heritage Interpretation International and Alberta Culture and Multiculturalism, Edmonton, Canada, 215–228.

Corsane, G. and Holleman, W. (1993) 'Ecomuseums: a brief evaluation'. In De Jong, R. (ed.) *Museums and the Environment,* Southern African Museums Association, Pretoria.

Countryside Commission (1995) *The Country Park Welcome Charter; Visitor Welcome Checklist,* Countryside Commission, Cheltenham.

Curry, D. A. (1969) 'Warleigh Wood Field Museum and Educational Nature Reserve', *Museums Journal,* September, 70–71.

Davies, M. (1987) 'Wildlife as a visitor attraction'. In Talbot-Ponsonby, H. (ed.) *Recreation and Wildlife: Working in Partnership,* Proceedings of the 1987 Countryside Recreation Conference, Countryside Recreation Research Advisory Group, Bristol.

Deane, C. D. (1965) 'A Field Museum and Nature Trail in Northern Ireland', *Museums Journal,* 65 (2), 97.

Duclos, J-C. (1980) 'The Camargue Museum, Mas du Pont de Rousty, Arles, France', *Museum,* 32 (1/2), 22–32.

Edington, J. M. and Edington, M. A. (1986) *Ecology, Recreation and Tourism,* Cambridge University Press, Cambridge.

Endacott, A. (1992) 'Change in the countryside', *Museums Journal,* 92 (4), 32–33.

Engstrom, K. (1985) 'The ecomuseum concept is taking root in Sweden', *Museum,* 37 (4), 206–210.

Everhart, W. C. (1972) *The National Park Service,* Praeger Publishers, New York.

Evrard, M. (1980) 'Le Cruseout – Montceau-les-Mines: The life of an ecomuseum, assessment of ten years', *Museum,* 32 (4), 226–234.

Flores, F. P. (1980) 'The Montoso Ecomuseum of the Rainforest and the Natural History of Puerto Rico', *Newsletter of the ICOM Natural History Committee,* No 6, Paper/texte No 8.

Fortier, J. (1988) 'Heritage interpretation in Canada'. In Lunn, J. (ed.) *Proceedings of the First World Congress on Heritage Presentation and Interpretation, Banff, Canada, 1985,* Heritage Interpretation International and Alberta Culture and Multiculturalism, Edmonton, Canada, 37–46.

Gilbert, D. L. (1964) *Public Relations in Natural Resource Management,* Burgess Publishing Company, Minneapolis.

Greenwood, E. F. (1976) 'Smithills Hall trailside museum, Bolton', *Museums Journal,* 76 (2), 66.

Grodwohl, M. (1987) 'L'Ecomusée de Haute-Alsace', *Musées et Collections Publiques de France,* 177, 39–43.

Harrison, R. (ed.) (1994) *Manual of Heritage Management,* Butterworth-Heinemann and the Association of Independent Museums, Oxford.

Himsworth, K. S. (1969) 'The Lake District National Park Centre at Brockhole, Windermere', *Museums Journal,* 69 (2), 65–66.

Hodgson, R. W. and Fritschen, J. A. (1984) 'Evaluating Interpretation'. In *Supplement to a Guide to Cultural and Environmental Interpretation in the US Army Corps of Engineers, Instruction Report R-84-1,* US Army Engineers Waterways Experiment Station, Vicksburg, MO, 57–85.

Hubert, F. (1985) 'Ecomuseums in France: contradictions and distortions', *Museum,* 37 (4), 186.

Hudson, K. (1992) 'The Dream and the Reality', *Museums Journal,* 92 (4), 27.

Jacobsen, B. and Harris, S. (1988) 'Heritage presentation and interpretation in the South Western Pacific'. In Lunn, J. (ed.) *Proceedings of the First World Congress on Heritage Presentation and Interpretation, Banff, Canada, 1985*, Heritage Interpretation International and Alberta Culture and Multiculturalism, Edmonton, Canada, 57–60.

Jullien, R. (1989) 'Le Musée hors les murs: les tendances contemporaines en histoire naturelle', *Musées et Collections Publiques de France*, 182–183, 45–47.

Kavanagh, G. (1990) *History Curatorship*, Leicester University Press, Leicester, 16–19.

Konare, A. O. (1985) 'Ecomuseums for the Sahel: a programme', *Museum*, 37 (4), 230–236.

Kool, R. and Trayner, K. (1990) 'The JASON Project – an evaluation', *Curator*, 33 (4), 261–271.

Kusel, U. (1993) 'Museums without walls (A holistic approach to conservation)'. In De Jong, R. (ed.) *Museums and the Environment*, Southern African Museums Association, Pretoria, 137–142.

Lawes, G., Sekers, D. and Vigurs, P. F. (1992) 'Defining the undefinable – Ecomuseums: a Cinderella or another Ugly Sister?', *Museums Journal*, 92 (9), 32.

Le Doaré, J. (n.d.) *The Armorique Regional Nature Park*, [Guidebook].

Lewis, R. H. (1993) *Museum Curatorship in the National Park Service, 1904–1982*, Department of the Interior, National Park Service, Curatorial Services Division, Washington DC.

Longland, J. L. (1964) 'Education'. In The Nature Conservancy (ed.) *The Countryside in 1970: Proceedings of the Study Conference held at the Fishmongers Hall, London, EC4, 4–5 November 1963*, HMSO, London, 52–55.

Lowenthal, D. (1988) 'Heritage and its interpreters'. In Lunn, J. (ed.) *Proceedings of the First World Congress on Heritage Presentation and Interpretation, Banff, Canada, 1985*, Heritage Interpretation International and Alberta Culture and Multiculturalism, Edmonton, Canada, 7–28.

Machlis, G. E. (1986) *Interpretive views*, National Parks and Conservation Association, Washington DC.

Marsh, J. S. (ed.) (1986) *Natural and Cultural Heritage Interpretation Evaluation*, Interpretation Canada, Ottawa.

Marsh, J. S. (1988) 'Heritage presentation and interpretation in the oriental realm'. In Lunn, J. (ed.) *Proceedings of the First World Congress on Heritage Presentation and Interpretation, Banff, Canada, 1985*, Heritage Interpretation International and Alberta Culture and Multiculturalism, Edmonton, Canada, 47–55.

Marsh, J. S. (1988a) 'Heritage interpretation evaluation: needs and methods'. In Lunn, J. (ed.) *Proceedings of the First World Congress on Heritage Presentation and Interpretation, Banff, Canada, 1985*, Heritage Interpretation International and Alberta Culture and Multiculturalism, Edmonton, Canada, 285–294.

Mayrand, P. (1985) 'The new museology proclaimed'. *Museum*, 37 (4), 200– 201.

Moniot, F. (1973) 'The Ecomuseum of Marqueze, Sarbres: part of the Regional Natural Park of the Landes de Gascogne'. *Museum*, 25 (1/2), 79–84.

Nabais, A. (1985) 'The development of ecomuseums in Portugal'. *Museum*, 37 (4), 211–216.

Naisbitt, J. (1982) *Megatrends*, Warner Books, New York.

O'Brien, C. (1988) 'Current status of interpretation in Australia'. In Lunn, J. (ed.) *Proceedings of the First World Congress on Heritage Presentation and Interpretation, Banff, Canada, 1985*, Heritage Interpretation International and Alberta Culture and Multiculturalism, Edmonton, Canada, 61–79.

Peart, B. (1988) 'Beyond Freeman Tilden'. In Lunn, J. (ed.) *Proceedings of the First World Congress on Heritage Presentation and Interpretation, Banff, Canada, 1985*, Heritage Interpretation International and Alberta Culture and Multiculturalism, Edmonton, Canada, 107–112.

Pennyfather, K. (1975) *Guide to Countryside Interpretation; Interpretive Media and Facilities*, HMSO, Edinburgh.

Querrien, M. (1985) 'Taking the measure of the phenomenon'. *Museum*, 37 (4), 199.

Rivard, R. (1985) 'Ecomuseums in Quebec'. *Museum*, 37 (4), 202–205.

Rivière, G. H. (1973) 'Role of museums of art and of human and social sciences', *Museum*, 25 (1/2), 26–44.

Rivière, G. H. (1985) 'The Ecomuseum: an evolutive definition', *Museum*, 37 (4), 182–183.

Roggenbuck, J. W. and Propst, D. B. (1981) 'Evaluation of Interpretation', *Journal of Interpretation* 6 (1), 13–23.

Schuylkill Center (1993) *The Schuylkill Center: Information Pack*, The Schuylkill Center for Environmental Education, Philadelphia.

Stansfield, G. (1967) 'Museums and the Countryside', *Museums Journal*, 67, 212–218.

Stansfield, G. (1970) 'The 1969 Countryside Centres Conference', *Museums Journal*, 70, 17–18.

Stansfield, G. (1983) 'Heritage and Interpretation', *Museums Journal*, 83 (1), 47–51.

Stevenson, S. (1982) 'The territory as Museum; New Museum Directions in Quebec', *Curator*, 25 (1), 5–16.

Tilden, F. (1957) *Interpreting Our Heritage*, University of North Carolina Press, Chapel Hill.

Torkildsen, G. (1986) *Leisure and Recreation Management*, E. & F. N. Spon, London.

Tynan, A. M. (1969) 'The Border Forest Park Museum, Northumberland', *Museums Journal*, September, 69–70.

Vander Stoep, G. A. (ed.) (1991) *Interpretation; A Resource and Curricula Guide for the United States and Canada*, National Association for Interpretation, Fort Collins, CO.

Van Mensch, P. J. A. (1993) 'Museology and the management of the natural and cultural heritage'. In De Jong, R. (ed.) *Museums and the Environment*, Southern African Museums Association, Pretoria, 57–62.

Varine, Hugues de (1973) 'A fragmented museum: the Museum of Man and Industry, Le Creuseot – Montreau-les-Mines', *Museum*, 25 (4), 292–9.

Wasserman, F. (1989) 'Les Ecomusées, ou comment une population reconnait, protège, met en valeur les richesses naturelles et culturelles de son territoire', *Musées et Collections publiques de France*, 182–183, 53–55.

Wilson, A. (1992) *The Culture of Nature; North American Landscape from Disney to the Exxon Valdez*, Blackwell, Cambridge: MA.

Wood, W. (1991) 'The Cowichan and Chemainus Valleys Ecomuseum; The ecomuseum concept as a successful planning tool for government and the governed', *Muse*, VIII (4), 10–11.

6. *Biodiversity – issues and opportunities*

> The last word in ignorance is the man who says of an animal or plant 'What good is it?' If the land mechanism as a whole is good, then every part of it is good, whether we understand it or not. If the biota, in the course of aeons, has built something we like but do not understand, then who but a fool would discard seemingly useless parts? To keep every cog and wheel is the first precaution of intelligent tinkering.
>
> Aldo Leopold, *The Round River*

> When you have seen one ant, one bird, one tree, you have not seen them all.
>
> Edward O. Wilson, *Time*, 13 October 1986

Wilson (1992) defines biodiversity as:

> The variety of organisms considered at all levels, from genetic variants belonging to the same species through arrays of species to arrays of genera, families and still higher taxonomic levels; includes the variety of ecosystems, which comprise both the communities of organisms within particular habitats and the physical conditions under which they live.

In essence it is the variety of life which has sustained and fascinated humankind; the variety of life which has been collected so avidly to stock museums and of which museums have taken stock.

The threat posed to the myriad life forms on Earth by a growing and ever-demanding human population has been labelled the 'biodiversity crisis'. Although the acceptance and usage of the term 'biodiversity' is a product of the late 1980s, (and more widely used since the Earth Summit), the threat to habitats and individual species is one of the major causes which has exercised the mind and energies of the environmental lobby since the 1960s. Campaigns to aid individual species (for example, Operation Tiger) and groups of animals such as Greenpeace's fight to protect cetaceans, have been supported by large scale publicity efforts which have drawn public attention to the loss of larger species. Such campaigns are important, but demands for more widespread action have increased as the scale of habitat and species loss has been recognized by the scientific community. Speaking at the ICOM Conference in 1992 Des Griffin, Director of the Australian Museum, Sydney reminded delegates that in the one week of the Conference '1.77 million people were added to the world's population, 950 species of plants and animals became extinct, 400,000 hectares of rainforest were destroyed' (Griffin, 1992). There is now widespread concern for the loss of species, and 'the academic community of biology now sees the biodiversity crisis as a very real phenomenon meriting our closest scrutiny.' (Eldredge, 1992).

The loss of animal and plant species is not a new phenomenon. Five large-scale extinctions have occurred in the last half-billion years of geological time, the best-known being that recorded at the Cretaceous–Tertiary boundary some 65 million years ago which saw the extinction of the dinosaurs. This was not the most severe – the Permian extinction 265 million years ago resulted in the loss of between 77 and 96 per cent of all marine animal species – 'if these estimates are reasonably accurate, global biology (for higher organisms at least) had an extremely close brush with total destruction' (Raup, 1988).

So is the current biodiversity crisis (which differs from the others in being caused by the unthinking actions of one species) the sixth major extinction phase in the history of life on earth? Wilson (1992) leaves us in little doubt that this is the case as he charts the long history of man's destruction of wildlife. He cites the colonization of the Pacific islands, where the Lapita:

> subsisted ... especially in the early days of settlement ... [on] whatever edible animals they encountered. They ate fish, turtles, and a profusion of bird species that had never seen a large predator and were easily caught .... The voyagers ate their way through the Polynesian fauna. On Eua, in present-day Tonga, twenty-five species [of birds] lived in the forests when the colonists arrived around 1,000 years ago, but only eight survive today.

The depredations of Europeans added to the destruction of the Pacific fauna caused by the first colonizers; referring to Hawaii, Wilson (1992) notes that:

> When European settlers arrived after Captain Cook's visit in 1778, there were approximately fifty native species of landbirds. In the following two centuries, one-third disappeared. Now we know from bone deposits that another thirty-five species identified with certainty, and very likely twenty other species less-well documented, had already been extinguished by the native Hawaiians.

Humans have been cited as a causal factor of the extinction of species (mammoth and ground sloths for example) as far back as the Pleistocene; there is strong circumstantial evidence to link the collapse of diversity of late-Pleistocene faunas with the influx of man in North America, Madagascar and New Zealand. The first Maoris, colonizing New Zealand around AD 1000 found about 13 species of large flightless birds – the moas – which had evolved to fill the niches normally taken by mammals, which were absent on these remote islands. All moa species had been hunted to extinction by about 1300. This is just one well-known example of species extinction – other notable fatalities include the Dodo, Great Auk, Passenger Pigeon, Carolina Parakeet and Quagga – specimens of which are now treasured by many museums (see Plate 21). Although all these species were hunted to extinction, other factors equally culpable were habitat destruction, the introduction of exotic, competitive species, and the spread of

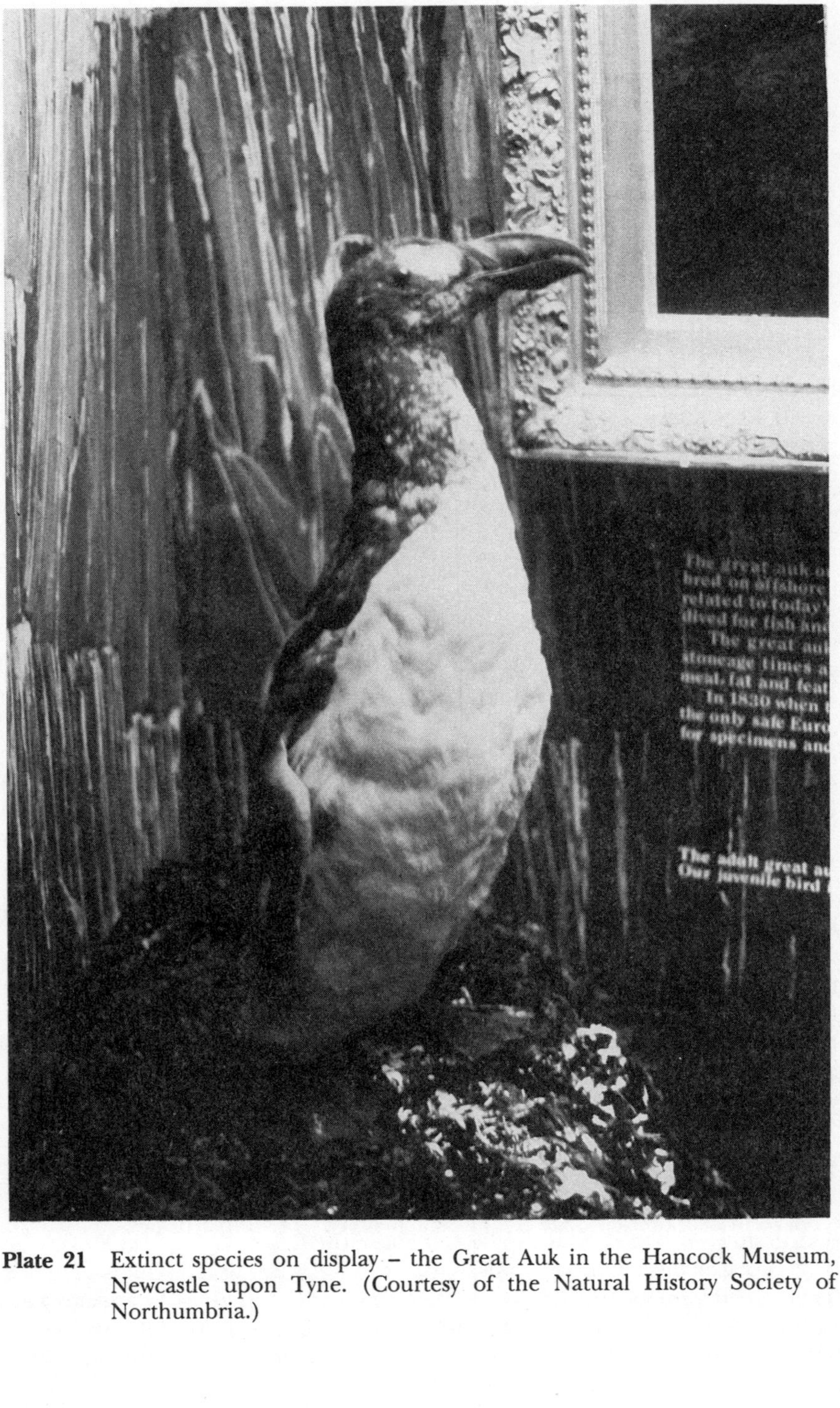

Plate 21 Extinct species on display – the Great Auk in the Hancock Museum, Newcastle upon Tyne. (Courtesy of the Natural History Society of Northumbria.)

disease carried by such exotics – the effect was dramatic, on the landscape and on habitats, as well as on species. If hunting was the primary cause of extinctions up until the end of the nineteenth century, there can be little doubt that habitat destruction has taken its place as the root cause of the biodiversity crisis – not through human malevolence, but simply the growing demands for living space and natural resources. Much of the loss of habitat is recent – Wilson (1992) cites a number of examples including that of Madagascar, with its spectacular endemic animals, including 30 primates, (all lemurs), and two-thirds of the world's chameleons and an estimated 10,000 plant species about 80 per cent of which are endemic. 'In 1985 the forest remaining intact was down to a third of the cover encountered by the first colonists fifteen centuries ago. The destruction is accelerating along with population growth, with most of the loss having occurred since 1950.' It is inevitable that the greatest loss of species will be felt in parts of the world, like Madagascar, with greatest diversity of species – the so-called 'hotspots' – it is here that biodiversity research and conservation efforts must be concentrated.

At what rate is the world losing species? No precise estimates can be made – it is not known how many species there are, habitats are frequently fragmented, and the distribution range of most species is unknown. Despite these problems, some calculations have been made, particularly with reference to the most biologically diverse habitats, including the tropical rainforests, which cover only 7 per cent of the earth's land surface, yet are estimated to contain 50 per cent of all species (Myers, 1988). Remote sensing surveys have done much to provide objective data on the scale of deforestation; Norman Myers (1989) estimated the loss in area of tropical rainforest as 1.8 per cent each year; from these figures, and using a model derived from island biogeography it is possible to determine the rate of species loss.

The model (MacArthur and Wilson, 1967) states that $S = CA^z$, where S is the number of species, A is the area of the place the species live, and C and z are constants; C can be effectively ignored, and the value of z depends on the kind of organism and the habitat under consideration, but a typical z value is 0.30. As a general rule of thumb, the model states that when an area is reduced to one tenth of its original size, the number of species is halved; knowing this Wilson (1992) concluded that 'reduction in the area of tropical rainforest at the current rate can be expected to extinguish or doom to extinction about half a percent of the species in the forests each year' and:

> if deforestation continues for thirty more years at the present rate, one tenth to one quarter of the rainforest species will disappear. If the rain forests are as rich in diversity as most biologists think, their reduction alone will eliminate 5 to 10 percent or more – probably considerably more – of all the species on earth in thirty years.

Taking the lowest z value possible (0.15) and assuming that the number of rainforest species is in the order of 10 million (certainly an underestimate), and knowing the rate of forest destruction, Wilson concludes that 27,000 species become extinct each year, equivalent to 74 species per day – 'clearly we are in the midst of one of the great extinction spasms of geological history'.

Why does biodiversity matter?

The assumption is continually made that high biodiversity is essential for the good of the planet – surely it is common sense that more species mean a healthier ecosystem? Yet ecological and energetics research had been unable to prove that this was the case; the costs of re-creating ecosystems in a meaningful way (i.e. so that all inputs and outputs can be measured) had, until recently, proved prohibitive. As a consequence the argument for maintaining high biodiversity has centred on its utilitarian value – as a supply of genes to enhance our crops, as a source of medicines, as a regulator of rainfall and water supply, or simply for aesthetic reasons. However, recent research carried out at London's Imperial College, using an ecosystem simulator called the 'Ecotron' (a facility which cost £1 million) has now answered the fundamental question – what does biodiversity *do*? It is simply that the more species you have, the more efficiently organisms can harness energy from sunlight – more diverse ecosystems are more productive (Cherfas, 1994).

Other recent research (e.g. McGinley and Tilman, 1993; Tilman and Pacala, 1994) has revealed that more complex ecosystems are more stable, are less affected by outside stress (for example, drought, fire or high grazing pressure impacts on grasslands) and recover more quickly from it. Biodiversity is therefore a safety net against adverse environmental conditions, as well as a means of maximizing productivity. Two broadly opposed models have been proposed for the link between species diversity of an ecosystem and its stability – Cherfas (1994) refers to these as 'rivet' and 'passenger' models:

> In one of the models, each species in an ecosystem can be likened to one of the rivets that holds an aircraft's wings to the fuselage: every rivet plays its part in keeping the plane intact, and the loss of even one or two will weaken it. Lose more than a few and the system is in for a catastrophic decline. The alternative ... [the passenger model] ... sees the species as the people on the plane – crew as well as passengers. You could remove most of them, but as long as a few key individuals remain the system will still fly. On the other hand, removing those key individuals will lead to disaster, no matter how many other passengers remain.

Which is the correct model is a major question of significance for human survival – how many species can we afford to lose before our life-support system begins to collapse? If it is the rivet model then we are already in trouble, if the passenger model then we need to identify the crew as a matter of urgency. Little wonder then that biodiversity issues were a major topic of debate in Rio.

Biodiversity – a global political issue

The 'Earth Summit' in Rio brought together well over 100 heads of state and government, and involved no less than 178 countries. The end result was five major agreements – *Forest Principles* (a failed attempt to negotiate a forest convention), *The Rio Declaration* (covering environment and development), *Agenda 21* (an action plan for sustainable development), the *Framework Convention on Climate Change* (seeking to reduce the emission of 'greenhouse' gases) and the *Convention on Biological Diversity*, which focuses on the preservation of species diversity and seeks to establish guidelines for the use of biological resources and biotechnology. Biological diversity has been the focus of natural history museums from the eigtheenth century; consequently the Biodiversity Convention is of immense importance to museum biologists, especially those engaged in taxonomic research. Particularly relevant is the fact that making biodiversity a political issue, and providing international legislation, may release funds for the urgent taxonomic work which is required in order to document the world's biological resources. *Agenda 21*, stating that 'substantial new and additional funding ... will be required' estimated that a global programme to conserve biological diversity would total US$3,500 million per year in the developing countries (Grubb *et al.* 1993).

It has been estimated that there are about 30 million species on the Earth, and that about a quarter of these face extinction within the next 30 years. Approximately one-and-a-half million species have been described (UNEP, 1992). The scientific case outlining the erosion of biodiversity is overwhelming, and the environmental ethic now so strong in developed countries that action is urgently required; it is therefore unfortunate (and potentially disastrous) that negotiations to draft a legal instrument became enmeshed in an argument between countries about the exchange of biological resources in return for biotechnological expertise. Consequently, the Biodiversity Convention reflects compromise between the developed countries, who view biodiversity as common heritage, and the developing countries' insistence on sovereign rights over genetic resources. The Objectives of the Convention are stated as:

> conservation of biological diversity, the sustainable use of its components and the fair and equitable sharing of the benefits arising out of the utilisation of genetic resources, including by appropriate access to genetic resources and by appropriate transfer of relevant technologies.

This emphasis on the utilitarian reasons for conserving biodiversity is far removed from the deeper philosophical concerns of most environmentalists. Despite this divergence of views, the Convention, by recognizing that genetic resources are important, stresses the need for developed and developing nations to work together to maintain biodiversity. Whilst taking this pragmatic approach, the Convention preamble takes a broader view, stressing the 'ecological, genetic, social, economic, scientific, educational, cultural, recreational and aesthetic' values of maintaining species diversity.

From the museum-based taxonomists' point of view, Article 6 is of particular relevance, stating each Party:

> shall, in accordance with its particular conditions and capabilities, develop national strategies, plans or programmes for the conservation and sustainable use of biological diversity or adapt for this purpose existing strategies, plans or programmes which shall reflect, *inter alia*, the measures set out in this Convention relevant to the Contracting Party concerned.

As part of these plans, under Article 7, each country is required '. . . as far as possible and as appropriate' to 'identify components of biological diversity important for its conservation and sustainable use' whilst 'paying particular attention to those requiring urgent conservation measures and those which offer the greatest potential for sustainable use'. As Grubb *et al.* (1993) have identified 'This may encourage governments to neglect the conservation of those species with aesthetic or cultural value only' – into which categories the majority of species fall. Despite the rather weak character of Article 7, (and indeed of the Convention as a whole) there is a prerequisite for producing inventories of species and to assess which are rare or endangered; from this data appropriate conservation strategies can be formulated. Museums, and museum biologists, must be at the cutting edge of such taxonomic and biogeographical studies; the science of systematics and taxonomy carried out in the world's great natural history museums now has increased relevance and purpose. The results of that research, and the collections upon which it has been based, are the fundamental resource on which to launch a renewed effort to document and conserve the world's biota.

Biological collections – some inherited problems and the consequences for the museum's role in sustaining biodiversity

Interestingly, within the museum world of the 1960s little reference was made to the loss of species, or to the museum role in conservation (e.g. Oliver, 1969); taxonomic research took centre stage, with only passing reference to the potential benefits of that research for environmental protection (Netting, 1962). Yet the impact of environmentalism was such that in the USA a major review of the nation's biological collections carried out in the early 1970s pivoted around the theme of biological conservation:

> the very cornerstones to studying, understanding and managing natural ecosystems are the systematic collections of the plant and animal species of the world . . . the major systematic collections are essential complementary components of a system which catalogs the world biota, and which deserves strong support by this nation to assure our gaining essential insights into our own relationship to that tortured biota (Conference of Directors of Systematic Collections, 1971).

This emotive statement is itself a strong reflection of the mood of the times, a statement of the impact that environmentalism had made. The realization of the central role of biological collections in biodiversity studies has continued to gain ground since the 1970s, and many museum directors (especially in the USA) have publicly stated their institutions' commitment to the role of preserving the Earth's biological diversity (e.g. Nicholson, 1991; Novacek, 1990). It is salutory to compare the readiness of larger institutions in North America to adopt the environment to the situation in Britain, where despite the best efforts of museum biologists, progress is constantly hampered by funding crises and conflicting priorities.

How well equipped are museums in the mid-1990s to become involved in the fight to save the planet's plants and animals? The legacy of past collecting policies, the perceived low status of natural history collections (especially in multidisciplinary museums), pose many problems which need to be solved if museums are to become key players in maintaining biodiversity. The downgrading of systematics as an academic discipline, and a lack of understanding of the significance of its role, is of equal concern, and is addressed in the next chapter.

The urge to collect natural history specimens has resulted in collections of tremendous scale and diversity in the museums of the world's developed countries, and the history of the emergence and development of many of the more outstanding museums has been well documented (e.g. Stearn, 1981; Strahan, 1979; Winsor, 1991). Large-scale natural history collections were made for two main reasons – first to provide education and enjoyment for the general public (often focusing on large and attractive organisms, especially birds and mammals) and secondly, for taxonomic and biogeographic research. The latter frequently had little or nothing to do with any educational or conservationist aim – it was scientific research for its own sake. Past collecting efforts in both these categories, despite their often seemingly random nature, can only now be appreciated, as the collections begin to realize new meaning and significance – for example, the well documented use of biological collections in determining the increase in pollutants in the environment (e.g. Johnels, 1973). In this, and many other ways, (see Pettitt, 1991 and 1994; Danks, 1991; Wiggins *et al.*, 1991) historical collections housed in museums and other institutions chart our understanding of the Earth's fauna and flora and its past distribution – our collections and their associated data give us a unique view of past worlds, and a yardstick with which to measure change.

It is unlikely that the curator of a mid-nineteenth century museum, engrossed in collection-building, identifying and describing new species, and preparing named examples for display, would have acknowledged that his work had any significance in terms of conservation. It was the heyday of natural history, and despite some small voices of protest about over-collecting being raised, huge collections of plants, animals and fossils were being made by individuals and institutions. Some specialist private collections rivalled those of the national museums – the shell collection of Hugh Cuming (1791–1865) was estimated to contain 52,789 specimens, including many types in 1846 (Barber, 1980); Walter Rothschild (1868–1937) used his fortune to establish his personal museum at Tring in Hertfordshire, the largest collection in natural history ever assembled by one man. Rothschild employed more than 400 collectors throughout the world – the resulting collections being used to describe more than 5,000 new species and subspecies (Purcell and Gould, 1992). These (and many other) great collections are well-known, and made an enormous contribution to our understanding of the diversity of life, yet throughout Britain, and much of the world, other individuals were making collections on a smaller scale – most of the important individual collections found their way into museums. The number of such 'named' nineteenth- and early twentieth-century collections held in museums is still not yet known, but the preparation of individual catalogues of natural history collections held in museums and other institutions in the various regions of Britain (e.g. Davis and Brewer, 1988) have done much to reveal their size, strengths and diversity. There is no doubt that it is an enormous resource for the understanding of biodiversity; collections of international significance can be found in the most unsuspected institutions, and even the smallest local museum may hold collections which help to reveal the status of species and the significance of biological and geological sites in its area – the most important prerequisite to any conservation effort.

Often led by charismatic and individual directors, there can be little doubt that personal, civic, national and institutional pride fuelled by a sense of competition, spurred museums to finance substantial collecting expeditions (as in the United States) or encourage the purchase and donation of collections (as in the UK). There were few thoughts of collaboration between museums on collecting policies in the nineteenth and early twentieth centuries, no 'National Plans' for systematic collections and collecting emerged until the 1970s (see Irwin *et al.*, 1973). As a consequence there has been little specialization by individual institutions, and every major museum has collections which cover 'popular' taxa such as Lepidoptera, Mollusca or Mammalia. Similarly, geographical boundaries of collecting activity received only lip-service, and hence collections in most major museums reflect a haphazard and sometimes bizarre geographical distribution pattern. The lack of a clearly defined collecting policy has meant that provincial museums have frequently accepted as donations material which lies well beyond their geographic sphere of influence and which they frequently have neither the expertise nor funding to curate and conserve, let alone actively research. This might, to the outsider, seem an unprofessional approach. However, as donations frequently come from nearby research institutions or individuals who have some association with the museum, but whose collecting area (geographically and taxonomically) falls well outside that of the museum, they are difficult if not impossible to refuse – curators are regularly compromised by such links. The Hancock Museum in Newcastle upon Tyne, England perfectly illustrates the wayward nature of past collecting, holding, for example, a collection of Australian land-shell (including type specimens) collected by George French Angas (1822–1886), the C. H. E. Adamson (d. 1930) collection of Burmese butterflies, and a world-wide collection of micro-crustacea of international significance made by George Stewardson Brady (1832–1921). There seems to be little logic in such collections being curated in the north of England, but the reason is simply that all three collectors had strong links with the city of Newcastle and its Natural History Society, which ran the museum.

This situation is repeated world-wide; for example, in France's 187 provincial natural history museums '... sont conservées plusiers dizaines de millions d'echantillons et de spécimens. Ils sont des temoins non seulement de l'histoire naturelle de notre pays mais aussi du monde entier ... ' (Leclaire, 1989). This world-wide scatter of collections has prompted some individuals (e.g. Haas, 1993) and organizations to suggest repatriation as a solution, in particular when collections are poorly curated or underutilized – The Royal Society in its evidence to the Dainton inquiry (Dainton, 1991, 6.23) suggested the cost of curation might be mitigated by 'relocating specific collections via long-term loan to relevant user groups in other countries. ... ' Some evidence of this actually occurring is the transfer of a collection of Irish Lepidoptera to the Ulster Museum, Belfast, from the Smithsonian (John Wilson, *pers. comm.*). The arguments for maintaining the integrity of the UK collections (the ability to compare material, concerns about repatriation to developing countries) are much stronger, yet there is little doubt that as a consequence of historical collecting a huge amount of information vital to the biodiversity cause is locked away, seemingly inaccessible. However, in the UK for example, the increasing number of individual museums which have generated computer databases of their collections, the work of regional collections research units (CRUs), and the pioneering work of the Federation for Natural Science Collections Research (FENSCORE) in generating a national database for natural science collections (Pettit, 1993) gives cause for some optimism. Couple these advances with the ability in future to access the databases through the World Wide Web, and there is a real revolution of great significance, a start has been made on cutting a key to unlock data which is currently inaccessible.

The biodiversity debate at the Madrid Conference – the implications for the care of biological collections

The International Symposium and First World Congress on the Preservation and Conservation of Natural History Collections held in Madrid in 1992 was primarily a forum to discuss the need better to prepare, conserve, curate and store biological and geological collections, yet a number of important contributions focused on the role of collections as they relate to the biodiversity debate. The published accounts (e.g. May, 1993; Lovejoy, 1993) hint at the underlying conservation ethic (that is, the conservation of species and biological and geological sites in the field) which has pervaded all aspects of work within natural history museums since the 1960s and the beginning of environmentalism. It is clear (although not always admitted even by museum biologists and geologists themselves) that the reasons for collecting, preserving and curating natural history collections go beyond simply the pursuit of strictly scientific or even curatorial goals. They are to ensure the survival of species and sites of natural history interest, and hence ultimately to serve society (Davis, 1994).

The Preamble to the Resolutions from the Madrid Conference (Anon, 1992) includes reference to human impact on the environment, and in particular the way in which increasing demands are reducing biodiversity – all the resolutions and their accompanying recommendations are focused back to this prime issue:

> Knowledge and understanding of biodiversity is essential for the conservation, management, and sustained use of ecosystems. Thus, natural history collections are important for the study of biodiversity. Recent research demonstrates that our catalogues of biodiversity are seriously inadequate (as much as 90% of species diversity is unknown to science) and urgently needs to be improved, so that science can place proper values on our natural resources and provide models to predict the consequences of lowered biodiversity.

The concept that museums are 'libraries of life and supporting geological structures' makes them a 'priceless archive.' According to Howie (1993) the priceless archive (an estimated 2 billion specimens) has:

> been slowly deteriorating over the past two centuries. The extent and understanding of the processes involved are as yet largely undocumented and unknown ... recent surveys ... reveal that some areas are at crisis point through the unchecked growth of collections, poor training in conservation and collection management procedures, and pure indifference.

The condition of many natural history collections prompts Howie to the conclude that ' ... a third of the world's natural history collections is in an extremely poor state with possibly as many as thirty million specimens per year deteriorating to the extent that they are of no future benefit.' The growing number of orphaned collections supports this view. Emery (1993) has itemized the threats posed to collections by museum managers, curators, collections managers, economic pressures, use of the collections, the cost implications and currently available technologies for preservation and conservation, and the type of specimen itself.

Clearly there is a real problem here if the collections held in museums are to be of any value to the needs of the biodiversity crisis. There is a real need for better collections management and improved conservation facilities, for funding which would enable research on conservation problems and techniques, and above all a commitment of management to the preservation of collections. The formation of specific organizations such as the National Institute for the Conservation of Cultural Property (NIC) (1973), the Society for the Preservation of Natural History Collections (SPNHC) (1985) in the USA and the Natural Sciences Division of the United Kingdom Institute for Conservation (UKIC) (1993), now (1996) known as the Natural Sciences Conservation Group, has meant a growing voice for the needs of collections, and a focus for improving standards of collection care.

One of the major resolutions of the Madrid Conference was the formation of the World Council on Collections Resources (WCCR). The Council has a wide-ranging remit, but some specific aims relating to collections include:

- to promote exchange of information and technology relating to the preservation and conservation of natural history collections;
- to promote high professional standards and establish educational programmes for collection managers and conservators;
- to work with national organizations to help society understand the mission and value of natural history collections and the needs for professional care of collections, large and small, in perpetuity.

Patterns of biodiversity and the distribution of natural history museums

A simplified view of the pattern of biological diversity is that it is greatest in the equatorial regions, especially in areas of tropical forest, becoming less diverse with increasing latitude. This pattern seems to hold true for larger vertebrates and higher plants, but has been questioned for some invertebrate groups, for example, spiders (Platnick (1992)). This generalized pattern of distribution suggests that as a consequence the greatest loss of biodiversity is occurring in equatorial, often developing countries, and it is here that systematic collecting of biological material is most urgently required in order to determine which areas are of greatest significance and to promote active conservation policies for them.

Those developing countries are facing a real dilemma as a result – many lack the systematics skills required, many do not have a national museum of natural history of any standing. The very real economic pressures such countries face mean that insufficient resources can be put into scientific education, museums or environmental protection. If biodiversity is to be considered as a global resource, as the Rio declaration has suggested, then equally the global scientific community has an obligation to work with developing countries, to develop scientific research programmes and to implement systematics training. Natural history museums are well placed, indeed possibly the most important organizations available to guide and conduct research and to influence public opinion in developing countries. There is a real need to promote museum developments, and to ensure that staff are trained in systematics and collections management. If the larger perspective promoted by environmentalism – the image of a small blue, fragile planet is still a strong one – is to have any real effect, then it will require a major commitment by the large, well-financed museums, which are primarily in the northern hemisphere.

Mares (1993) has made a comparative study of the development of museums in the USA and in the countries of South America; he notes that by 1992: 'there were 1500 natural history related organizations in the United States (1176 actual natural history museums) compared with only 412 organizations in all of Latin America (326 actual museums)' and that:

> without the prodigious efforts of thousands of research biologists from throughout the world, there is little doubt that the taxonomic and ecological information available on the biota of the Americas would be poor indeed. Today, standing on the verge of possible cataclysmic losses of species and habitats, this research base allows a reasonable chance to circumvent an ecological disaster.

However, the end result of this scientific (often museum-based) activity in developing countries is that the majority of the material evidence of their biodiversity is now housed in remote institutions; information vital to the conservationists is inaccessible. There is a need to develop a global network of museums and information exchange, and a number of initiatives have been proposed, including the development of the NBIC at the Smithsonian (see Chapter 7).

The museum community needs to assist the growth of natural history museums in developing countries if any hope of retaining fragments of the rich biodiversity there are to be realized. The disproportionate distribution of the 6,294 natural history museums in the world in 1992 has been described by Mares (1993), who states:

> The number of museums located within any country is a function of the economic well-being of that country . . . there are more museums in the United States than in all other American countries combined. In the Old World, the pattern is similar. Indeed, countries with high per capita incomes have 2.5 times as many museums as countries with low per capita incomes, although there are only twenty-seven developed countries versus 137 developing nations.

He has demonstrated convincingly that not only is there is a direct relationship between the number of natural history museums and per capita income within a country, but also a significant relationship between population size and number of museums once a threshold of personal income has been passed (see Figures 6.1 and 6.2).

However, a solution to the loss of habitats and species is a matter of real urgency – the biodiversity crisis will not wait for an improvement in the economic conditions in developing countries, for that vital threshold of per capita income to be reached, and for the development of new museums. The mantle of responsibility must fall on the global museum community – it is the world's natural heritage (and perhaps ultimately, survival) which is at stake. There is clearly a need to develop a world-wide network of natural history museums which, wherever they are, face the same essential challenges of collecting, preserving and interpreting the natural heritage. The promotion of co-operative scientific and educational programmes between institutions in developed and developing countries must be seen as a matter of some urgency, and was recognized at the Madrid Conference, when the WCCR was mandated to 'promote efforts to establish regional training centres in one or more developing countries, particularly in tropical regions, to train natural history museum collection managers and conservators to properly maintain collections in tropical regions'. (Anon, 1992). In addition:

> Industrialised and developing countries should be encouraged to form partnerships to build programs in collections management and research on materials and methods for preservation. All countries should share information and improve access to scientific and worldwide cultural and natural resources.

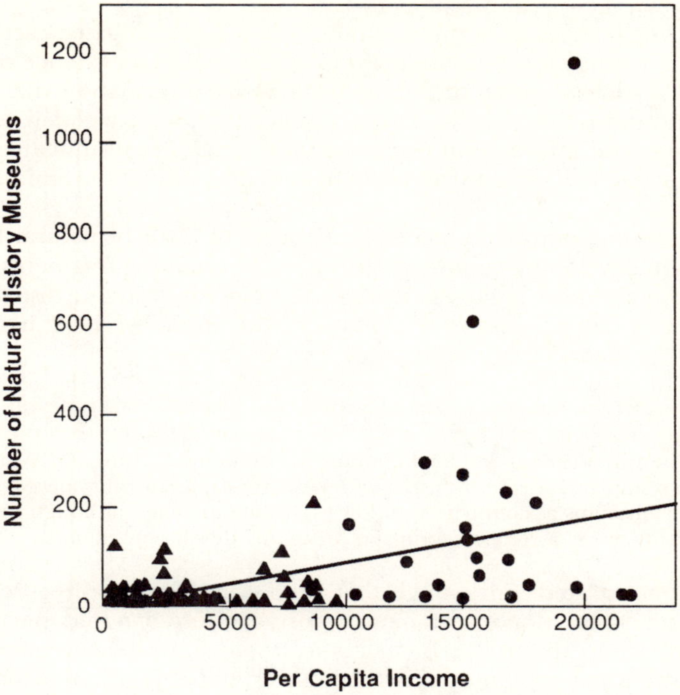

Figure 6.1 The relationship between the per capita income and the number of natural history museums for different countries: • developed countries ▲ developing countries. (Based on Mares, 1993)

Ethical issues – to collect or not to collect?

Tattersall (1992) commented that:

> in a world where both ecological communities and large numbers of systematic groups are under threat everywhere, museums will in some areas at least find it increasingly difficult, or often impossible, to continue collecting the kinds of material of which they have built up their collections over the past couple of centuries. And indeed, no responsible museum professional would wish to add to the stress upon populations already on the brink of disappearance. Thus a change of emphasis seems not only desirable but mandated.

The museum biologist has always faced an ethical dilemma with regard to collecting specimens – yet if we are to document fully an area's biological diversity, there is no alternative but to collect for most taxa.

Ethical considerations are now compounded by practical problems. In many developing countries the enforcement of new laws relating to the collection and export of natural history material (usually to limit wildlife trade as a national response to the demands of CITES) is already influencing the way in which museums can collect, sometimes to the detriment of

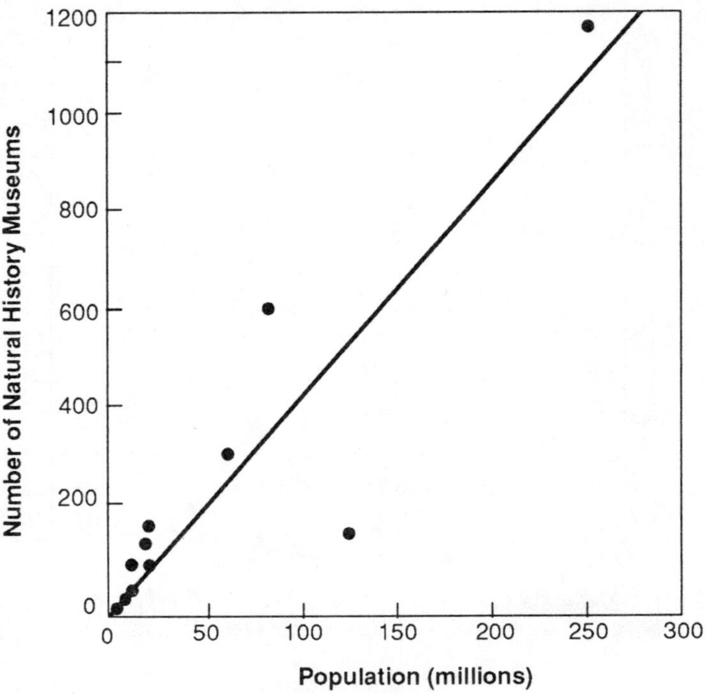

Figure 6.2 The relationship between the number of natural history museums and population size in developed countries. (Based on Mares, 1993)

our understanding of the biodiversity of those countries. Similarly, the Biodiversity Convention has resulted in the need for greater control over the movement of specimens – botanical gardens, with their propensity to move plants and seed, have been especially affected, with many gardens using documentation rigorously to control the exchange of living material (Nelson, *pers. comm.*). It is quite probable that these activities, when compared to the impact on biodiversity of illegal (and legal) trade in animals, plants and derivatives for non-scientific purposes are trivial (Braun and Mares, 1991).

Mahan (1980) proposed an 'International Ethics Code for Natural History Museums' which sets out broad guidelines for the museum curator, and explores the ethics of field collecting. It is unfortunate that this document has not been more widely circulated, as it sets practical and attainable standards, including the prerequisite for careful planning of expeditions and the disclosure of findings, as well as stressing the need for appropriate conduct when in other countries. As he notes 'Being a member of a scientific team and having authority to collect, does not absolve a museum worker from observing the principles of good conservation ... nor permit him to behave in a callous or inhumane fashion.' An important part of ethical collecting is being aware of wildlife legislation, particularly as it relates to endangered species, and complying with it (Saito, 1993). It is interesting to note that the Madrid Conference Resolutions called for 'Rapid and *focused* surveys and inventories of the Earth's biota' (Anon, 1993), a strong indication that the international museum community has recognized the need for selective collecting, and that the conservation ethic is firmly established.

References

Anon. (1992) 'Resolutions from the International Symposium and World Congress on the Preservation and Conservation of Natural History Collections, Madrid, 1992', *ASC Newsletter*, August, 20 (4), 132–136.

Barber, L. (1980) *The Heyday of Natural History*, Jonathan Cape, London.

Braun, J. K. and Mares, M. A. (1991) 'Natural History Museums: working toward the development of a conservation ethic'. In Mares, M.A. and Schmidly, D.J. (eds) *Latin American Mammalogy: History, Biodiversity and Conservation*, University of Oklahoma Press, Norman, 431–454.

Cherfas, J. (1994) 'How many species do we need?', *New Scientist*, 6 August, 36–40.

Conference of Directors of Systematic Collections (1971) *The Systematic Biology Collections of the United States: An Essential Resource*, A Report to the National Science Foundation by the Conference of Directors of Systematics Collections, January, 1971.

Dainton, Lord (Chairman, Select Committee on Science and Technology) (1991) *Systematic Biology Research*, Vol 1 – report; Vol 2 – Oral Evidence and Written evidence received after 21 May 1991, HMSO, London.

Danks, H. V. (1991) 'Museum collections – fundamental values and modern problems', *Collection Forum*, 7 (2), 95–111.

Davis, P. (1994) 'Curators and the conservation ethic', *The Biology Curator*, Issue 1, September, 13–16.

Davis, P. and Brewer, C. C. (1988) *A Catalogue of Natural Science Collections in North-East England, with Biographical Notes on the Collectors*, North of England Museums Service, Durham.

Eldredge, N. (ed.) (1992) *Systematics, Ecology and the Biodiversity Crisis*, Columbia University Press, New York.

Emery, A. R. (1993) 'Changing philosophies, roles and responsibilities'. In Rose, C. L., Williams, S. L. and Gisbert, J. (eds) *Current Issues, Initiatives, and Future Directions for the Preservation and Conservation of Natural History Collections*, International Symposium and First World Congress on the Preservation and Conservation of Natural History Collections, 1992, Madrid, Spain, Vol 3, 111–126.

Griffin, D. (1992) 'Museums and the Environment: A challenge for us, not them'. In Saucier, S. (ed.) *Museums: Re-thinking the Boundaries*, Proceedings of the 16th ICOM General Conference, Quebec, Canada, 1992.

Grubb, M., Koch, M., Munson, A., Sullivan, F. and Thomson, K. (1993) *The Earth Summit Agreements; A Guide and Assessment*, Earthscan Publications and The Royal Institute of International Affairs, London.

Haas, J. (1993) 'Repatriation in a global context for natural history museums'. In Rose, C. L., Williams, S. L. and Gisbert, J. (eds) *Current Issues, Initiatives, and Future Directions for the Preservation and Conservation of Natural History Collections*, International Symposium and First World Congress on the Preservation and Conservation of Natural History Collections, 1992, Madrid, Spain, Vol 3, 71–81.

Howie, F. M. (1993) 'Natural science collections: extent and scope of preservation problems'. In Rose, C. L., Williams, S. L. and Gisbert, J. (eds) *Current Issues, Initiatives, and Future Directions for the Preservation and Conservation of Natural History*

Collections, International Symposium and First World Congress on the Preservation and Conservation of Natural History Collections, 1992, Madrid, Spain, Vol 3, 97–110.

Irwin, H. S., Payne, W. W., Bates, D. M. and Humphrey, P. S. (1973) *America's Systematics Collections – A National Plan*, Association of Systematics Collections, Washington.

Johnels, A. G. (1973) 'Natural History Museum Collections – A basis for future research'. In Engstrom, K. and Johnels, A. G. (eds) *Natural History Museums and the Community*, Universitetsforlaget, Oslo.

Leclaire, L. (1989) 'Les Musées D'Histoire Naturelles en France: Un Patrimoine Scientifique et Culturel Méconnu', *Musées et Collections publiques de France*, no. 184–5, 38–90.

Lovejoy, T. E. (1993) 'The role of Natural History Museums in a changing world'. In Rose, C. L., Williams, S. L. and Gisbert, J. (eds) *Current Issues, Initiatives, and Future Directions for the Preservation and Conservation of Natural History Collections*, International Symposium and First World Congress on the Preservation and Conservation of Natural History Collections, 1992, Madrid, Spain, Vol 3, 27–34.

MacArthur, R. H. and Wilson, E. O. (1967) *The Theory of Island Biogeography*, Princeton University Press, Princeton, NJ.

Mahan, H. (1980) 'An International Ethics Code for Natural History Museums', *ICOM Natural History Committee Newsletter*, No 6, [unpaginated].

Mares, M. A. (1993) 'Natural History Museums: Bridging the past and the future'. In Rose, C. L., Williams, S. L. and Gisbert, J. (eds) *Current Issues, Initiatives, and Future Directions for the Preservation and Conservation of Natural History Collections*, International Symposium and First World Congress on the Preservation and Conservation of Natural History Collections, 1992, Madrid, Spain, Vol 3, 367–404.

May, R. M. (1993) 'Global change; the need and concern for collecting and preserving'. In Rose, C. L., Williams, S. L. and Gisbert, J. (eds) *Current Issues, Initiatives, and Future Directions for the Preservation and Conservation of Natural History Collections*, International Symposium and First World Congress on the Preservation and Conservation of Natural History Collections, 1992, Madrid, Spain, Vol 3, 35–42.

McGinley, M. A. and Tilman, D. (1993) 'Short-term responses of old-field plant communities to fire and disturbance', *American Midland Naturalist*, 129 (2), 409–413.

Myers, N. (1988) 'Tropical forests and their species – going, going . . . ?' In Wilson, E. O. (ed.) *Biodiversity*, National Academy Press, Washington, DC.

Myers, N. (1989) *Deforestation Rates in Tropical Forests and their Climatic Implications*, Friends of the Earth, London.

Netting, M. G. (1962) 'Objectives of Museum Research in Natural History', *Museum News*, 41 (4), 30–43.

Nicholson, T. D. (1991) 'Preserving the Earth's Biological Diversity: The Role of Museums', *Curator*, 34 (2), 85–108.

Novacek, M. J. (1990) 'Research and Education in Natural History Museums – The need for commitment', *Museum Management and Curatorship*, 9, 352–358.

Oliver, J. A. (1969) 'Remarks made at the Centennial Convocation of the American Museum of Natural History', *Museum News*, May 1969, 28–30.

Pettitt, C. (1991) 'Putting "Bloody mice" to good use', *Museums Journal*, 91 (8), 25–28.

Pettitt, C. (1993) 'Building on success: Update on the FENSCORE initiative'. In Roberts, D. A. (ed.) *European Museum Documentation Strategies and Standards*,

Proceedings of an International Conference held in Canterbury, England, 2–6 September, 1991, 227–230.

Pettitt, C. (1994) 'Using Natural History Collections'. In Stansfield, G., Mathias, J. and Reid, G. *Manual of Natural History Curatorship*, HMSO, London, 144–166.

Platnick, N. I. (1992) 'Patterns of Biodiversity'. In Eldredge, N. *Systematics, Ecology and the Biodiversity Crisis*, Columbia University Press, New York, 15–24.

Purcell, R. W. and Gould, S. J. (1992) *Finders, Keepers; Eight Collectors*, Hutchinson Radius, London.

Raup, D. M. (1988) 'Diversity crises in the geological past'. In Wilson, E. O. and Peter, F. M. *Biodiversity*, National Academic Press, Washington DC, 51–57.

Saito, K. (1993) 'Animal Rights and Wrongs', *Museum News*, 72 (3), 72–74.

Stearn, W. T. (1981) *The Natural History Museum at South Kensington*, Heinemann, London.

Strahan, R. (1979) *Rare and Curious Specimens; An Illustrated History of the Australian Museum*, The Australian Museum, Sydney.

Tattersall, I. (1992) 'Systematic versus ecological diversity: the example of the Malagasy primates'. In Eldredge, N. *Systematics, Ecology and the Biodiversity Crisis*, Columbia University Press, New York, 25–39.

Tilman, D. and Pacala, S. (1994) 'The maintenance of species richness in plant communities'. In Ricklefs, R. E. and Schluter, D. (eds) *Species Diversity in Ecological Communities: Historical and Geographical Perspectives*, University of Chicago Press, Chicago, 13–25.

UNEP (1992) 'The state of the global environment', *Our Planet*, 4 (2), 7.

Wiggins, G. B., Marshall, S. A. and Downes, J.A. (1991) 'Importance of research collections of terrestrial arthropods. A brief prepared by the Biological Survey of Canada (Terrestrial Arthropods)'. *Bulletin of the Entomological Society of Canada*, 23 (2), Supplement, 16 pp.

Wilson, E. O. (1992) *The Diversity of Life*, Penguin, London.

Winsor, M. P. (1991) *Reading the Shape of Nature; Comparative Zoology at the Agassiz Museum*, University of Chicago Press, Chicago.

7. *Museums, biodiversity and systematics*

I don't know if it is a characteristic only of British people but so often it seems that a really worthy cause only gets the interest and support it needs when it is virtually lost. It must be something to do with not noticing your possessions until you haven't got them or not recognising artists until they're dead.

HRH Duke of Edinburgh, 1963

We cannot even estimate the number of species of organisms on earth to an order of magnitude, an appalling situation in terms of knowledge and our ability to affect the human prospect positively. There are clearly few areas of science about which so little is known, and none of such relevance to human beings.

Peter Raven, 1992

This isn't like naming your dog Spot

Richard Melville, Secretary,
International Commission on Zoological
Nomenclature, 1984

The terms 'systematics' and 'taxonomy' are often regarded as synonyms, although systematics is usually defined as the broader discipline – in its evidence to the Dainton inquiry (see below) the Royal Society noted in its evidence that 'Systematic biology builds on taxonomy (the systematic giving of names to specimens) to investigate relationships between organisms.' Novacek (1992) attempted to redefine the terms, describing systematics as 'the science that deals with the organization, history and evolution of life. It ultimately asks, how did life forms originate? How did they diversify and how are they distributed both in space and time?' Taxonomy is regarded as 'the practice of describing and naming life forms and arranging them in classifications that reflect patterns of relationships. It provides the language for systematics. It is part of systematics, not the whole of systematics.' Some confusion is also created by the overlaps which exist between the three major categories of systematic research – descriptive (alpha-taxonomy), revisionary (beta-taxonomy) and interpretive (gamma-taxonomy) – yet the best systematic research probably combines elements of all these.

Ecology versus systematics?

The activities of the early naturalist–conservationists were devoted to finding and describing new species, producing comprehensive lists of plants and animals for specific sites of natural history interest, and identifying patterns of distribution and indexes of rarity. In contrast, the new breed of British ecologists of the 1920s and 1930s were more interested in finding out how ecosystems worked. Lowe (1976) has suggested that ecology in Britain had arisen as a discipline in conscious reaction to the obsession of the field naturalist and systematist with novelty and rarity; the emergence of other specializations in the biological sciences, particularly the more fashionable disciplines of physiology or molecular biology, have further served to isolate and marginalize the systematist. The division between the ecologist and the systematist is still evident today, and Eldredge (1992) has suggested that this is particularly noticeable in their different perceptions of the term 'diversity', which:

> to the systematist is the number of species within a monophyletic taxon, [genealogical diversity] rather than the number of species represented by local populations within an ecosystem [ecological diversity] .... The mechanism of extinction may lie squarely within the province of ecology, but we *measure* extinction taxonomically, squarely within the realm of systematics.

The divide between ecologists and systematists in the United States was emphasized during the hearings on the Biological Diversity Bill, as the Nature Conservancy and the largely museum-based systematic community sought to claim priority for managing biodiversity databases (Hoagland, 1990a).

Such argument seems to be of little benefit to conservation. The truth is that sound ecological research demands accurate species identification, and hence ecology and systematics should develop together. Neither ecologist nor systematist can work alone if conservation of species and habitats is seen as the ultimate aim of their research. It is perhaps one of the saddest ironies that the rise of ecology has been implicated in the decline of the teaching of taxonomy and systematics in universities; the need for an understanding of procedures in systematics and taxonomy should have been identified as a building block in the training of ecologists. Yet the drift away from 'whole organism' studies to more fashionable and seemingly exciting spheres of biology has continued until relatively recently. The continuing pressure of environmentalism, and the increased awareness of the biodiversity crisis following the publicity generated by the Rio summit seems to be slowly reversing this trend; for example, the Department of Marine Sciences and Coastal Management at the University of Newcastle, which for many years had done little systematics teaching, launched a new course in 'Biodiversity' with a considerable emphasis on taxonomy and systematics in 1994 and the Natural History Museum, London, is cooperating with University College London to launch a Master's degree in Taxonomy and Biodiversity.

The status of taxonomy, systematics and biological collections

The status of taxonomy in Britain was reviewed in the late 1970s by the Advisory Board for the Research Councils (Smith, 1979) when taxonomy was recognized as the:

> essential basis of all biological sciences because it enables the enormous diversity of organisms, living and fossil, to be recognised and described, named according to an internationally accepted code, and systematically classified. A good taxonomic classification reflects the genealogical and evolutionary relationships of organisms, and by defining the nature, attributes and relationships of individual species, facilitates the identification of specimens by workers in other branches of biological science to whom it is therefore indispensable.

This review recognized the intricate weave of taxonomy in Britain – a mix of specialist institutions (including national and provincial museums, and botanic gardens), universities and amateur naturalists – and recommended specific support for taxonomic work on 'the floras and faunas of changing and/or disappearing ecosystems, and in support of nature conservation policies'. The report struck an optimistic note for taxonomy, regarding it as 'emerging from the shadows'. However, the publication of a more recent investigation into the state of systematic biology in Britain carried out by the Systematic Biology Research Subcommittee of the Select Committee on Science and Technology, (Dainton, 1991) revealed that 'activity in teaching and research in systematic biology had undergone a decline, and that the curation of collections in museums left something to be desired' (Dainton, 1993).

Curation of collections was recognized as a major underpinning element to research in systematics, indeed curation and research in museums was integral to the Subcommittee's work – it had originally been instigated following the outcry from scientists around the world to proposed changes in systematic research (and in staffing levels) at the Natural History Museum, London. Its brief was of necessity widened, and its findings were disturbing. It discovered that despite the rich systematic resources held in the nation's museums (for example 33 million specimens in the Natural History Museum, 9.4 million specimens in Oxford, 7 million specimens in the National Museum of Scotland) active research has continued to decline, not simply within the museums themselves, but also generally. Despite this, Dainton notes that:

> in 1990 museums represented 34 percent, both of total expenditure on systematic biology in the United Kingdom and of manpower expressed in man years. If we separate curation from research we see that museums accounted for 27 percent of the total United Kingdom spend on systematic biology research. The pre-eminent position of the NHM in research is readily apparent – it alone accounted for 24 percent of the United Kingdom total.

Dainton also neatly summarized the reasons why collections are important (Dainton, 1991, 6.2, p 32) to the study and understanding of biodiversity:

– they preserve 'type' material, the specimens to which scientific names are permanently linked. These are the irreplaceable basis of biological nomenclature and essential in determining the names to be applied;

– they are necessary to verify published research on particular organisms. Such research which cannot be so verified is of limited value;

– they are a database of the geographical distribution, range and seasonality of species. Years and dates of collection can be used to monitor changes in distribution;

– they provide an important source for taxonomic research and enable the variation in and the limits between species to be determined;

– as new techniques are developed for characterising species, type collections offer the only means of verifying the presence of newly recognised features in the specimen on which a species was based.

Hence museums and botanic gardens clearly have a moral and scientific obligation to preserve and curate their collections; yet Dainton clearly identified the financial hardships which affected such institutions and frequently prevented them from carrying out those obligations. University collections were identified as being especially at risk, whilst local authority collections revealed 'a picture of gentle decay'. The Museums Association survey of biological collections in provincial museums (Williams, 1987) confirms the latter picture, demonstrating (for example) that 190 museums housing natural history collections did not have a full-time biologist, and that 80 per cent of the museums surveyed reported that biological collections had been lost through neglect, poor curation or disasters. These disturbing findings give an indication not only of the low status of taxonomy in all but a few of the larger museums, but an almost total lack of appreciation of biological collections by the governing bodies of provincial museums. This is particularly ironic when in the majority of cases the founding collections of such museums, their initial *raison-d'être*, was based on the natural sciences. Although some additional resources had been made available following the publication of this survey, (in particular the Museums and Galleries Commission's Natural History Incentive Fund) they proved to be far from adequate – low levels of funding remain the root cause of the threat to biological collections. Dainton's principal recommendation was that:

> in future core funding for research and curation of the collections in the major systematics institutions be maintained in real terms; that ABRC establish a special fund for five years of £1m. a year for systematic biology research; that OAL set up a rolling programme of up to £0.5m a year to assist systematics collections outside the grant-in-aid institutions ... (Dainton, 1991, 9.7, p 50).

Systematics needs a new impetus and a higher profile in order to attract research funding and to change the current downward trend – the biodiversity crisis, backed by the demands of the Biodiversity Convention, could be the catalyst required. A national policy for systematics research, which can release funds to museums and universities, is urgently required in Britain. Some progress has been made following the formation of the UK Systematics Forum, which is funded (for three years in the first instance – up until March 1996) by the Office of Science and Technology, in response to the Dainton Report. The main aim of the Forum is to promote co-ordination and communication amongst the major taxonomic institutions and the wider systematics community and to develop priorities for systematic work. Based in the Natural History Museum, the Forum sees the creation of a Directory of Current UK Systematic Expertise and Research as a necessary first step in this process (Anon, 1994). A recognition of the role of systematics is also given within the framework of *Biodiversity: The UK Action Plan* (HMSO, 1994), which has as its goal 'To conserve and enhance biological diversity within the UK and to contribute to the conservation of global biodiversity through all appropriate mechanisms'. The plan recognizes the significance of Britain's taxonomic institutions ... 'some of the finest collections in the world ... a systematic, geographical and historical perspective for present-day activities in ex-situ conservation' and identifies a need to 'conduct a comprehensive survey of the nature, extent and location in the UK of preserved reference specimens and the associated staff expertise' – a need now being addressed by the Systematics Forum.

Collections-based research in the United States is primarily centred on museums, and especially on the major institutions such as the Smithsonian, the Field Museum Chicago, the Academy of Natural Sciences in Philadelphia, the principal botanic gardens and the great museums associated with the Universities of Yale, Harvard and Berkeley. Although the pattern of taxonomic research is very different from the UK, the problems being faced are much the same. Much of the federal funding for the support of systematics and collections is directed towards the major museums (Novacek, 1990), but increasing demands on resources have had implications even for these well-known academic institutions. For any museum, grants from the National Science Foundation are more difficult to obtain, and competition is keener; more and more collections have been orphaned by debilitated programmes, especially in university museums. Molecular systematics research laboratories for DNA sequencing, installed to ensure a museum's role at the cutting edge of taxonomic research, incur huge running costs which have had a knock-on effect on other research fields. Systematics research in museums has been undervalued by government, and there has been little effective lobbying and representation to politicians, despite the considerable efforts within the scientific community of the Association of Systematics Collections (ASC). Attitudes to political activism have changed in recent years, partly as a result of the Biological Diversity Bill (HR 1268), and the growing awareness of the museum systematics community that they would have to be more aggressive to maintain their position. As an example, 'The Politics of Biodiversity' was the theme for the 1990 ASC annual meeting, a new agenda for museum-based scientists, not known for their radical attitudes (see Anon, 1990).

Internationally, museums and other taxonomic institutions face a tremendous challenge. The number of species is far greater than imagined, (the conservative estimate is 10 million, but 30 million has been suggested), and of the 1.4 million species described, about 1 million are from temperate regions rather than the tropics (Peter Raven – evidence to the Dainton inquiry) – in other words we know least about the parts of the world where biodiversity is greatest. This is not to say that we know everything we need to know about temperate regions – Canada has an estimated 66,000 species of arthropods, half of them undescribed (Danks, 1979); New Zealand has about 20,000 species, again only half have been described, but 90 per cent of them are endemic (Crosby, 1990).

The museum role in conserving biodiversity

A new direction for scientific survey, collecting and research is required if museums are to fulfil a meaningful role in preserving biological diversity and there can be little doubt that in the global biodiversity arena it will be the major national museums in Europe, and the well-financed museums of North America that will be the key players. Their pre-eminent place in terms of existing collections, but also their highly developed taxonomic and systematic skills place them at the cutting edge of biodiversity studies. If these museums are to take on the mantle of environmental protector, (and the majority of them now have mission statements which suggest this is a major goal) then it will mean significant change in the ways that their scientists work. No longer is it possible for museum scientists to pursue their own dedicated studies in a secluded corner – biodiversity research needs to be focused geographically and taxonomically, and teamworking will be an essential element of change. Museums also need to move away from being simply a service industry – supplying information for others to use in implementing conservation action – and become advocates of conservation in their own right. Making such changes is not a simple matter – the Natural History Museum in London has identified 'sustainable management of biological diversity' as one of its key roles, and has had to reorganize its scientific base. Its Director, Neil Chalmers records that:

> It has not been easy, and I would not suggest that any strategic changes in great institutions will be. But we must face up to the need to change . . . . If we fail to do so, taxonomy will wither and decline. If we succeed, taxonomy and natural history collections will return to the centre of modern biology, as they deserve (Chalmers, 1993).

This is a significant point – that through their involvement with biodiversity museums will not only play a significant social role, but flourish once again as scientific institutions on centre stage.

It is important to consider what major changes in policy and practice in natural history museums have already occurred, what other steps need to be taken, and how these might be achieved in order to make their collections and scientific expertise central to the biological diversity debate. Environmentalism has ensured a change in philosophy, most museums recognize that their collections are a major resource for conserving the world's natural heritage, that they have relevance to modern problems (Emery, 1993), and have thus acquired a strong conservation ethic. The following factors are especially pertinent to museums wishing to build on that ethical stance and play an active role in conserving biodiversity.

Improving collections management and conservation procedures

In the preceding chapter reference was made to the resolutions concerning collections care made at the Madrid Conference in 1992, in particular the scale of the problem posed by the needs of curation, storage and preservation. These resolutions are important – and can be quoted to good effect by museum curators attempting to draw the attention of governing bodies to the needs of scientific collections in their institution. Collections are central to all aspects of the natural history museum's role and their role in establishing the nature of biodiversity, and the implications for individual museums needs to be further explored.

Hoagland (1989) referred to the important role that collections can play in biological conservation, and points to the growing need to refer to older specimens to obtain comparative data (for research on the accumulation of persistent toxins in animal tissue for example, or to assess changes in geographical distribution of species), as well as referring to newly collected material. Such conservation-oriented work may include systematics, biogeographic research, ecological research, establishing the range of geographic variation, and environmental assessment. This practice of utilizing the wide spectrum of information associated with specimens has been referred to as 'museum ecology' (Cambray, 1990). Ecologists and conservationists can learn a great deal from museum collections, and as species become rarer in the wild, the ethical argument for maintaining collections and rescuing orphaned ones, so ensuring the well-being of irreplaceable resources collected in the past, becomes overriding.

Because of the growing significance of collections, whether they be recently collected or historical, a first priority must be adequate preservation and curation (see Plate 22), and yet it is apparent that for many natural history museums:

> A large proportion of the specimens in collections are uncatalogued, and . . . with present staff resources, the situation will improve very little or may even get worse. Everywhere, there is a shortage of resources for the management and conservation of collections in museums (Griffin, 1992).

This major problem for museums can only be resolved by preparing for the growth of collections by the use of collection impact statements (McGinley, 1993 gives a typical list of questions for a CIS) and setting priorities for collections care by implementing collection management plans. Such techniques are becoming more commonplace, and are as essential to the small museum (see, e.g. Silvy and Cato, 1991) as they are to the major institutions (see, e.g. McGinley, 1993). Collections management has been narrowly defined by Roberts (1988) as all activities pertaining to information management, but for most museum curators it also encompasses all aspects of collection-related activities other than research. Herholdt (1990) divided the major areas of collections management under the headings of Physical Aspects (collection growth, specimen preparation, storage, fumigation, collection security, repair, research), Documentation and Information Management, Policies and Procedures, Funding and Literature (using

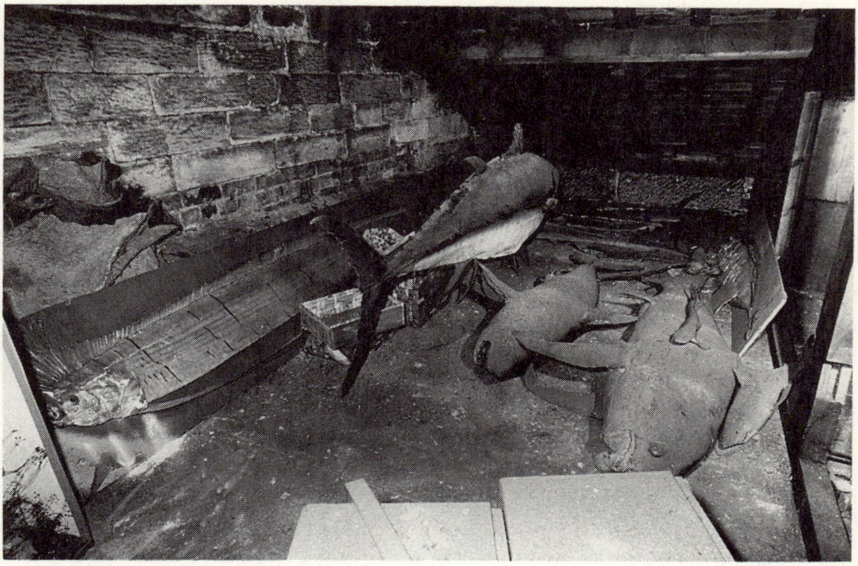

Plate 22 A forgotten collection. This situation is (sadly) repeated in natural history museums around the world, as resources for collections care continue to dwindle. The importance of natural history collections remains to be fully recognized and understood by funding bodies.

published information to best effect). In her review of the needs of collections management the need for additional training for collection managers (especially in her country, South Africa) is a recurring theme.

The conservation of biological and geological material demands a professional approach, and recognition of the skills of the preparator and conservator. The number of specimens in natural history collections is so daunting that remedial conservation is impractical. For example, the Natural History Museum, London, houses 25 million insect specimens representing some 800,000 species (Danks, 1991). Consequently much emphasis is now correctly placed on preventative conservation as the only realistic approach to the scale of the problem, but which can only be achieved '... through a long-range conservation plan that incorporates preservation policies and procedures into the daily activities of the museum, and develops strategies for addressing conservation problems in a logical, feasible manner' (Rose, 1991). Such a plan should be preceded by a conservation assessment, for which published guidelines are available (National Institute for Conservation, 1990, summarized in Wolf, 1993). A conservation plan needs to also include statements relating to documentation and preservation research; the latter is particularly important as commonly practised preparation techniques are still frequently based on traditional methods developed during the nineteenth century which may be totally inappropriate for the needs of the modern researcher.

There needs to be a commitment to the conservation of collections in all collecting activities to ensure that necessary support and funding is available, hence incorporating a strategy for conservation within a collections management plan must be seen as an integral step in making collections secure. It is important that curators receive good advice from conservators, yet in the UK, there are only about 20 conservators dealing with natural history collections, compared to 1,400 trained individuals caring for humanities collections (Horie, 1993). In other countries in Europe, with the exception of Germany, the situation is similar – too few conservators, still using out-of-date techniques. Clearly there is a need for more conservators in natural history, more research and more training courses. One of the key tasks for the World Council on Collections Resources is to 'work with international organisations such as the International Council for Museums (ICOM) to promote high professional standards and establish educational programs for collection managers and conservators' (Anon, 1992).

How many collections should be maintained? Many museums, often the smaller, less well-financed institutions, house collections which lack data and are frequently in poor condition. Even with the greatest imagination, it is hard to consider that such collections are going to be of any relevance to the biodiversity debate. There are, of course, other uses to which such collections can be put, but should they receive the devoted attention of biologist and conservator at the expense of other collections and activities – which might include biological recording and the making of new collections of conservation significance?

Research and collection building with a conservation purpose

Collecting and conservation, to anyone but a museum curator, would be seen as a paradox; collecting implies killing animals or plants and removing them from their habitat. There are therefore ethical issues which need to be faced, but there are also the practicalities of identifying and quantifying the biodiversity of an area, which in many cases can only be resolved by collecting in an appropriate manner. There can be no doubt that the activities of naturalists have had a detrimental effect on natural populations in the past, yet the impact of collecting, including research activities has been little studied. The few references (Ratcliffe, 1967; Ellis, 1967) are somewhat anecdotal and largely inconclusive, though it is suggested that habitat destruction is a far more serious risk to populations than any collecting effort.

If biodiversity issues are going to be addressed then there is a requirement to refer to newly collected specimens, collected for a specific purpose, and stored in new ways – but how, where and what do we collect? Hoagland (1989) notes that:

> Efforts are now being made to return to original localities to recollect species – to confirm their continued existence or to obtain fresh material for new studies; the whole range of life history of a species is often collected today, including larvae and juveniles; a series of collections are made over time at the same site to record

recent shifts in the biota, including extinctions, related to climatic change; specimens from a single species and locality are preserved in several ways, including freezing and chemical preservation, so that they can be used for studies of nutrition, genetics, and so on.

This implies that museums need to apply new strategies for collecting. Danks (1991) has correctly identified that such changes in strategy have tremendous potential simply to lead to more and more collecting. Interests in biodiversity means maintaining large collections, but making new ones too; the invertebrate taxonomist now uses methods (including the 'fogging' of the rainforest canopy with insecticides) to collect enormous numbers of species with very little effort – Danks refers to a routine one-season sampling trip for invertebrates in southern Alberta resulting in a haul of about 3.5 million specimens. Bulk sampling on this scale could overwhelm any museum and its staff. Given that museums are also acquiring orphaned collections, voucher specimens associated with ecological assessments, 'living' collections of frozen tissue or seed banks, and making and maintaining regional collections, then there is potential for collecting on a massive scale. Unless properly financed and staffed, such an approach is not simply impractical, but also unethical. It is when faced with these problems that collection impact statements become invaluable – what can the museum afford to collect, and where do the priorities lie? Should museums evaluate their systematic expertise and storage and conservation facilities before making the decision on how and what to collect? Such suggestions are reinforced in the resolutions from the Madrid Conference which included the appropriately worded statement that 'This congress and the WCCR encourage institutional and societal policies that promote *selective collecting* of organisms, recognizing *legitimate research needs* and both the problems of the *impact of collecting on species survival,* and the problems of appropriate *space for storage of specimens* in perpetuity' [my emphasis] (Anon, 1992).

With the constraints of limited finance and manpower, how do the major natural history museums determine where geographically to put their scientific efforts? As Platnick (1992) notes:

> So many different species and so many different habitats are threatened in so many different countries, by so many different human economic needs, that it is clearly impossible to preserve every species, and every habitat, that still exists today. Is there a rational way to decide where to concentrate our efforts?

He states that there is no easy answer to which parts of the world have greatest biodiversity, arguing that the concept of greater diversity in the tropics is in part due to 'boreal bias' (which assumes that the pattern of diversity in the Holarctic is also true for the southern temperate zones) and 'megafauna bias' (i.e. that what is true for vertebrates is true for all other groups of organisms). He sums up the latter succinctly – 'Let me be blunt: speaking of biodiversity is essentially equivalent to speaking about arthropods. In terms of number of species, other animal and plant groups are just a gloss on the arthropod theme.' As a spider expert with an interest in southern temperate areas, Platnick concludes that ' ... for ... spiders, and

I suspect for most non-vertebrate groups as well, the picture of a species-rich tropics and species-poor north and south temperate zones is inaccurate.'

There is another problem to be faced when deciding which geographical areas deserve investigative effort and preservation. It cannot be assumed that when comparing two or more sites that the one with the most species, or a larger number of rarer species should automatically be considered the most important site; we need to assess which species are there, and their total distribution pattern. It is not simply the number of species that counts, but also the proportion of endemic species or those which have a restricted distribution.

These factors impinge on the question of where a museum should put its taxonomic effort. It is essential that sites are selected which are under some threat, yet at the same time it is important within those sites there are also areas of high endemism. And, as Platnick (1992) notes:

> it is equally important that we stop portraying the problem as one that exists only in the tropics. We must admit that threatened areas in South America, in Southern Africa, New Caledonia, and even Australia and New Zealand, as well as some in the north temperate zone, may imperil significantly large and (more importantly) unique components of biodiversity.

Evidently then there is a need for selective yet intensified collecting; the concept of 'triage' – the selective choice of one site or a specific taxon for study at the expense of another – is now widely accepted as the only pragmatic approach. If conservation activities are based on systematics priorities, (for example preserving the tuatara taxa because they provide information about early divergence – May (1990)), the distribution of effort might not conform to widespread preconceptions about the significance of biodiversity. Clearly there is a need to take more than systematics into consideration when devising conservation strategies, but it must be seen as a significant element in decision-making.

Because species and habitats are being lost at an alarming rate, there is a need for conservation strategies to be put in place quickly, for reserves to be designated and appropriate mechanisms put in place to maintain them. Despite the demands of some scientists it will never be possible fully to document all the elements of a complex ecosystem and determine their interrelationships before acting to conserve that ecosystem. Although it has been suggested that 'to attempt . . . [conservation management] . . . without a full understanding of what species exist in the environment and how they relate is arrogance of epic proportions' (Darling and Thomson-Delaney, 1992), one must ask how long we can afford to wait – a degree of arrogance, and indeed a little emotional concern, is almost certainly necessary.

May (1993) has highlighted the need for quick and crude surveys to provide ' . . . inventories of the fauna and flora of a few appropriately chosen sites, partly so that researchers can make better estimates of how many species there are in total and monitor extinction rates.' Such 'quick and dirty' surveys have been a feature of Costa Rica's conservation programme following the creation of INBio, the country's National Biodiversity Institute, which has the remit of determining the biodiversity of conservation areas, and looking at how that biodiversity might be used in a sustainable fashion for society, intellectually and economically. Realizing the urgency of the situation, and having limited resources, INBio turned to an under-utilized resource:

> An army of local rural residents, called 'parataxonomists,' ... [has been trained] ... in the philosophy and technology required to understand and conduct the process of collection and preparation of ... specimens in the field. The parataxonomist program has attracted wide national and international attention because of its effectiveness and potential applicability to other countries (Gamez, 1993).

Parataxonomists work from offices located in conservation areas and their specimens are identified by curatorial staff and visiting international experts – the Natural History Museum, London, is closely involved with the project, teaching taxonomic skills in Spanish (Mound, 1991). The preparation of an inventory of species is allied to a search for species or species products with economic potential – such 'economic prospecting' is carried out in partnership with a major drugs company. This indigenous answer to solve the complex problems of integrating conservation of biotic resources through facilitation of sustainable use must be seen as a model for other developing countries.

Developing new technologies and scientific facilities

Bonhomme (1993) in describing the methodology and use of the *in vitro* DNA amplification technique known as the Polymerase Chain Reaction (PCR), suggests that it is applicable to many museum specimens, even historical ones, so opening up a new field which potentially gives renewed significance to collections. The investigation of species relationships through the study of DNA and other molecules has been carried out using a variety of museum specimens, including dried material such as herbarium specimens, alcohol-preserved items and bones. Ethanol is regarded as the best stabilizer of DNA; formalin however tends to bind proteins and create an insoluble matrix, and should be avoided as a museum preservative. The resolutions at the Madrid Conference included a call for ' ... all curators, preparators, and collections managers to prevent the use of treatments that would irreversibly alter or destroy DNA content. Guidelines must be established to preserve useful DNA (and other genetic and biochemical materials) in new collections.' (Anon, 1992.) For many years curators have reflected on the loss of potential data as soft parts of molluscs and internal organs of vertebrates have been discarded in line with nineteenth-century musems practice. There is an overriding need to preserve and store such materials, and the development of 'frozen collections' must be viewed as a viable alternative. Hackett and Zink (1989) discussed the implications of collecting tissues in relation to birds, and have suggested that they are best frozen (in liquid nitrogen or dry ice) within half to one hour in order best to preserve mitochondrial DNA.

DNA investigation has not only caused a reappraisal of preservation techniques, but the provision of laboratory facilities at some major museums to carry out such work. Hoagland (1989) cites the Academy of Natural Sciences, Philadelphia, the National Museum of Natural History, Washington, and the American Museum of Natural History, New York, as examples. Although in the long term such facilities will undoubtedly make a real contribution to the conservation of biodiversity, the cost of such installations ($1.8 million in 1990 at the AMNH – Novacek, 1990) and their running costs is prohibitive for most museums.

Co-operative research

In the past most of the great scientific museums mounted their own collecting expeditions alone, with no perceived need to co-operate with other institutions or individuals. Small-scale field collecting by provincial museum biologists in Britain is still done in this way, as individuals decide their geographical and taxonomic priorities and set their own work programmes. However, in order to investigate broader issues (either taxonomic or geographic), greater co-operation between ecologists, evolutionary biologists and systematists, especially in relation to field expeditions and the collection of more elaborate ecological field data, is vital. Such expeditions also need to involve conservationists to ensure that their research is correctly oriented towards conservation objectives.

The formation of new, stronger coalitions between museums and other organizations has been suggested by Griffin (1992), particularly where large-scale, global issues are paramount. He cites Diversitas, a consortium of IUBS-SCOPE-UNESCO-IUMS whose aim is to promote a network of systematists and their institutions, and stresses that co-operative research 'will provide exciting results which will illuminate the way museum science really does contribute to understanding. Secondly it will forge closer alliances with ecologists and other biological scientists, groups with whom collaboration is not always as active as it might be.' May (1993) has also called for more co-ordination, citing the network of museum-based entomologists, the Entomology Collections Committee (ECC) as an example. As he notes 'Time is too short, and the tasks too large, to continue addressing biodiversity on a piece-meal basis. It is too easy to see the past superimposed on the present, when in fact the present is a revolution.' The formation of the World Council on Collections Resources (WCCR) at the Madrid Conference in 1992, with a specific goal to foster collaborative collections-based research, as well as access to collections, training, data exchange and technology transfer, must be seen as a significant step for the scientific community working in museums.

The most recent, and most comprehensive call for co-operation is *Systematics Agenda 2000*, an ambitious programme initiated in 1994 by the American Society of Plant Taxonomists, the Society of Systematic Biologists, the Willi Hennig Society and the Association of Systematics Collections, and aided by financial support from the US National Science Foundation. Its first documents (Systematics Agenda 2000, 1994a and 1994b) set out three major missions, which, it is predicted, will require 25 years of intensive effort at a cost of US$3 billion annually. They are:

> Mission 1 – To discover, describe, and inventory global species diversity.
>
> Mission 2 – To analyze and synthesize the information derived from this global discovery effort into a predictive classification system that reflects the history of life.
>
> Mission 3 – To organize the information derived from this global program in an efficiently retrievable form that best meets the needs of science and society.

All three missions will produce information which will enable decision-makers in a number of scientific fields, but especially relevant is data which will benefit biological conservation, wise use of resources, and ecotourism.

The need for new museums in tropical countries

The paucity of natural history museums in countries with significant biodiversity has been discussed earlier (p.139) , but in listing potential action for the future it is worth emphasizing the point again. The establishment of natural history museums can have a dramatic impact on conservation policies in developing countries, for example Taboada (1992) has described the impoverishment of the fauna and flora of Cuba, and notes that '... Cuba did not have a natural history museum ... until the Felipe Poey Museum was opened in Havana in 1964. It was designed then merely as a set of exhibit halls with neither research collections nor staff qualified to conduct research.' This museum is now designated as the National Museum of Natural History and has a team of 15 research workers and technicians actively involved in biological survey and systematics. It is evident that there is a need to develop museums and natural history collections in those countries with high biodiversity, where possible through collaborative ventures with major scientific institutions in North America and Europe. Such programmes should include elements of staff training in systematics, collections management and conservation techniques.

Training for systematists – a role for museums

In a report on research priorities in tropical biology by the US National Research Council (quoted in Raven, 1981) it was estimated that no more than 1,500 systematic biologists in the entire world are able to deal at a professional level with the identification of one or more groups of tropical organisms – ' we clearly need many more specialists if we are to be able to accelerate the pace of classifying and providing information about the vast array of organisms that is present in the tropics of the world'. Fifteen years later, it is hard to imagine that the situation has improved – the reverse is more likely to be true – a decline in systematists is matched only by a decline in species world-wide.

A number of initiatives have been proposed by the Association of Systematics Collections to counter the decline in numbers of trained systematists in the United States, including joint international training programmes

(Hoagland, 1990b). A major report on research needs in biological diversity by the National Science Board of the NSF in 1990, and the ASC's 1989 report on Systematics Resources for the 1990s both lamented the decline in the number of systematists employed in universities in the USA and the lack of training and internment opportunities for systematists.

There is a growing need for the development of training programmes as collaborative ventures between major museums and universities, not just in systematics, but also in collections management. As Barrowclough (1992) and others have pointed out, the present distribution of taxonomists and systematists does not accurately reflect the abundance of organisms. Some taxa, especially vertebrates, are studied by a disproportionate share of taxonomists, whilst others, notably arthropods, are relatively neglected. Taboada (1992) has noted the impact of the lack of training for systematists in his native Cuba, where:

> Owing to this deficiency, the technical work in the area of management and protection of faunal resources is usually done by professionals from other fields such as physicians, geographers, economists and agricultural or forestry engineers, a situation that sometimes has led to unsound decisions.

The increased demand for Environmental Impact Statements in the USA (in response to NEPA) and the UK (in response to European Legislation) has resulted in a marked upsurge in biological monitoring – a fundamental requirement of which is accurate identification of animal and plant groups. The importance of correctly identifying species, and maintaining voucher collections for EIAs has been identified by Pettitt (1991):

> Two large and expensive surveys, one for a river valley authority in America and one for an oil company, failed to preserve voucher material in a permanent collection. Both surveys were carried out by recent graduates with little taxonomic experience, and their findings have since been successfully challenged; without the voucher material these surveys were largely a waste of time.

With these scenarios in mind, due recognition of the skills of museum-based taxonomists might yet be given by enforcement authorities – Leicestershire Museums have already taken on a formal role of assisting in the enforcement of wildlife laws and regulations (Stansfield, *pers. comm.*). The growing need for taxonomic expertise has been addressed by the Natural History Museum, London, which in 1993 launched an initiative which promoted higher standards in identification, a new scheme which awards certificates in competence of identification called IDQs. The courses will not produce systematists, but will do much to increase the numbers of parataxonomists able to determine common species in a selection of major British habitats:

> IDQs will be geared towards habitats and natural assemblages of organisms, and to suites of indicator species, rather than towards narrow systematic groups. These may include, for example, woodland plants, sessile marine invertebrates, freshwater insects and intertidal and inshore crustaceans. (Natural History Museum, 1993).

If such schemes were repeated world-wide, they might have a significant impact.

The need for more systematists and additional training was identified as a key issue in the Resolutions from the Madrid Conference in 1992:

> as collections management for systematic research requires individuals who are authoritative concerning the systematics, taxonomy and nomenclature of genetically allied groups of species and genera, and as today there are too few experts for many groups, especially in taxon-rich developing countries, it is clear that there is an immediate need for the education and training of systematists with emphasis on studying the most poorly-known group of organisms with the greatest economic and cultural value (Anon, 1992).

It was suggested that universities might develop systematics training, perhaps in association with museums, and that a sustained effort was required to influence government agencies of the pressing need for more trained systematists.

Promoting the role of systematics and obtaining increased financial support

The majority of people have no understanding of what systematics is, including the scientific community. Indeed, Ricklefs (1981) suggested that museum professionals have failed generally to convince other biologists that collections and the insights gained from specimen-based research have any value. It is vital to raise the profile of collections-based research; convincing fellow museum workers of the value of natural history collections is important, convincing ecologists of the need for museums is urgent, convincing funding organizations is paramount. Campaigning for additional resources to maintain biological and geological collections and their associated information must now be seen as an essential role for the natural history curator. Griffin (1992) has made a plea for museums to take 'an engaged and public role', participating in debates and decision-making about the natural environment, as one of the means of promoting the museum role in the biodiversity debate. The recent 'Sunflower campaign' mounted by the Biology Curator's Group in the UK is an excellent example of how awareness of collections and systematics can be raised, especially in political circles. Only if politicians realize the importance of scientific collections to society will the necessary extra funding be obtained – current levels of financial support are so pitifully low for natural history museums that the Madrid Conference urged a tenfold increase in resources before the end of the century (Romero-Sierra, 1993). If the major museums are to play a role in the science of biodiversity then the costs will be considerable – the millennium appeal launched by the American Museum of Natural History set a total target of $250 million; $113 million 'to foster exploration and discovery', which would include 'curatorships, laboratories, collections, expeditions and library services'. In 1992 the work carried out by the

Division of Systematics cost the Academy of Natural Sciences of Philadelphia \$2.3 million (Thomson, 1992); the National Museum of Natural History at the Smithsonian spent 78 per cent of its 1992 budget on research and collections management – some \$35.5 million (Kennedy, 1992).

The impact of new information technology – using the data

The principal role played by the world's major museums of natural history in the biodiversity debate is to be responsible for classifying and identifying species from sites of specific wildlife interest, and especially those which are under threat. Major sources of such biogeographic information – in the form of published scientific papers and books – have been a major feature of natural history from the nineteenth century. It is now evermore important that data held in this form, and information associated with the material evidence of biodiversity, the specimen collections, are made widely available. It is vital that this knowledge, representing years of scientific endeavour, can be transferred from one museum to another, but perhaps even more important that a means of communicating data between museums and nature conservation organizations be established.

The amount of data associated with museum specimens is on such a scale that it can only be handled effectively using powerful computer software; used correctly, such data can provide vital information to help make decisions about biological conservation. The Natural History Museum, London has developed *Worldmap* (see Fig 7.1) to produce maps, from a local to a global scale, which indicate those areas of greatest genetic and taxonomic diversity – the more information that is input from collections around the world, the more sophisticated and meaningful *Worldmap* becomes (Vane-Wright *et al.*, 1991). *Worldmap* charts biodiversity using an equal area projection map of the world which cover a range of spatial scales from grid cells of less than 100 km^2 to over 611,000 km^2. These cells can indicate biodiversity in terms of richness in species, (the most simple measure of biodiversity), richness in characters and character combinations (an indication of taxon diversity and cladistic dispersion) and richness in endemic species (a measure of range-size rarity). Although designed initially as a mapping tool, *Worldmap* enables explicit choices to be made about conservation using these criteria; particularly useful is the ability to detect 'hotspot' areas; deselecting such an area then enables the 'best next area' to be detected. This in turn can be deselected, and so on, so drawing up a priority list of geographical areas for species conservation (Williams, 1994). However, care needs to be taken with the 'eco-taxa hotspot' approach, as McAllister (1993) has warned:

> Examination of hotspot maps shows that different taxa may have different hotspots. ... Conservation needs to take into account that butterflies and birds may have different needs. The old saw about a reserve for tigers protecting many other 'lesser' species needs to be reconsidered.

Another biodiversity database is ERIN (Environmental Resource Information Network), an extensive GIS system developed in Australia, which can be

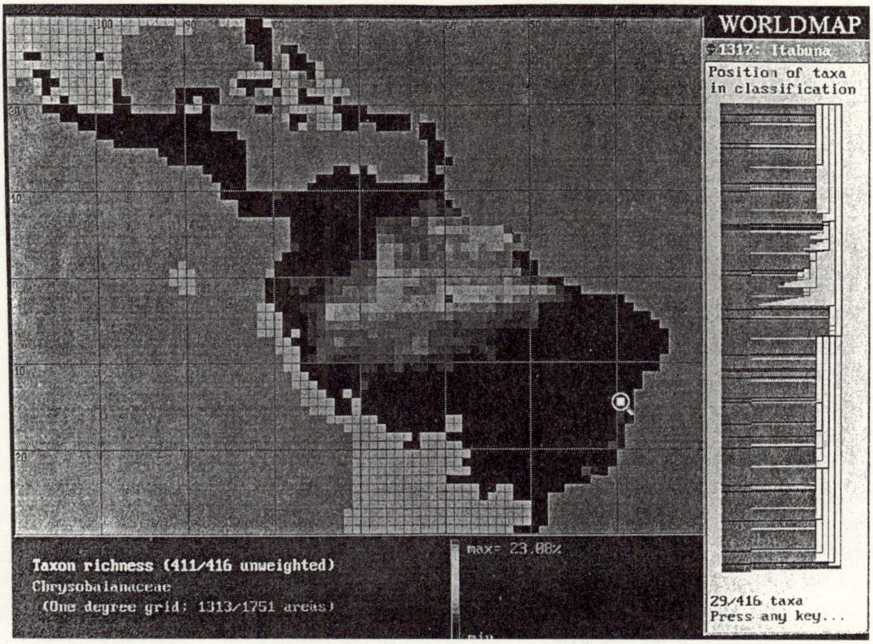

Figure 7.1 *Worldmap* – an example of output from the Natural History Museum's software which can produce a global view of the pattern of biodiversity. (Reproduced with permission. P. H. Williams, The Natural History Museum, London.)

updated by satellite imagery. Griffin (1992) notes that ERIN can receive 'information on the distribution of animal or plant species – available from already computerized collections ... correlations can be drawn ... with the principal factors in the environment critical to the species' distribution. ERIN is predictive: it shows where species ought to occur even if they haven't been found there.'

The computerized documentation of biological and geological collections in individual museums is at best uneven. However, the relatively low price of hardware, and the increasing sophistication and availability of suitable software means that comprehensive databases of biological collections are becoming more commonplace. The Information Superhighway is already doing much to facilitate the exchange of data between museum-based biologists – through JANET in the UK and NSFNET in the USA – but the ability to search museum databases through the Internet is currently extremely limited. The development of new links, and the ability to subscribe to major databases of biological collections such as Taxacom (based at Harvard) through the network must be seen as a priority. According to Wallace and Jones-Garmil (1994) the current list of US institutions with collections of interest to the museum biologist available on the Internet is confined to Harvard, the University of California at Berkeley, the National Museum of Natural History (Smithsonian Institution) and the Museum

Computer Network. The list produced by Miller (1994) is much more extensive. He estimates that collections available world-wide on Internet give access to over 2 million records. However, such is the pace of developments in 'Highway' technology, that this number will grow dramatically.

The exchange and compilation of data from museum collections must be of tremendous significance for future, positive action towards biodiversity issues. In response to the Rio Biodiversity Convention, a number of initiatives have already begun, including the Biodiversity Information Network (BIN21) which aims to ' ... become a comprehensive and authoritative source of information on world biodiversity without holding all useful information on-site' (Canhos, *pers. comm.*) – museums are seen as an integral part of this network. Another important step has been taken by the National Museum of Natural History in Washington, with a proposal to establish a United States National Biodiversity Information Centre (NBIC), where the Smithsonian would act as a clearing house for biodiversity data. The project, with an estimated start-up cost of $1.7 million, will catalogue, link to, complement and strengthen existing databases rather than duplicate them. In a letter accessible via the Smithsonian gopher in October 1994, Thomas Lovejoy commented that ' ... the concept of a NBIC is simple: a focal point where the many parties can generate, manage, or use data on biological resources, can collaborate and make decisions leading to broader access to that information.' Morin (1994) has restated this approach, and quotes Stan Shetler's (NMNH) view that:

> museums should view the National Biodiversity Information Center as the main entry point for the effective mining of information found in museum collections and should use the Center in a way to ensure that museum collection information and systematics information are a very prominent part of the mix of data being offered.

If such national databases can be further linked to other networks, especially the Conservation Information Network (CIN), then real progress might be made. The cost of carrying out such work must not be underestimated – it has been suggested that to create a single database for all known species, which included ecological and distribution data, some £60 million sterling might be necessary (Anon, 1992).

Making collections databases and other sources of informaton widely available on the Internet has tremendous potential for the protection and sustainable use of biodiversity. However, it is not without its risks, and embraces questions of academic and legal ownership, the responsibilities of owners and users, and not least, those of payment for the use of data. ASC has argued that collections databases are a national resource and require public funds; access to a federally supported Collections Database Network would be free to data providers and reciprocal users, but charges made to consultants or other product and profit-oriented users. The ASC Guidelines for Institutional Database Policies provide a sound basis for future progress, and address primary isues such as database definitions, ownership and stewardship, ethics and obligations (Anon, 1993). The 1995 AGM of the ASC featured 'Collections on the Information Highway'.

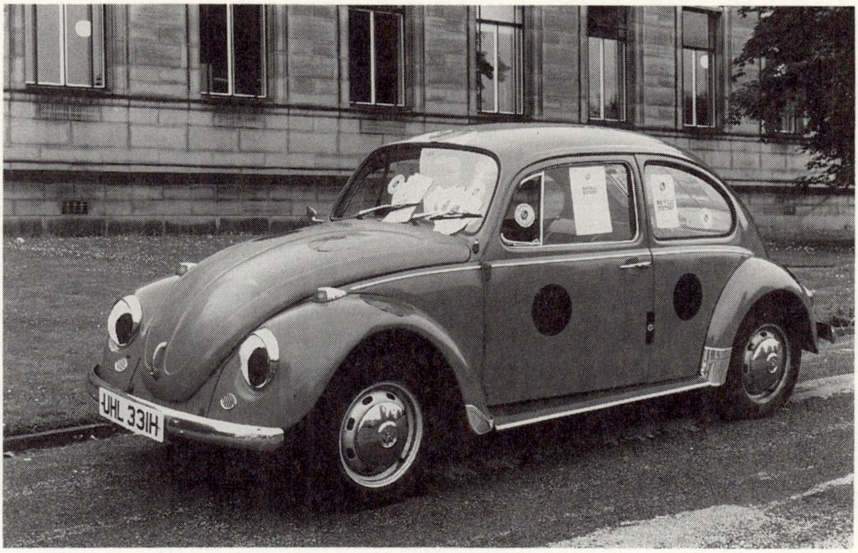

Plate 23 The 'Beetle-down' campaign launched by the Biology Curator's Group to increase awareness of the role of natural history museums included this unusual promotional vehicle. (Photographed by Paul Richards, Sheffield City Museums, 1988.)

Promoting a public understanding of biodiversity and the role of the natural history museum

The role of museums in displaying the natural world to the public is described in Chapter 4. However, it is worth noting here how ineffective museums are at explaining what they do – and especially their role in systematics and conservation. Museum-based biological research rarely hits the headlines, yet ' ... we need to be seen, in public, joining in the debate' (Griffin, 1992). In the UK the timely 'Beetle down' (to your local museum) campaign first mounted by the Biology Curators' Group in 1987, did much to break down the mystique that surrounds biological collections and the role of the museum biologist. Natural historians in museums throughout Britain participate regularly in a variety of educational activities (workshops, guided tours, music and drama events) during 'Beetle-down week', gaining much-needed publicity for their museum, the services they provide, and for their specialism. Backed by 78,000 leaflets, window stickers and T-shirts, the campaign was initially aimed at younger museum visitors, but has acted as ' ... a springboard to tackle the more serious issues of service losses, collection deterioration, lack of resources and loss of natural history staff' (Whiteley, 1991). Amongst Britain's biologists, the 'Beetle-down' campaign (see Plate 23) is remembered in particular for the subversive alternative leaflet which circulated the desks of curators emploring visitors to 'B****r-off' (to someone else's museum).

Museum-based biologists have played a major role in the early history of nature conservation in many countries, often establishing significant databases of information long before a formalized conservation network, or major conservation organization existed. For example, 'By the time the need for nature conservation in Southern Africa was realized, museum taxonomists had established basic classifications and keys that allowed conservationists to scientifically identify species in conservation areas, including those taxa considered for conservation' (Rautenbach and Herholdt, 1990). By undertaking comprehensive faunal and floral surveys as part of their mandate, individual curators acquire a deep knowledge of their geographical area. With a deeply rooted conservation ethic, individual curators have an obligation to use this knowledge, and the information associated with their collections, to promote more effective nature conservation strategies. This personal dedication is one of the key ways in which museums have made a significant contribution towards the environmental movement – there are very few natural history curators who do not readily give their time to a charitable nature conservation organization, from national to local level.

Conclusions – priorities for natural history museums

Tattersall (1992) remarked that:

> natural history museums, while priding themselves on being libraries of the diversity of the living world, are in danger of simply becoming history museums, containing records of a diversity that has vanished. Clearly, it is our responsibility to do everything possible to prevent this from happening.

Few museum biologists would argue with this premise, yet also recognize that they do not, as yet, have the resources to cope with such a massive responsibility. It was recognized by UNESCO in 1992 that:

> The preservation of ecosystems is linked to exploration, collection, museology, botanical and zoological gardens, archives and libraries. The channelling of resources to technology has left inventories and collections of biological materials of developing countries in low priority. Adequate funding is essential for scientific research and education.

Museums and their collections are essential to researching and hence conserving the Earth's diverse ecosystems, – 'It is systematics that will define the parameters of yet undiscovered biological communities in the tropical basins or the deep sea floor' (Novacek, 1992), – yet until the fundamental role of natural history museums is recognized, and they are properly funded, this most important of roles will never be achieved.

References

Anon. (1990) 'The politics of biodiversity', *ASC Newsletter*, 18 (5), 65–69.

Anon. (1992) 'Resolutions from the International Symposium and World Congress on the Preservation and Conservation of Natural History Collections', Madrid 1992, *ASC Newsletter*, August, 20 (4), 132–136.

Anon. (1993) 'ASC Guidelines for Institutional Database Policies: Executive Summary', *ASC Newsletter*, 21 (6), 65–71.

Anon. (1994) 'Note on the UK Systematics Forum'. In *The Biology Curator*, Issue 1, September, 2.

Barrowclough, G. F. (1992) 'Systematics, Biodiversity and Conservation Biology'. In Eldredge, N. *Systematics, Ecology and the Biodiversity Crisis*, Columbia University Press, New York, 121–143.

Bonhomme, F. (1993) 'Museum collections and the new molecular techniques'. In Rose, C. L., Williams, S. L. and Gisbert, J. (eds) *Current Issues, Initiatives, and Future Directions for the Preservation and Conservation of Natural History Collections*, International Symposium and First World Congress on the Preservation and Conservation of Natural History Collections, 1992, Madrid, Spain, Vol 3.

Cambray, J.A. (1990) 'The role of natural history museums in aquatic conservation'. In Herholdt, E. M. (ed.) *Natural History Collections – Their Management and Value*, Transvaal Museum, Special Publication No 1, 129–137.

Chalmers, N. (1993) 'Achieving strategic change: natural history collections for the twenty-first century'. In Rose, C. L., Williams, S. L. and Gisbert, J. (eds) *Current Issues, Initiatives, and Future Directions for the Preservation and Conservation of Natural History Collections*, International Symposium and First World Congress on the Preservation and Conservation of Natural History Collections, 1992, Madrid, Spain, Vol 3, 143–6.

Crosby, T. K. (1990) 'The New Zealand Arthropod Collection', *Insect Collections News*, 4, 22–23.

Dainton, Lord (Chairman, Select Committee on Science and Technology) (1991) *Systematic Biology Research*, Vol 1 – Report; Vol 2 – Oral Evidence and Written evidence received after 21 May 1991, HMSO, London.

Dainton, Lord (1993) 'Systematic Biology Research; an outsider's visit to the secret garden'. In Rose, C. L., Williams, S. L. and Gisbert, J. (eds) *Current Issues, Initiatives, and Future Directions for the Preservation and Conservation of Natural History Collections*, International Symposium and First World Congress on the Preservation and Conservation of Natural History Collections, 1992, Madrid, Spain, Vol 3, 345–361.

Danks, H. V. (ed.) (1979) *Canada and its Insect Fauna, Memoirs of the Entomological Society of Canada*, 108, 573 pp.

Danks, H. V. (1991) Museum Collections: fundamental values and modern problems, *Collection Forum*, 7(2), 95–111.

Darling, C. and Thomson-Delaney, J. (1992) Bluebirds, blowflies and parasitic wasps: Scientific rather than emotional responses are needed to protect the diverse life on Earth', *Rotunda*, 25(1), 37–40.

Eldredge, N. (ed.) (1992) *Systematics, Ecology and the Biodiversity Crisis*, Columbia University Press, New York.

Ellis, A. E. (1967) 'Are research activities always compatible with conservation?' In *The Biotic Effects of Public Pressures on the Environment*, NERC, Monks Wood Experimental Station, 35–37.

Emery, A. R. (1993) 'Collections of biotic diversity: relevance to modern problems'. In Palacios, F., Martinez, C. and Thomas, B. *First World Congress Sessions and Contributions on the Function and Management of Natural History Collections*, International Symposium and First World Congress on the Preservation and Conservation of Natural History Collections, Madrid, May 1992, 59–68.

Gamez, R. (1993) 'Costa Rica's National Biodiversity Institute: Linking biodiversity, conservation and development'. In Rose, C. L., Williams, S. L. and Gisbert, J.

(eds) *Current Issues, Initiatives, and Future Directions for the Preservation and Conservation of Natural History Collections,* International Symposium and First World Congress on the Preservation and Conservation of Natural History Collections, 1992, Madrid, Spain, Vol 3, 363–366.

Griffin, D. (1992) 'Museums and the Environment: A challenge for us, not them'. In Saucier, S. (ed.) *Museums: Re-thinking the Boundaries.* Proceedings of the 16th ICOM General Conference, Quebec, Canada, 1992.

Hackett, S. J. and Zink, R. M. (1989) 'Summary of techniques to preserve frozen tissues'. In Rogers, S. P. and Wood, D. S. (compilers) *Notes from a Workshop on Bird Specimen Preparation held at the Carnegie Museum of Natural History, Pittsburgh, August 1985,* Pittsburgh.

Herholdt, E. M. (1990) 'Vertebrate collections management: a personal perspective with special reference to Southern Africa'. In Herholdt, E. M. (ed.) *Natural History Collections – Their Management and Value,* Transvaal Museum, Special Publication No 1, 1–11.

HMSO (1994) *Biodiversity: The UK Action Plan,* HMSO, London.

Hoagland, K. E. (1989) 'Socially responsible – in the 1990s natural history museums will focus their efforts on maintaining the diverse life of a healthy planet', *Museum News,* September/October, 50–52.

Hoagland. K. E. (1990a) 'Hearings on Biological Diversity Bill point to need for redefinition of the relationship between systematists and the Nature Conservancy', *ASC Newsletter,* 18 (1), 3–4.

Hoagland, K. E. (1990b) 'Biodiversity Initiatives abound in Washington', *ASC Newsletter,* April, 18 (2), 21–26.

Horie, V. (1993) 'Conservation in Europe', in Rose, C. L., Williams, S. L. and Gisbert, J. (eds) *Current Issues, Initiatives, and Future Directions for the Preservation and Conservation of Natural History Collections,* International Symposium and First World Congress on the Preservation and Conservation of Natural History Collections, 1992, Madrid, Vol 3, 189–197.

Kennedy, L. A. (1992) *National Museum of Natural History Report 1992,* Smithsonian Institution, Washington DC.

Lowe, P. D. (1976) 'Amateurs and professionals; the institutional emergence of British plant ecology', *Journal of the Society for the Bibliography of Natural History,* 7, 517–535.

May, R. M. (1990) 'Taxonomy as destiny', *Nature,* 347, 129–130.

May, R. M. (1993) 'Global change: the need and concern for collecting and preserving'. In Rose, C. L., Williams, S. L. and Gisbert, J. (eds) *Current Issues, Initiatives, and Future Directions for the Preservation and Conservation of Natural History Collections,* International Symposium and First World Congress on the Preservation and Conservation of Natural History Collections, 1992, Madrid, Spain, Vol 3, 35–41.

McAllister, D. E. (1993) 'The eco-taxa hotspot approach to the conservation of biodiversity: evolving tools', *ASC Newsletter,* 21 (4), 47–48.

McGinley, R. J. (1993) 'Where's the management in collections management?' In Rose, C. L., Williams, S. L. and Gisbert, J. (eds) *Current Issues, Initiatives, and Future Directions for the Preservation and Conservation of Natural History Collections,* International Symposium and First World Congress on the Preservation and Conservation of Natural History Collections, 1992, Madrid, Spain, Vol 3, 309–333.

Miller, S. (1994) 'Biological Collections Databases Available on Internet', *ASC Newsletter,* 22 (4), 57.

Morin, N. R. (1994) 'Report to ASC on Biological Survey efforts and US Federal Agencies', *ASC Newsletter,* 22 (5), 67–70.

Mound, L. (1991) 'Note on the INBio project', *Museums Journal*, 91 (8), 22.

National Institute for Conservation (1990) *Conservation Assessment Book*, Getty Conservation Institute and National Institute for Conservation, 45 pp.

Natural History Museum (1993) 'Announcement: IDQs – Identification qualifications', Press Release, The Natural History Museum, London, May 1993.

Novacek, M. J. (1990) 'Research and Education in Natural History Museums – The need for commitment', *Museum Management and Curatorship*, 9, 352–358.

Novacek, M. J. (1992) 'The meaning of systematics and the biodiversity crisis'. In Eldredge, N. *Systematics, Ecology and the Biodiversity Crisis*, Columbia University Press, New York, 101–108.

Pettitt, C. (1991) 'Putting "Bloody mice" to good use', *Museums Journal*, 91 (8), 25–28.

Platnick, N. I. (1992) 'Patterns of Biodiversity'. In Eldredge, N. *Systematics, Ecology and the Biodiversity Crisis*, Columbia University Press, New York, 15–24.

Ratcliffe, D. A. (1967) 'Conservation and the collector'. In *The Biotic Effects of Public Pressures on the Environment*, NERC, Monks Wood Experimental Station, 16–19.

Rautenbach, I. L. and Herholdt, E. M. (1990) 'Curators and the research collections of natural history museums – their contribution to nature conservation in South Africa'. In Herholdt, E. M. (ed.) *Natural History Collections – Their Management and Value*, Transvaal Museum, Special Publication No 1, 145–151.

Raven, P. H. (1981) 'Natural Heritage'. In *The World's Heritage; The Museums Responsibilities. Proceedings of the 12th General Conference of ICOM, Mexico City, 1980*, International Council of Museums, Paris.

Ricklefs, R. E. (1981) 'Response: It's time for museums and biology departments to get back together', *Auk*, 98, 196–198.

Roberts, D. A. (1988) Introduction to Roberts, D. A. (ed.) *Collections Management for Museums*, Museum Documentation Association, Cambridge.

Romero-Sierra, C. (1993) Closing comment. In Rose, C. L., Williams, S. L. and Gisbert, J. (eds) *Current Issues, Initiatives, and Future Directions for the Preservation and Conservation of Natural History Collections*, International Symposium and First World Congress on the Preservation and Conservation of Natural History Collections, 1992, Madrid, Spain, Vol 3, 429–430.

Rose, C. L. (1991) 'The conservation of natural history collections: addressing preservation concerns and maintaining the integrity of research specimens'. In Cato, P. S. and Jones, C. (eds) *Natural History Museums: Directions for Growth*, Texas Tech University Press, Lubbock, 51–59.

Silvy, V. and Cato, P. S. (1991) 'Collections care in a small natural history museum'. In Cato, P. S. and Jones, C. (eds) *Natural History Museums: Directions for Growth*, Texas Tech University Press, Lubbock, 61–67.

Smith, E. (1979) *Taxonomy in Britain: Report by the Review Group on Taxonomy set up by the Advisory Board for the Research Councils*, London, HMSO, 126 pp.

Systematics Agenda 2000 (1994a) *Systematics Agenda 2000 – Charting the Biosphere – A Global Initiative to Discover, Describe and Classify the World's Species*, New York, 20.

Systematics Agenda 2000 (1994b) *Systematics Agenda 2000 – Charting the Biosphere – A Global Initiative to Discover, Describe and Classify the World's Species. Technical Report*, New York, 34 pp.

Taboada, G. S. (1992) 'The conservation of animal diversity in Cuba. In Eldredge, N. *Systematics, Ecology and the Biodiversity Crisis*, Columbia University Press, New York, 169–177.

Tattersall, I. (1992) 'Systematic versus ecological diversity: the example of the Malagasy primates'. In Eldredge, N. *Systematics, Ecology and the Biodiversity Crisis*, Columbia University Press, New York, 25–39.

Thomson, K. S. (1992) *The Academy of Natural Sciences Annual Report, 1991–1992*, Academy of Natural Sciences, Philadelphia.

UNESCO (1992) *Declaration of Belém.* Report on the III UNESCO Forum on Science and Culture 'Towards Eco-ethics: alternative visions of culture, science, technology and nature', SECULT, Belém, Brazil, 8 pp.

Vane-Wright, R. I., Humphries, C. J. and Williams, P. H. (1991) 'What to protect? – systematics and the agony of choice', *Biological Conservation*, 55, 235–254.

Wallace, B. and Jones-Garmil, K. (1994) 'Museums and the Internet: A Guide for the Intrepid Traveller', *Museum News*, 73 (4), 32–37.

Whiteley, D. (1991) 'Note on the "Beetle-down" Campaign', *Museums Journal*, 91 (8), 23.

Williams, B. (1987) *Biological Collections U.K.*, The Museums Association, London.

Williams, P. H. (1994) *Using Worldmap – Priority Areas for Biological Diversity. Version 3*, October 1994, Natural History Museum, London.

Wolf, S. J. (1993) 'Conservation Assessments and long-range planning'. In Rose, C. L., Williams, S. L. and Gisbert, J. (eds) *Current Issues, Initiatives, and Future Directions for the Preservation and Conservation of Natural History Collections*, International Symposium and First World Congress on the Preservation and Conservation of Natural History Collections, 1992, Madrid, Spain, Vol 3, 289–307.

8. *Museums and environmental recording I – History and framework*

> Now more than ever in recorded history we are faced with enormous losses of species and degradation of the environment. Natural History museums must respond to these global changes by documenting biota before they and the information they contain are severely altered or destroyed.
>
> ICOM Committee for Conservation, Copenhagen, 1984

The debate about the role of museums in conserving biodiversity has centred on the world's major taxonomic institutions. The collections and systematics expertise of museums such as London's Natural History Museum, the Smithsonian in Washington or the Musée d'Histoire Naturelle in Paris are on such a scale that one might get the impression that smaller, less well-known museums can play no part in assessing biological diversity. It is true that the resources required to address the status of biodiversity on a global scale, to research the major issues of systematics and taxonomy, can only be found by the major institutions, either working alone, or through their co-operation within a framework such as that proposed by Systematics Agenda 2000. It is those museums and botanic gardens which have the scientific expertise and can create the databases and mapping software such as *Worldmap* which will determine priority areas for species conservation world-wide. However, to make the assumption that other museums can do little towards furthering the cause of biological conservation would be wrong, indeed nothing could be further from the truth. The maxim 'think global, act local' holds true for museums as well as individuals, and every natural historian in every museum can make an important contribution by surveying and recording sites of wildlife interest which fall within their geographic area. One of the greatest impacts of environmentalism on museums has been an increase in the time devoted by museum natural history staff to the documentation of biological and geological sites within their region. The dramatic development of museum-based 'biological record centres' or 'environmental record centres' in Britain from the early 1970s provides convincing evidence of the growth of the conservation ethic. The natural history curator has changed – no longer is (s)he inward-looking and collections-based. The emphasis has shifted to biological and geological survey – the documentation of environmental change. The progressive curator is now actively involved in conservation-oriented field work and becoming more central to the network of conservation organizations which flowered during the rise of environmentalism.

Museum curators have always been involved in documenting the environment through their collecting activities. However, environmentalism has led naturalists to re-evaluate the significance of the data attached to existing collections, their revelations about past environments. The emphasis of collecting has changed too, it is now not simply collecting specimens, but the collection of new information which can be used for the benefit of geological and biological conservation. Sometimes this has meant collecting new material to add to museum collections, but more often a field visit is devoted to completing recording cards, which now exist (in Britain) for the majority of taxa. Biological recording has been defined as 'the collection, collation, storage, dissemination and interpretation of records, both in space and time, of kinds and numbers of wildlife, assemblages of organisms, and their habitats, especially when the records are related to specific, localized sites' (Harding, 1991). As a discipline it is arguably more museum-based in Britain than elsewhere; sadly, the ideal of environmental records centres which would take a holistic approach to the environment has not been achieved, and the mainstays of museum-based recording activity are still the natural scientists. However, for most provincial scientists in Britain this does not mean greater involvement in pure systematics, (which is still largely the preserve of the major museums), but it does require biological and geological curators to have the ability to identify species they are observing – or if they lack expertise in a particular group they need to be able to get competent help. Networking between museums, other professional biologists, geologists and respected amateurs is essential to ensure the accurate recording of the distribution of species and the ranking of the significance of sites of biological and geological interest in their geographical region.

These co-operative efforts to document Britain's natural resources are able to build on the country's long tradition of field natural history, which is nowhere better demonstrated than in botany. The following account of the exploration of the flora of County Durham, England provides an example of how past and present recording efforts combine, and illustrates both the role of the museum and the significance of museum collections.

The flora of County Durham, England

In 1671 the British naturalist John Ray (1627–1705), on his final 'simpling' journey through Britain, saw and recorded the presence of shrubby cinquefoil *Potentilla fruticosa* on the Yorkshire bank of the River Tees near Eggleston Abbey – he had earlier published the first record for this plant from a specimen sent to him (Ray, 1670). Often regarded as the first true field naturalist, Ray had links with many other early botanists, including Thomas Lawson (1630–1691), a schoolmaster from Penrith in Cumbria, who forwarded a list of plants considered rare in northern England to him in 1688 which was incorporated into the second edition of the *Synopsis methodica stirpium Britannicarum* (Ray, 1693). Lawson's records include plants from

County Durham, for example *Bryonia dioica* near Darlington and *Glaux maritima* from Fryer's Goose near Gateshead. These records mark the beginning of a number of detailed investigations into the distribution of vascular plants in Durham, one relatively small area of England. The Quaker botanist Edward Robson (1763–1813) of Darlington produced the first definitive lists of Durham plants in 1794 – the manuscript *Plantae Dunelmenses* and the privately printed *Plantae rariores agro Dunelmensi indigenae.* Robson exchanged specimens with other botanists, and forwarded a list of his *Desiderata* to Nathaniel John Winch (1768–1838) in Newcastle; it was the efforts of Winch and his network of local botanists that resulted in the publication of the first definitive works on the flora of County Durham and neighbouring Northumberland (Winch *et al.*, 1805; Winch and Thornhill, 1807; Winch, 1831). Winch's *Flora of Northumberland and Durham* (1831) has stood the test of time, and is still a remarkable source of information on plant localities, being acknowledged by writers of later *Floras* (Baker and Tate, 1868; Graham, 1988; Swan, 1993) as the major pioneering work. A review of the number of botanists involved in the study of the plants of County Durham from Ray to the present day (Davis, 1988) not only gives an indication of the enormous co-operative efforts made by local naturalists, but also details the end result, namely comprehensive published accounts of plant distribution and localities of botanical significance and the major herbaria which act as the material evidence for the published records. The co-operation which was so necessary to the success of these ventures was due not only to the enthusiasm of individuals, but also to the established scientific societies which flourished in the region, notably the Newcastle Literary and Philosophical Society (1793), the Natural History Society of Northumberland, Durham and Newcastle upon Tyne (1829) and the Tyneside Naturalists' Field Club (1847). Their aims, to encourage research, disseminate knowledge through their publications and build significant collections did much to create an environment which stimulated the exploration of the area.

The collections resulting from these botanical excursions are now stored in the region's museums, in particular the Hancock Museum in Newcastle upon Tyne, Sunderland Museum and the Yorkshire Museum, York. All three museums had their origins in early scientific societies, which acted as meeting places for local naturalists and as repositories for collections. Hence individual naturalists, societies, their collections and museums are inextricably linked in the history of the documentation of the distribution of plants in one county of England. It is through the legacy of this network – the herbaria which are the results of past collecting efforts, often accompanied by other evidence such as field notebooks and correspondence – that museums in the north of England are now able to provide a comprehensive historical account of the plant diversity of the region.

The significance of these plant collections and archives was realized during the most recent efforts to update knowledge of the local flora (Graham, 1988; Swan, 1993). Both authors made extensive use of the herbaria in Sunderland and the Hancock Museum to check the validity of records, and gained detailed information from the comprehensive data recorded by the nineteenth-century naturalists on their herbarium sheets. For example, the relict late-glacial flora of Upper Teesdale was of particular interest to the early botanists, and herbarium specimens frequently convey details of the discovery of new plants there (see Plate 24). James Backhouse (1794–1869) of York, an early explorer of the area, credited the discovery of the Teesdale Sandwort *Minuartia stricta* to his friend G. S. Gibson who accompanied him on a trip to the dale on 29 June 1844; the actual specimen is in the Yorkshire Museum and is labelled 'When in summer with my son and some other friends we found on Widdybank Fell by an alpine rill the enclosed plant. It is *Alsine stricta* ... previously only known as a native of Lapland.'

Herbaria also reveal the extent of collecting activities in particular localities; the fern craze of the 1840s led to the wholesale destruction of many botanical sites, and the extinction of a number of species. The Oblong Woodsia, *Woodsia ilvensis* is a scarce fern that was found on dolerite rock scree at Falcon Clints in Teesdale; it is represented by specimens from this locality in most of the local museums, but has been extinct in the county since about 1895 (Graham, 1988). A similar fate befell the Holly Fern, *Polystichum lonchitis*, which has always had a precarious existence in Teesdale as herbarium records reveal; one or two plants survive in the Teesdale National Nature Reserve, a remnant of the 'dozen or so clumps which must have grown there during the last century' (Graham, 1988). Castle Eden Dene, a deep magnesian limestone valley on the Durham coast, now a National Nature Reserve, was also subject to attack by the 'moss troopers' seeking its 16 species of fern, but at this site orchids were equally sought by collectors and gardeners. Burdon (1856) recalls that:

> The Lady's Slipper [*Cypripedium calceolus*] is now so nearly extinct, that I vainly endeavoured to exhibit a native of it to the Society [Tyneside Naturalists' Field Club] when they last visited the Dene. The *Epipactis ensifolia*, the *Pyrola rotundifolia*, the Fly Orchis and the Bird's nest Orchis have all been taken up and none are likely to survive in a garden.

Despite the pleas and creditable behaviour of some visiting societies:

> (four plants [of *Cypripedium*] ... were seen, but as great treasures and with a due regard to that excellent rule of the [Tyneside Naturalists' Field] Club which pledges itself to prevent as far as possible the extirpation of rare species – they were left undisturbed (Carr, 1848))

the Lady's Slipper was extinct in Durham by the 1920s. The only evidence of its existence are the dried specimens, collected between 1831 and 1882, in the Hancock Museum and other herbaria.

This example of the history of botanical exploration of a northern England county provides an indication of the significance of museum collections to the study of the distribution of species. It is a story which could be repeated, not just for flowering plants, but for other taxa, throughout Britain and in much of continental Europe and North America. It is important to note that museums were not always central to the production of floras, or other accounts of species distribution – but they (or the

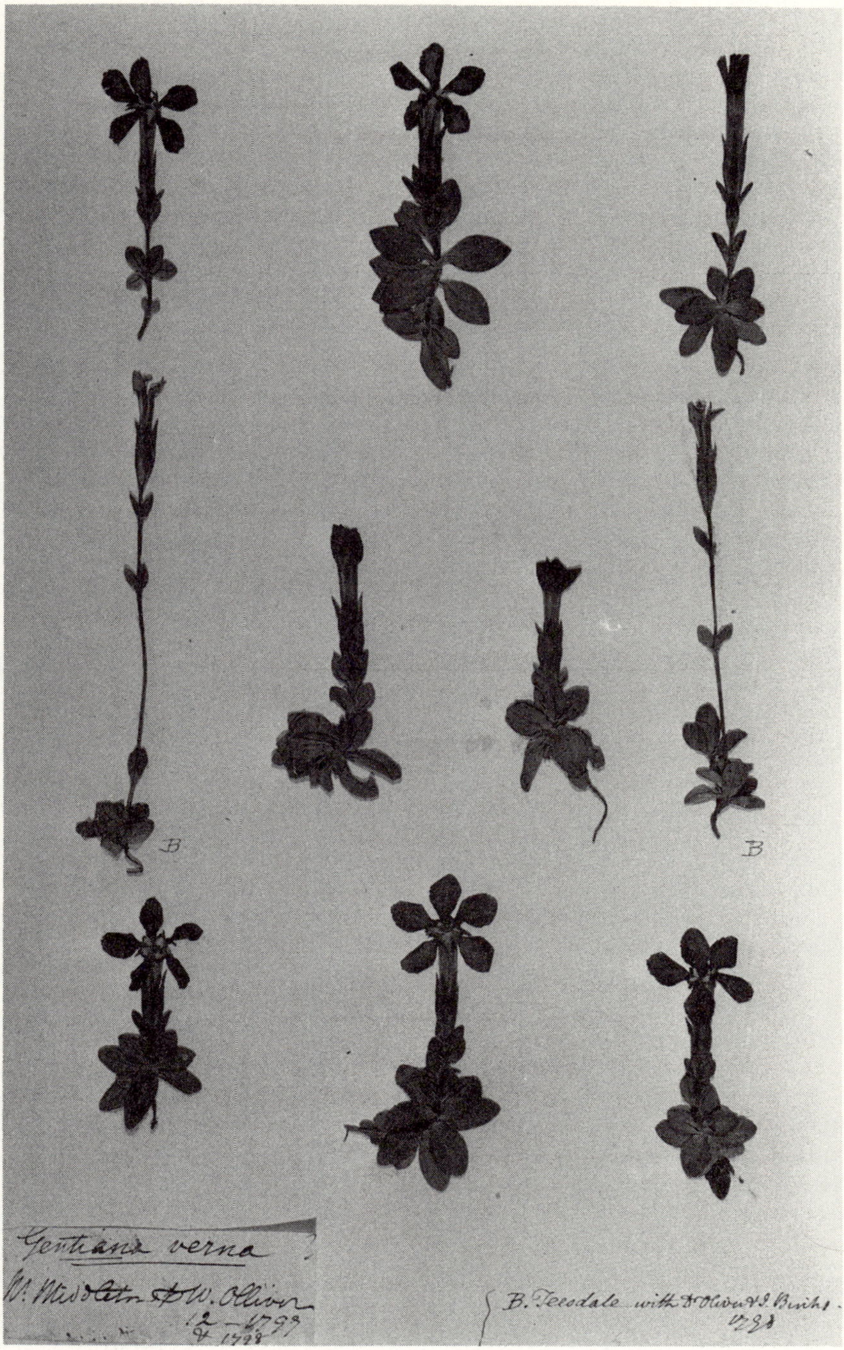

Plate 24 Specimens of *Gentiana verna* from Teesdale, Co. Durham collected in 1797–8. From the herbarium of the botanist Edward Robson of Darlington held in Sunderland Museum. (Courtesy of Tyne and Wear Museums.)

societies which ran them) did play a significant role in co-ordination of effort, and as a store for the material evidence, the so-called 'voucher specimens'. The tradition of natural history in Britain is so strong that this remains the case with many taxonomic groups – for example the field work for the most recent *Flora of Durham* (Graham, 1988) was carried out mainly by amateur botanists, co-ordinated and written by a clergyman. The role of the local museum was limited – Sunderland Museum did help with administration, organizing meetings and fund-raising for the publication, but its major role was still as a repository for voucher material and as a link to specialist taxonomic expertise. Interestingly *The Flora and Vegetation of County Durham* (Graham, 1988) was seen into press by Durham County Conservation Trust (now Durham Wildlife Trust). All previous accounts of the local flora were seen as academic exercises which explored patterns of distribution – they were scientific endeavours published by scientific societies. Environmentalism has changed that view. It is now recognized that distribution data provide a powerful conservation tool, a means to give objective assessments of species rarity or site significance. A number of museums have been involved in the research and publication of County Floras, including Hitchin Museum (Bedfordshire) and Leicestershire Museums (Rutland). Such Atlases emphasize the need for conservation to stop the erosion of species – for example, Graham (1988) records that 111 species which had been listed in previous floras were not seen in Durham during his survey – of these '14 ... have been extinct for more than 150 years; 29 were extinct or very rare at the beginning of this century; 50 disappeared between 1900 and 1950 and another 18 or so have not been recorded since 1950.'

Mapping the distribution of species – the British experience

The new-found interest in natural history in the early nineteenth century prompted the formation of a number of specialist societies, which included, in 1836, the Botanical Society of London and the Botanical Society of Edinburgh. Both societies encouraged the exchange of plant specimens between members, and became aware of the enormous quantity of information to be had relating to the distribution of species. In 1838 William Brand (1807–1869), a member of the Edinburgh Society, divided the British Isles into 42 numbered 'districts' of roughly equal size, and in a loose-leaf register of the Society's collections entered the presence of species in columns numbered from 1 to 42 – 'the presence or absence of species in each district could be pinpointed at a glance' (Allen, 1976). A map showing the boundaries of Brand's districts was exhibited by Edward Forbes at the British Association that year, and the suggestion was made that a base-map could be usefully established to record the distribution of plants and animals in the British Isles – an agreed version was available for use in 1842.

However, in 1843, Hewett Cottrell Watson (1804–1881), driving force behind the Botanical Society of London, published provisional distribution maps for 39 species of flowering plants based on 18 broad subdivisions of the British Isles; Watson's system took precedence over Brand's proposals,

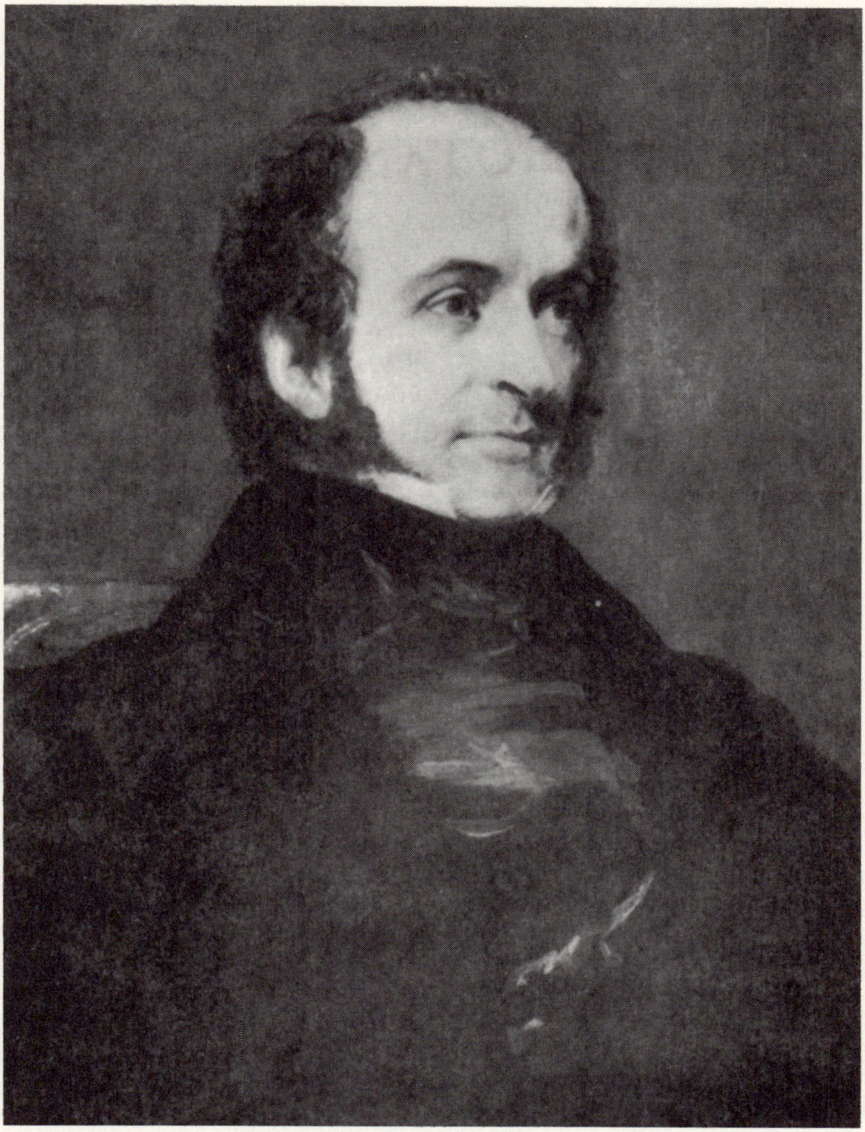

Plate 25 Hewett Cottrell Watson (1804–1881), the man who devised a recording system for British flora. (Courtesy of the Royal Botanic Gardens, Kew, and St Pauls' Bibliographies, Winchester.)

perhaps due to his authority as a writer of key botanical works. Watson (see Plate 25) subsequently moved to progressively smaller units, so that by the time of the publication of *Topographical Botany* (Watson, 1873) he had established 112 'vice-counties' – this 'Watsonian Vice-County System' became firmly established as the basis for biological recording in Britain.

David Allen (1976) explains why the vice-county system had such an appeal, and why it still lingers on today:

> For the first time a unit had been sanctioned which was small enough to offer almost inexhaustible scope for new records – and thus by his own reckoning, for demonstrable scientific usefulness – to the ordinary, comparatively static local worker. From now on a naturalist need not leave his own estate or his parish in order to feel that his hunting served a purpose in the national scheme of things. A man could share pride in his county: with his 'vice-county' he could identify individually. For those who wanted a relatively undemanding pretext for combing the countryside immediately to hand Watson had provided the perfect answer; and throughout the final quarter of the [nineteenth] century the task of filling in the gaps that his preliminary cataloguing had disclosed supplied the study with a greater part of its momentum.

Mapping the distribution of species using 'dot-maps' allied to a national grid system had been in use in Germany as early as 1860, and had been widely accepted in Europe, but largely due to the influence of the vice-county system, failed to make an appearance in Britain until 1936. Even in the early 1950s the *Biological Floras* of the British Isles published in the *Journal of Ecology* adopted distribution maps using Watsonian vice-counties. However, *Flora of the British Isles* (Clapham, Tutin and Warburg, 1952), by providing the botanist with the means to identify all the species of British plants in one volume, proved vital to the success of a new project to map the distribution of the British flora using the Ordnance Survey 10 kilometre grid square as a basis. Proposed by Roy Clapham at a meeting of the Botanical Society of the British Isles in March 1950, the scheme generated some 1.5 million records contributed by more than 1,500 volunteers, (Harding and Sheail, 1992) which were transformed into 1,700 distribution maps and published as the *Atlas of the British Flora* (Perring and Walters 1962) 12 years later. The change to recording species distribution based on the national grid had been effected, and has proved to be adaptable to all levels of recording from $10km^2$ for national maps to tetrads ($2km^2$) at the local level.

It is interesting to note the subtle changes in attitude which were taking place during the 1960s. Perring (1992) recalls:

> What is almost unbelievable to us now, perhaps, is that nowhere in the 1950 [BSBI] Conference Report is there a reference to conservation . . . yet by the time the *Atlas* and the *Critical Supplement* had been published the threat to our flora, particularly from modern agriculture, was only too apparent, and Monks Wood Experimental Station, to which we had now moved, had been set up in 1961 to study the threat to our flora and fauna. . . . It was only then, I believe, that the full potential of the *Atlas* as a conservation tool was appreciated, though the fact that we had data of immediate value was only as a by-product of the method.

The method involved the automatic sorting of data held on punch card, and the use of a 'tabulator' which produced the dot maps – these amazing machines which served British naturalists so well (and paved the way for much more sophisticated data-handling computers) are now on display in the Digital Computer Museum at Marlboro, Massachusetts.

A number of factors had come together at the right moment – a growing interest in the environment, a dedicated group of amateur botanists, new technology which could cope with the data, and a scientific demand to investigate species distribution and species loss – to inspire the production of the *Atlas of the British Flora.* What was revealed was a useful conservation tool – a means of establishing which species were rare or threatened, and which sites were botanically significant. A new organization was needed to develop this idea, and much of the credit for the establishment of the Biological Records Centre (BRC) at Monks Wood near Huntingdon in 1964 must go to Max Nicholson, then Director General of The Nature Conservancy. During the 1960s and 1970s BRC began to play a major role in helping scientific societies (including the Bryological Society, British Conchological Society and Deer Society) process data on the distribution of species. The growth in various recording schemes was phenomenal – six existed before the foundation of BRC; from 1964 to 1973 another 18 commenced, and a further 31 have begun since then (Harding and Sheail, 1992) – a number of *Atlases* (e.g. Heath, 1978; Heath *et al.*, 1984) have been published as a result. In 1973 BRC became part of the Institute of Terrestrial Ecology, and in 1989 the largest part of its Environmental Information Centre (Wyatt, 1992). The data have been used effectively to aid conservation, influencing the species schedules of the Wildlife and Countryside Act 1981, and providing the necessary information for the WWF's *Red Data Books* (Perring and Farrell, 1977, 1983).

A framework for biological recording in the British Isles

The Biological Records Centre had become established as a focal point for biological recording in Britain by the early 1970s. Yet the numbers of organizations and individuals taking part in biological surveillance were considerable, and included individual naturalists, local groups (e.g. natural history societies), and national groups (BSBI, BTO) representing amateur interests, and a significant number of professional biologists attached to academic institutions, local authorities, commercial organizations and conservation agencies. The quantity of available data was such that some network was desirable to ensure that the flow of information was maintained – and perhaps more importantly that it was accessible, yet safe.

Museums, already working closely with many conservation bodies, and being central to the work of the local naturalists, seized the opportunity to expand their role as centres for stored biological and geological information, and to formalize their situation as locally based information centres on wildlife. In 1971 Liverpool Museum (now the National Museums and Galleries on Merseyside) established the North West Biological Field Data Bank, one of the first museum-based records centres. Other pioneer museums were Leicester and St Albans, and many other city and county museums followed their lead, establishing their own 'local biological records centres'. A Conference held in Leicester in 1973 acted as a forum for 23 established and proposed centres when it was suggested that 'Environmental Records Centres should be set up and paid for by local authorities to cover areas based upon the existing vice-county system' (Stansfield, 1973). A flurry of activity followed; the Museums Association convened a Standing Committee on Environmental Records Centres in 1977 (Stewart, 1980); the same year saw the establishment of the Biological Recording in Scotland Committee (BRISC) (Sommerville, 1977) and a meeting of Records Centre organizers at Monks Wood resulted in the publication of the *Handbook for local biological records centres* (Flood and Perring, 1978). By 1980, there were some 60 centres, mainly museum-based, covering most English counties and parts of Wales and Scotland (Harding and Greenwood, 1981); 70 centres were listed as being operational in February 1987 (Berry, 1988). In museums, the day-to-day organization of such centres has been devolved to natural sciences staff; in some instances (e.g. Sheffield) the authority has based its ecologists within the local museum to organize the centre and its work. The user-friendly nature of museums, their accessibility to the public and specialist groups, their independence, the complementary skills of curators and their high profile are all seen as valid reasons for promoting the museum as a base for ecological skills (Rotherham, 1992).

Co-ordination of the work of the centres, (which still have no formal link with BRC), avoiding duplication of effort and ensuring data standards were the subjects for a major conference on biological recording convened by the Biology Curator's Group in 1984. The seminar concluded (Wyatt, 1992) that:

- existing arrangements for recording, storing and retrieval of biological data were unsatisfactory and under-funded;
- agreed standards for biological recording should be set, bearing in mind the needs of both amateur naturalists and professional scientists;
- museums should, where possible, manage local biological databanks which should provide a range of information services to the general public.

Two major initiatives resulted from this meeting, first the formation in 1985 of an umbrella group, the Biological Recording Forum (Copp and Harding, 1985) from which evolved the National Federation for Biological Recording (NFBR) in 1986, secondly, the formation of the Working Party on Biological Surveillance under the auspices of the Linnean Society. The latter produced a report (Berry, 1988) which included amongst its recommendations proposals for a national operating system for biological recording under the direction of a Co-ordinating Commission for Biological Recording, which would consist of the recording community and the users of data. This body was established in 1990 and has prepared a statement of intent (Co-ordinating Commission for Biological Recording, 1990); funding has been identified as a key factor in any proposals for a national system, which must address the application of standards and their adoption, and provide software support and documentation.

These changes, which bode well for the future, have been accompanied by two significant factors. There is now real awareness of the significance of stored data for environmental monitoring and protection; the demand for biological information as part of the planning process and a prerequisite for Environmental Impact Statements have made data valuable – both scientifiically and financially. The second factor is in the method of data handling and its effect on biological recording; the computerization of environmental data requires strict control of terminology and data structure. Collaboration between the Nature Conservancy Council, the RSNC, WWF and natural history curators enabled the development of a new package for data storage and handling – RECORDER – which is now being used a standard by most local records centres and Wildlife Trusts (see Figure 8.1). This sophisticated software does more than simply produce lists of species or distribution maps – it can produce maps showing the distribution and overlap of related species, can determine flowering times of plants within a local area or estimate sex ratios of particular species (Yeates, 1991). The ability to store and analyse information, and to disseminate it widely in a recognized form, will do much to enable a national recording network, and provide information in a form which planners and land managers can use. However, some concerns have been voiced about the impact of RECORDER on the strategic development of biological recording. As more organizations seek to acquire and use it, so biological data becomes increasingly fragmented, when what is really required for effective conservation is union catalogues and authoratative databases. Wyatt (1992) has pointed to other costs of computerization:

> if computer systems are to be exploited to their full potential, then it is necessary that the providers of data exercise discipline and adhere to basic minimum standards with regard to the form, the consistency and the accuracy of the data and that the users of such systems observe any restrictions which may be necessary, either because of characteristics of the data (e.g. precision) which constrain their use or because of other restrictions on their dissemination (e.g. considerations of confidentiality). Such restrictions can be applied relatively easily in an environment where a single organisation is responsible for all aspects of the database. Biological recording rarely conforms to this pattern; by its nature biological surveillance tends to involve many different specialisms, from many different organisational backgrounds.

Despite these difficulties, there is little doubt that RECORDER has already done much to enable record centres to operate more effectively, and to provide biological data, correctly interpreted, to decision-makers. This is only a beginning however. Other avenues are being explored, including the possibility of carrying out multivariate analysis to some data (e.g. Luff *et al.*, 1992), and to the development of GIS analysis which links species distribution data with other environmental datasets, examining patterns of species richness and habitat-associated species assemblages.

Mapping species distribution in the USA

The mapping of the distribution of plants in larger countries poses difficulties. In the USA, with large areas to explore and document, producing meaningful maps becomes a difficult task, and as a consequence the development of a framework for recording in America has developed along different lines, and in contrast to the British experience, does not have the

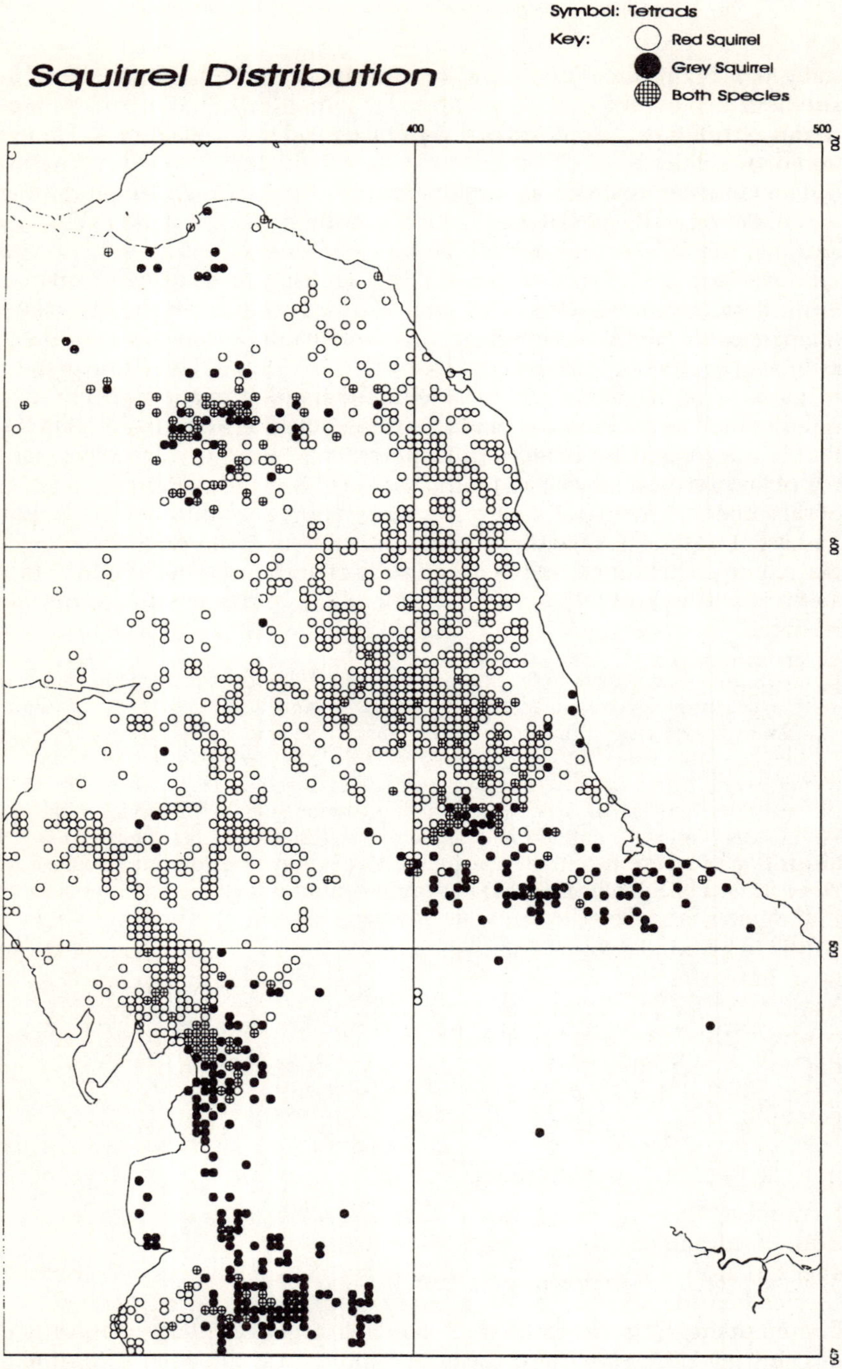

Figure 8.1 A distribution map of red and grey squirrels in northern England and southern Scotland, produced using the RECORDER software. (Reproduced by courtesy of Tyne and Wear Museums.)

same significant contribution from the amateur naturalist. The scale of the problem is so great that the production of state distribution maps demands highly efficient organization, and the involvement of professional scientists and organizations.

The Flora of Pennsylvania project provides an interesting example, and a marked contrast to the production of the Durham Flora. It is a joint venture between the Morris Arboretum (University of Pennsylvania) and the Academy of Natural Sciences of Philadelphia and began in the mid-1930s, as enthusiastic scientists began to incorporate collections of Pennsylvania plants into the herbarium, many of them collected on botanical field trips mounted by the University. From 1939–41 some 30 workers were employed to mount and label these field-collected specimens – some 50,000 specimens in the University of Pennsylvania collections were documented. As the Flora project gathered pace, data from other major collections of Pennsylvania plants were examined, including those from the Philadelphia Academy of Sciences, Penn State University, The Carnegie Museum of Natural History, Pittsburgh, Gray herbarium, Harvard and the State Museum of Harrisburg. The initial run of enthusiasm faltered under the weight of data, and it was only in the 1970s that the collecting and mapping programme was resumed – a resurgence of interest undoubtedly linked with the rise of environmental awareness. These original data were plotted by hand on a base map of Pennsylvania to produce the *Atlas of the Flora of Pennsylvania* (Wherry *et al.*, 1979). Since 1979, data have been transferred on to a computer database (Dbase) which now holds over 400,000 records of 3,318 taxa of flowering plants (Rhoads, *pers. comm.*). A new *Atlas* has been published (Rhoads and Kline, 1993), each dot on the maps representing a voucher specimen. This information has been used to good effect in affording protection to rare species; all data have been sent to the State Department of Environmental Resources, where they have been used in the Endangered and Threatened Species Programme.

It is interesting to note that the Flora of Pennsylvania is compiled only from herbarium specimens – there are no records of plant identifications made on-site, even for the commonest and readily recognized species. This contrast to the preparation of British Atlases, where field observations form the bulk of recording work is fascinating. Although this can be linked in part to the greater diversity of the American flora (and fauna), there is also a longer tradition in Britain for the publication of sight records of, for example, unusual species occurrences, or lists of species made on field outings of naturalists' clubs. From the earliest days of their founding the Naturalists Field Clubs published their record lists. There is, however, a distinct advantage in the American method – all identifications can be checked against a specimen. In Britain this may only be possible for those species considered difficult to identify, and voucher specimens may be limited to critical genera such as *Taraxacum*, *Hieracium*, *Rosa* or *Rubus*.

The study of the distribution of insects in the USA, poses major difficulties due to the sheer diversity of organisms. An example of how this problem has been addressed is the California Insect Survey, a cooperative venture based on the major insect collections held at the California Academy of

Sciences, Los Angeles County Museum, and the University of California at Berkeley (The Essig Museum) and Davis (R. M. Bohart Museum of Entomology). Specialists in these institutions work on a variety of arthropod taxa world-wide; the total number of specimens is estimated at some 20 million. Much of the early work on the distribution of Californian insects was carried out from the State Agricultural Experimental Station in the 1940s. The main objectives have been to publish identification keys and distribution maps for Californian insects; in the 44 publications produced since 1950 it is estimated that only 12 per cent of the insect fauna has been covered (Powell, *pers. comm.*). Major problems have been caused by the number of new species encountered, often resulting in the necessity for comprehensive taxonomic revisions, and by the lack of specialists. Although the project still continues, there are now fewer taxonomists interested in state mapping, as biodiversity studies concentrate entomological minds on tropical regions of the world.

Environmental recording activities in the USA are carried out by a range of statutory and voluntary bodies, covering a variety of taxa – the essential difference is the geographical scale of operation, as the two previous examples have demonstrated. Co-operation is essential within this framework, and the Nature Conservancy, founded in 1951 for educational and scientific purposes, is central to these efforts, being involved in the identification of sites of significance, their purchase and management, not only in the USA but also in other countries. A non-profit organization with 645,000 members in 1992, the Nature Conservancy maintains 1,300 nature reserves covering 6.2 million acres in the USA and 20 million acres in other countries. With a central administration and 'Chapters' at a state level, it is equivalent to the RSNC in Britain – though far more influential and better financed.

The Nature Conservancy identifies areas of significance through the State Natural Heritage Inventory Programs. These are usually administered by a state agency, and act as significant catalysts to the charting of biodiversity, through their encouragement of networking at a state level. In Latin America, these inventory programmes are called Conservation Data Centers (CDCs). Progress in collecting information has taken some important steps forward through these inventory programmes. For example, in 1991 an agreement was signed which will allow individual military installations to contract with Natural Heritage Programs to provide information from heritage databases and conduct field surveys on Department of Defense land (Chipley and Weiting, 1991). The New Mexico Natural Heritage Program is working with the New Mexico Ornithological Society to develop a bird record database which uses Advanced Revelation software (the basis of the UK's RECORDER); the natural heritage inventory of the California Department of Fish and Game has been recently made available to the California Academy of Sciences – many records in the database were drawn from the CAS collections (Anon, 1991); the northern Mexican state of Sonora joined the network in 1991, with the CDC being located in the Centro Ecologica de Sonora, a semi-autonomous agency of the state government that carries out research on rare species and threatened ecosystems

(Anon, 1991a). In 1992 the network of Natural Heritage Inventory Programs and Conservation Data Centers comprised 82 members throughout North, Central and South America.

The need for a more systematic approach to the collection and dissemination of biological data in the USA, especially in relation to effectively enforcing the provisions of the Endangered Species Act, led to the formation, in 1993, of the National Biological Survey (NBS). Researchers and support staff from seven US Department of the Interior bureaux were merged to create NBS, including staff from the Fish and Wildlife Service, National Parks Service and the Geological Survey. The mission of this new body is 'the consolidation and enhancement of existing biological research in the U.S. Department of the Interior for the wise stewardship of our nation's natural resources and for the promotion of understanding biological systems and their benefits for society'. In its circular of June 1993 the following points were identified as key areas of NBS operation:

- The NBS will provide information on where species and habitats occur and identify environmental declines before the crisis point of an 'endangered species' listing. We need a professional biological science organization, not to make regulatory decisions, but to help the Nation understand and protect threatened ecosystems and carefully use its land and natural resources.
- The NBS will ensure that biological science remains independent from management and regulatory applications of science. This separation of functions will enhance the integrity and objectivity of scientific results.
- The NBS will serve to improve the scope, efficiency and co-ordination of biological research. Through the NBS, scientific information will become more accessible and more credible for the benefits of scientists, resource managers, industry and public.

The budget for the financial year 1994 for the NBS was set at $179 million, with 46 per cent being devoted to ecological research, 17 per cent to inventory and monitoring, 8 per cent to information transfer and the remainder to administration. NBS became operational on 11 November 1993.

Prior to the acceptance of the idea of NBS, the museum community, through the Association of Systematics Collections (ASC) had suggested the formation of a National Centre for Biodiversity – a closely related but not identical concept, which ASC had suggested be run from the Smithsonian Institution (Hoagland, *pers. comm.*). The recent proposal for a National Institute for the Environment, (Committee for the NIE, 1993) with a mission 'to improve the scientific basis for making decisions on environmental issues' encompasses many of the ideas within the original suggestion for a National Centre for Biodiversity. The NIE, if approved, would create a centre for environmental statistics, and require a fundamental review of research efforts in environmental biology, including species and habitat survey.

The position of museums in this mêlée is interesting. The need for a co-ordinated approach to recording in the USA had long been recognized by museums, and during the 1980s considerable discussion took place to find a way to implement a National Biological Survey. The Association of

Systematics Collections (ASC) raised the issue at its 1985 Annual Meeting, the results of which were summarized in a Symposium volume (Kim and Knutsen (eds), 1986). Speakers represented a variety of interests, and wide-ranging issues relating to biological surveys, including systematics, ecological research, parasitology, tropical biodiversity, biotechnology, agriculture and input to multidisciplinary international research such as the Man and the Biosphere programme. However, the significance of such a survey for conservation purposes were recognized by a number of speakers (e.g. Hirsch, 1986), who identified the scattered and often incompatible nature of biological data held by a variety of organizations. However, the NBS proposals formulated by ASC (Association of Systematics Collections, 1986) drew the scientific community of systematics into conflict with The Nature Conservancy in the late 1980s. This occurred despite the long-held realization of the significance of systematics collections for environmental protection which had been evident at the ASC's meeting in 1974 (Barr, 1974). A sizeable museum-based taxonomic lobby, holding extensive databases, and seeking contracts to enable them to undertake field work, felt threatened by the Nature Conservancy, who wished to take over the biological recording role, using parataxonomists to collect the necessary data, so bypassing the systematics expertise in museums. The ill-defined relationship between museum collections and their attendant data and the Nature Conservancy's Heritage Program databases was another major reason for the deteriorating relationship – a sad situation when one reflects on the shared philosophies of the organizations. However, by 1991 representatives of the Nature Conservancy and ASC had proposed conciliatory talks, and ASC had prepared a statement on 'Collections use agreements' – that is, access to databases. The practical and ethical considerations regarding data sharing were addressed by ASC in 1992 (Anon, 1992). Although the Nature Conservancy had failed in its bid to run a national centre for biological information, the situation remained confusing, with independent conservation organizations, state agencies, private consultancies and museum-based biologists competing for funds to enable biological surveys to be carried out, and little co-ordination to aid practical conservation.

A breakthrough occurred in February 1994, when a group of natural history museum directors met Bruce Babbit, Secretary of the Department of the Interior, to discuss relationships between museums and the National Biological Survey. A Memorandum of Understanding (MOU) was signed with ASC at this meeting, with Babbit declaring the co-operative agreement 'a milestone – bringing the substantial resources of our nation's museums and botanical gardens into partnership with the National Biological Survey. The support of these scientists and their collective information on America's biological resources will be tremendously helpful to the NBS.' The MOU recognizes four areas of common interest: biological collections, scientific expertise in taxonomy and related areas, biodiversity information and education (Anon, 1994) and working groups have been established to examine these key issues (Anon, 1994a).

Earth Heritage Conservation

Earth Heritage Conservation has been defined as being 'concerned with sustaining the part of the physical resource of the Earth that represents our cultural heritage, including our geological and geomorphological understanding, and the inspirational and aesthetic response to the resource' (Wilson, 1994). George Perkins Marsh and other supporters of wilderness in America, with their love of landscapes and geological wonders would have understood the definition and supported the idea of conserving geological and geomorphological sites. Yet just as biological resources are threatened and are in decline, our geological heritage is equally at risk.

In 1991 the University of Nebraska State Museum carried out a survey of the palaeontological resources of 25 square miles of the Oglala National Grassland in Sioux County Nebraska, and assessed the impact of unauthorized fossil collecting in areas frequented by tourists and collectors (Lagarry-Guyon, 1992). Of the 39 sites of special interest, which included exposures which yielded complete vertebrate skeletons, skulls and groups of individual specimens, and microfossil sites, 11 (28 per cent) showed evidence of illegal collecting. The survey report suggested the implementation of a permit system for collecting, more regular patrolling by rangers and stronger prosecution measures in order to conserve the geological resources of the site. This example serves to illustrate the nature of the threat to geological sites and the need for action to conserve them. Just as for biological sites, geological exposures should be identified and ranked in terms of their significance, and here museum geologists have a significant role to play. Similarly, geological conservation requires a framework within which survey and assessment of geological sites can take place, as well as appropriate legislation.

In the USA the protection of geological sites appears at best piecemeal, and a proposal to pass new legislation to protect vertebrate fossil sites – 'The Vertebrate Palaentological Resources Protection Act' – proposed by Senator Max Baucus have caused confusion and concern amongst state authorities, landowners, fossil collectors and the museum community (Garfield, 1992). The Bill seeks to safeguard palaeontological resources by preventing commercial exploitation of sites – fossil exports from the USA were estimated to be worth some $22 million in 1991 – by limiting collecting activities to persons or organizations affiliated to scientific institutions, and ensuring that collections are deposited in museums or universities. Enforcing such legislation is fraught with difficulties, and the proposal within the Bill that the government will retain ownership of specimens, even after they have been deposited in museums and accessioned, runs contrary to the usual curatorial practice of museums establishing and maintaining legal title to collections. The Association of Systematics Collections has been central to discussions about these issues, lobbied to amend the legislation (Anon, 1993) and considerable changes in the Bill have been agreed as a result, including the idea of federal ownership; ASC collections standards must be met for a museum to hold vertebrate fossils from federal lands (Anon, 1995).

In the UK, protection for geological and geomorphological sites began on a voluntary basis in the 1940s, with an initial list of some 400 geological sites being prepared by the Society for the Promotion of Nature Reserves in 1945. The Nature Conservancy Council was able to capitalize on this list when designating geological sites as Sites of Special Scientific Interest under the legislation which brought it into being in 1949 – The National Parks and Access to the Countryside Act, and subsequently the Wildlife and Countryside Act (1981). In 1977 The Geological Conservation Review, (GCR) was launched by the Nature Conservancy Council, whilst the National Scheme for Geological Site Documentation was initiated by the Geological Society and the Geological Curators' Group. More recently (1990 onwards), the Regionally Important Geological/Geomorphological Sites (RIGS) initiative has been developed by English Nature. All of these initiatives have contributed greatly to the identification of sites of particular value. The GCR was completed in 1988, and provides a systematic review of 2,200 Earth Heritage Sites of which some 1,450 have been designated as Sites of Special Scientific Interest (SSSIs); seven are designated as National Nature Reserves. One of the last major achievements of the Nature Conservancy Council before it was broken up into the new National Conservation Agencies was the publication of *Earth Science Conservation in Great Britain – A Strategy* (Nature Conservancy Council, 1990) This sought to consolidate the SSSI network, extend the RIGS scheme, develop new conservation techniques, improve site documentation and public awareness, and develop international links. Museums, museum-based organizations and Local Geological Records Centres were identified as key participants in all the major strategies proposed. The Geological Curators' Group (GCG) has been central to discussions about geological site conservation, has promoted good practice in collection and curation of specimens, and through *The Geological Curator* been an important forum for debate.

The designation of museum-based record centres in the early 1970s saw the promotion of Local Geological Records Centres, working along similar lines of those devoted to biology – and frequently run together and by the same group of individuals. A *Handbook* (Cooper *et al.*, 1980) provided useful guidelines, and GCG did good work in promoting site documentation and co-ordination via these centres. The National Scheme for Geological Site Documentation was based on over 50 such centres, virtually all of them based in museums. The RIGS scheme, with financial support from a number of organizations, including the Nature Conservation Agencies, the RSNC and the Geologists Association provided an additional focus for them. As informal, voluntarily run groups, RIGS groups develop locally applicable criteria for site selection, and use them to designate a network of RIGS. Membership is drawn from societies, Wildlife Trusts, universities and museums, and normally cover a geographical area equivalent to a county. Enthusiasm for RIGS has continued to grow, and schemes now exist for 36 counties in England, with a further seven being given active consideration; museums continue to act as a focus for RIGS activities. Museum-based geologists play a significant conservation role in the UK, simply because they can provide expertise which is not widely available.

One of the principal outcomes of the discussions about environmental recording in the 1970s was a call for fully integrated 'Environmental Records Centres', which would pull together information relating not only to the natural environment, but also to that relating, for example, to archaeological sites, historic buildings and monuments. In terms of the planning process, such comprehensive data would be much more meaningful. As yet, few museums have set up such a system. Also, as far as is known, no provincial museums have capitalized on GIS software to provide digitized data relating to their geographical area, which on the one hand is surprising (considering GIS as a powerful tool), yet on the other understandable, as resources available for such development work are pitiably lacking.

References

Allen, D. E. (1976) *The Naturalist in Britain – A Social History*, Allen Lane, London.

Anon. (1991) 'New state initiatives', *Biodiversity Network News*, 4 (3), 6.

Anon. (1991a) 'Sonora, Mexico joins CDC Network', *Biodiversity Network News*, 4 (3), 7.

Anon. (1992) 'Report from the ASC workshop on data sharing and database ethics', *ASC Newsletter*, 20 (6), 157–166.

Anon. (1993) 'Vertebrate Fossil Collecting on Federal Lands: Views Aired on Pending Legislation', *ASC Newsletter*, 21 (2), 23–24.

Anon. (1994) 'Bruce Babbitt meets with Museum Directors; signs MOU with ASC', *ASC Newsletter*, 22 (2), 17–23.

Anon. (1994a) 'First Meeting of the Systematics Resources Working Group to Implement the ASC–DOI MOU held', *ASC Newsletter*, 22 (5), 61–64.

Anon. (1995) 'Major activities of the Association of Systematics Collections since moving to Washington DC', [Reference to Vertebrate Palaeontological Resources Protection Act] *ASC Newsletter*, 23 (1), 4.

Association of Systematics Collections (1986) *Foundations for a National Biological Survey*, ASC, Washington DC.

Baker, J. G. and Tate, G. R. (1868) 'A New Flora of Northumberland and Durham, with sketches of its climate and physical geography', *Transactions of the Natural History Society of Northumberland, Durham and Newcastle upon Tyne*, 2, 1–316.

Barr, S. W. (ed.) (1974) *Applications of Systematics Collections: The Environment*, Report from the Second Annual Meeting of the Association of Systematics Collections, Texas Tech University, 3 May.

Berry, R. J. (1988) *Biological Survey: Need and Network*, Linnean Society of London/ PNL Press, London.

Burdon, R. (1856) 'Presidential Address', *Transactions of the Tyneside Naturalists' Field Club*, 3, 48–51.

Carr, R. (1848) 'Presidential Address', *Transactions of the Tyneside Naturalists' Field Club*, 1, 212–213.

Chipley, R. and Wieting, H. (1991) 'Heritage Programs: Uncle Sam wants you!', *Biodiversity Network News*, 4 (3), 1–2; 7.

Clapham, A. P., Tutin, T. G. and Warburg, E. F. (1952) *Flora of the British Isles*, Cambridge University Press, Cambridge.

Committee for the National Institute for the Environment (1993) *A Proposal for a National Institute for the Environment: Need, Rationale and Structure*, Association for Systematics Collections, Washington DC.

Cooper, J. A., Stanley, M., Sedman, K. and Phillips, P. (1980) *Geological Record Centre Handbook*, Museum Documentation Association, Cambridge.

Co-ordinating Commission for Biological Recording (1990) *Programme for the Establishment of a National System for the Co-ordination of Biological Recording*, Institute for Terrestrial Ecology, Huntingdon, (unpublished).

Copp, C. J. T. and Harding, P. H. (eds) (1985) *Biological Recording Forum 1985*, (Biology Curators Group Special Report No 4), Biology Curators Group, Bolton.

Davis, P. (1988) 'A History of Botanical Recording in Durham'. In Graham, G. G. *The Flora and Vegetation of County Durham*, The Durham Flora Committee and Durham County Conservation Trust, 8–23.

Flood, S. W. and Perring, F. H. (1978) *A handbook for local biological records centres*, Institute of Terrestrial Ecology, Huntingdon.

Garfield, D. (1992) 'Bones of Contention', *Museum News*, 71 (6), 16–17.

Graham, G. G. (1988) *The Flora and Vegetation of County Durham*, The Durham Flora Committee and the Durham County Conservation Trust.

Harding, P. T. (1991) 'National species distribution surveys'. In Goldsmith, B. (ed.) *Monitoring for conservation and ecology*, Chapman and Hall, London, 133–154.

Harding, P. T. and Greenwood, E. (1981) 'Survey of local and regional biological records centres', *Biology Curators Group Newsletter*, 2, 468–478.

Harding, P. T. and Sheail, J. (1992) 'The Biological Records Centre; a pioneer in data gathering and retrieval'. In Harding, P. T. (ed.) *Biological Recording of changes in British Wildlife*, HMSO, London, 5–19.

Heath, J. (1978) *Provisional Atlas of the insects of the British Isles. Part 7, Odonata – dragonflies*, Biological Records Centre, Huntingdon.

Heath, J., Pollard, E. and Thomas, J. (1984) *Atlas of Butterflies in Britain and Ireland*, Viking, Harmondsworth.

Hirsch, A. (1986) 'The Role of a National Biological Survey in Environmental Protection'. In Kim, K. C. and Knutsen, L. (eds) *Foundations for a National Biological Survey*, Association of Systematics Collections, Lawrence, KS. 55–64.

Kim, K. C. and Knutsen, L. (eds) (1986) *Foundations for a National Biological Survey*, Association of Systematics Collections, Lawrence, KS.

Lagarry-Guyon, H. E. (1992) 'Unauthorized collecting of fossil vertebrates on federal lands in northwest Nebraska: a case study', *Programme and Abstracts, Society for the Preservation of Natural History Collections*, June 2–6, 1992, 17.

Luff, M. L., Rushton, S. P., Eyre, M. D. and Foster, G. N. (1992) 'Environmental and ecological applications of invertebrate distribution data from the Biological Records Centre'. In Harding, P. T. (ed.) *Biological Recording of changes in British Wildlife*, HMSO, London, 74–75.

Nature Conservancy Council (1990) *Earth Science Conservation in Great Britain: A Strategy*, Nature Conservancy Council, Peterborough.

Perring, F. H. (1992) 'BSBI distribution maps scheme: the first 40 years'. In Harding, P. T. (ed.) *Biological Recording of changes in British Wildlife*, HMSO, London, 1–4.

Perring, F. H. and Farrell, L. (1977) *British Red Data Books, I, Vascular Plants*, 1st edn, Royal Society for Nature Conservation, Lincoln.

Perring, F. H. and Farrell, L. (1983) *British Red Data Books, I, Vascular Plants*, 2nd edn, Royal Society for Nature Conservation, Lincoln.

Perring, F. H. and Walters, S. M. (1962) *Atlas of the British Flora*, Nelson, London.

Ray, J. (1670) *Catalogus Plantarum Angliae*, J. Martyn, London.

Ray, J. (1693) *Synopsis methodica stirpium Britannicarum*, 2nd edn, Smith, London.

Rhoads A. F. and Kline, W. M. (1993) *The Vascular Flora of Pennsylvania, an Annotated Checklist and Atlas*, American Philosophical Society, Philadelphia.

Rotherham, I. (1992) 'Beyond the museum: the role of information in environmental conservation', Paper presented at The Greening of Museums, The Scottish Museum Council's Annual Conference with the Museums Association, Edinburgh, 1992, Unpublished typescript.

Sommerville, A. (1977) *A Guide to Biological Recording in Scotland*, Scottish Wildlife Trust for the Biological Recording in Scotland Committee, Edinburgh.

Stansfield, G. (ed.) (1973) *Centres for Environmental Records*, Vaughan Papers in Adult Education No 18, University of Leicester, Leicester.

Stewart, J. D. (1980) 'A summary of local environmental records centres in Britain', *Museums Journal*, 80, pp 161–165.

Swan, G. A. (1993) *Flora of Northumberland*, The Natural History Society of Northumbria, Newcastle upon Tyne.

Watson, H. C. (1873) *Topographical Botany*, privately printed, Thames Ditton.

Wherry, E. T., Fogg, J. M. and Wahl H. A. (1979) *Atlas of the Flora of Pennsylvania*, Morris Arboretum, University of Pennsylvania.

Wilson, C. (ed.) (1994) *Earth Heritage Conservation*, The Geological Society and The Open University, Milton Keynes.

Winch, N. J. (1831) *Flora of Northumberland and Durham*, T. and J. Hodgson, Newcastle upon Tyne.

Winch, N. J., Thornhill, J. and Waugh, R. (1805) *The Botanist's Guide through the Counties of Northumberland and Durham*, Vol 1, Hodgson, Newcastle upon Tyne.

Winch, N. J. and Thornhill, J. (1807) *The Botanist's Guide through the Counties of Northumberland and Durham*, Vol 2, Marshall, Gateshead.

Wyatt, B. K. (1992) 'Resources for documenting changes in species and habitats'. In Harding, P. T. (ed.) *Biological Recording of Changes in British Wildlife*, HMSO, London, 20–26.

Yeates, C, (1991) 'Informed collecting', *Museums Journal*, 91 (8), 28–30.

9. *Museums and environmental recording II – roles and outcomes*

> We need to know more about the environment, to expand our knowledge of what's out there. The best way to get started on making ecosystem management work is to get the information together to make good, comprehensive resource management decisions.
>
> Bruce Babbitt
> US Secretary of the Interior

Museums can now be identified as part of a network of biological and geological recording activity; some of these networks are formal, others less so. Many museums are at the hub of recording schemes in their county, state or region. They have historical connections with recording, and house the supporting evidence, the voucher material, as well as other documentary evidence such as notebooks and correspondence. Museum staff are familiar with data-handling, have specialist identification skills, and operate within an environment which acts as a 'honeypot' for amateur naturalists, but which is also perceived as objective by inquirers. It is this mix of history, skills, credibility and objectivity that place museums in a unique position to become involved in biological and geological recording and a valuable asset in making decisions concerning the future of the environment.

The roles a museum takes within its environmental recording agenda include the following:

- As a repository for information and specimens – a safe haven for long-term storage
- As a provider of information to decision-makers, which may, for example
  - illustrate changing patterns of distribution
  - illustrate the status of a species or site
  - have associated expert interpretation
- Stimulating and inspiring local naturalists
- Involving the public in environmental recording, so stimulating a sense of community and guardianship
- Using data to support education programmes, from displays to inquiry services.

Vouchers and long-term storage

The significance of maintaining voucher collections of specimens as part of a vegetation survey has been indicated with reference to the Durham flora. The reasons for maintaining such collections – to check and verify records, and to provide historic evidence – are equally true for faunal surveys. Zoological voucher collections can be well used by a variety of specialists, even for comparatively easily identified groups such as mammals (see, e.g. Genoways *et al.*, 1979). As Pettitt (1994) has noted 'Field biology increasingly reflects the growing demand for information and prediction in support of ecological management decision ... but such work needs the objective backing of voucher specimens .... Vouchers are especially important for introduced species or those from limited habitats.' Museums are recognized as safe places for such material to be housed – they are permanent institutions, with scientific expertise, known to be able to care for material, and with links into the local (scientific) community. They are also without political bias or prejudice. It is for these selfsame reasons that museums have been chosen in the past as repositories for naturalists' field notebooks, species lists, photographs or correspondence, and why they are now recognized as suitable places to deposit and store biological and geological data. This public confidence is important, because the museum-based record centre is limited to what it can survey on its own, with limited staff and financial resources. This distinction between active recording by museum biologists and geologists and the passive collection of data via its networking activities is significant. For most centres, the latter is more important. The centre is essentially a catalyst, encouraging the recording activities of its local naturalists, and a place where collected information can be safely deposited and its future assured, the hub of a network of environmental information, a facilitator, collector and disseminator of information. It has been correctly argued (e.g. Ely, 1994) that passive collecting alone fails to provide adequate coverage, both geographically and taxonomically, and that museums need to be more than a repository – they must promote and encourage recording activities to be viable. In practice, however, insufficient funding and staff resources restrict museums to a passive role. Museums carry out an important function through the 'rescue' of voucher material which has been gathered by other agencies (universities, consultants, government agencies) when carrying out environmental assessment. For example, Santa Barbara's natural history museum instituted a 'voucher specimen collection program' for its area in the mid-1970s which has resulted in almost 100,000 specimens representing 3,000 different species, and has used the data to prepare distribution atlases (Sears, 1990).

The significance of maintaining collections and information in museums was realized in the late 1960s and early 1970s, following the publication of *Silent Spring*. In response to concerns about the use of persistent pesticides, scientists began to use museum specimens to determine historical levels of pollution (NMNH, 1976). Two particularly well-documented uses of collections and records in investigating this aspect of environmental change were the decline of avian eggshell thickness in response to pesticide use (e.g. Everett, 1990; Ratcliffe, 1970; Klaas *et al.*, 1979) and the variation in contamination of the tissues of fish (e.g. Kelly *et al.*, 1979) and birds (e.g. Johnels, 1973) by mercury. The thickness of eggshells in predatory and fish-eating birds was shown to have declined dramatically when comparing museum specimens and those collected in the late 1960s; this decline could be linked to a reproductive failure in some species, with populations declining where thinning persisted over a number of years. Museum specimens have enabled scientists to demonstrate that mercury concentrations in fishes vary markedly from one locality to another, and that mercury concentration is more closely linked to age and size of the fish rather than increases in pollution levels over time. Mercury contamination in predatory (and especially fish-eating) birds, on the other hand, increased dramatically in the 1960s. Voucher material is essential to help science unravel these environmental connections – the use of historical collections to establish norms of contamination by pollutants has strengthened the case for maintaining museum specimens and associated records. Without the availability of such data the cause of environmental conservation would undoubtedly have been set back many years – museums provided the material evidence of change, leading to new laws to mitigate against environmental degradation.

Using museum specimen-data, as part of a more extensive survey which also includes extensive literature research and field work to assess the changing distribution of species is commonplace and has been carried out by a variety of individuals and organizations. It has, for example been used for water beetles (Foster, 1992) where ordination analysis has been used to rank wetland sites in terms of conservation quality. Using species lists of water beetles recorded in field notebooks held in the Hancock Museum, Newcastle, and verifying the species lists using associated specimens, Newcastle-based entomologist Michael Eyre (unpublished, quoted in Foster, 1992) was able to detect changes in wetland habitat over a period of time, based on water-beetle assemblages. Based on these community types, Boldon Flats, a low-lying area in South Tyneside, was probably a large bare pool in 1850; in 1930 it was a large muddy pond; in 1988 it was a temporary pool with water beetles characteristic of that habitat. In 1850 Prestwick Carr in Northumberland was probably a *Sphagnum* mire, in 1970 it was a muddy pond, and in 1985 a lowland marsh.

The decline of epiphytic lichens and bryophytes due to air pollution, in particular sulphur dioxide, is a well-known phenomenon which has been extensively documented in Britain (e.g. Rose and Hawksworth, 1981). Evidence for the decline in species near to London is largely dependent on three major bryophyte collections: those made by Edward Forster from Epping Forest around 1800, Ezekial Varenne in the Kelvedon area from 1860 to 1876, and Frederick Brocas from the vicinity of Saffron Walden. Referring to these collections Adams and Preston (1992) note that:

> Forster's herbarium, in particular, is crucial to our understanding of the lethal effects of air pollution on bryophytes. Around 1800 he collected extensively in

the Epping Forest area, finding many species that have not otherwise been recorded in Essex; he effectively set a baseline for future observations.

These collections enable bryologists to chart the decline of individual species. The terrestrial moss *Rhytidiadelphus triquetrus* was regarded as common and collected in Epping Forest by Forster, yet had become scarce there by 1885, and must have become extinct well before the 1960s when its absence was first noted. Interestingly there is now growing evidence for the return of bryophyte and lichen species as pollution levels decline (e.g. Adams, 1990), demonstrating the continued need to resurvey areas. It should be a key objective of museum-based record centres to identify such changes and respond accordingly.

Providing information for environmental decision-makers – rarity and status

The historical information associated with museum specimens has already been referred to with regard to the project to map the distribution of the Durham flora – the presence of herbarium specimens allows the decline in some species to be detected; rare species in Durham have been identified and their localities pin-pointed. This information has been assessed by Durham Wildlife Trust, and used to promote the conservation of sites of particular significance in association with landowners and planning authorities. Yet this very practical use of data was not initially foreseen, nor was it a specific goal of the Flora project. It could therefore be argued that this approach, despite the lack of focus on conservation objectives, can act as a model in terms of survey technique, yet it is one which demands an enormous commitment, and is only made possible by a team effort over a considerable period of time.

How might record centres work best in order to provide 'rarity and status' data which will be of best use to decision-makers, which might include local authority departments, industry and commerce, statutory bodies or voluntary organizations? It has been suggested (Dawson and Game, 1988) that the principal prerequisites of data required to ensure site protection are completeness (total geographic and taxonomic coverage), consistency (using standard techniques), and precision (accuracy in geographic data and identification); any data must also be up to date, as well as providing a historical component. To achieve this for selected habitats is a reasonable objective for records centres, and one which in a planning context is especially useful. To be able to answer such questions as 'How many ponds are there in Rutland?' or 'What is the acreage of deciduous woodland in County Durham?' is especially important in terms of establishing the significance of sites in terms of habitat in its broadest sense. Many museums have successfully carried out comprehensive surveys of specific habitats, and are able to provide objective data about the past and present status of threatened habitats such as wetlands, hedgerows or hay-meadows. Fulfilling the same criteria for detailed inventories of specific sites, or for species recording is however much more difficult. The effort required is much more considerable, is primarily carried out by amateur naturalists, and can only be gained through co-operative ventures such as the BRC recording schemes. Species conservation also demands some assessment of the abundance of species rather than a simple presence/absence criterion, which demands a longer-term commitment – monitoring rather than simply recording.

Long-term research and monitoring programmes require dedicated effort, and significant levels of financial support. A good example of museum involvement in a major long-term research project to monitor environmental change is the Canadian Museum of Nature's research programme in the Arctic (Gray, 1979; Cumbaa, 1991). The museum (then known as the National Museum of Natural Sciences) established its High Arctic Research Station on Bathurst Island in the North West Territories in 1968, with a view to carrying out survey work. This involved mapping the distribution of species and habitats, but there was also emphasis on the study of behavioural adaptations of Arctic animals to their environment. Each year records are kept of the arrival, courtship, mating and breeding success of birds and mammals; since the early 1970s this research has been carried out against a backdrop of increasing human disturbance, primarily through seismic surveys as part of oil exploration activities. As a result the museum was in a commanding position to provide information to protect areas threatened by pipeline development, notably at Polar Bear Pass – the first site in the high Arctic to be selected for protection as a National Wildlife Area. Scientific data gathered by museum biologists proved vital to the campaign to protect other Arctic wilderness areas, which also benefited from a successful travelling exhibition 'An Arctic Oasis', mounted by the museum to increase public awareness and gain support for conservation measures.

The CMN provides a good example of a museum taking responsibility for recording at a national level, continuing the work to document Canada's natural heritage which began in 1841 through the Geological Survey of Canada (Birker, 1991). This work has had a considerable environmental impact; the Botany Division initiating the Canadian Rare and Endangered Plants Project in 1974, which yielded valuable information on the status of species that has been published as provincial and territorial lists of rare and endangered species.

In partnership with the Entomological Society of Canada, from 1977 the Zoology Division of the CMN has supported the Biological Survey of Canada (Terrestrial Arthropods). The museum, as the agent of government, provides a long-term base for the survey, whilst the Society provides scientific expertise through its advisory committee. The early pilot study for an insect survey established the methodology and led to a publication (Danks, 1979) which indicated that the arthropod fauna is poorly known.

> The adults of only about half the insects, and only one fifth of the mites have been described .... Groups of small insects of high diversity ... are less well known than groups that are of medical or veterinary importance (e.g. fleas), conspicuous (e.g. dragonflies) or composed of very few species (e.g. earwigs) (Danks, 1993).

With overwhelming biodiversity and a geographically diverse and huge country to survey, the CMN team has had carefully to focus its activities on key themes – specific habitats, regions, projects or taxa which will provide the most useful data – which will lead to a better understanding of Canada's arthropod fauna. Danks (1993) has summarized the key facets of the Canadian Biological Survey (CBS), which he defines under four major headings. The 'Clearing House' role includes maintaining databases about collections (it maintains links with another 100 entomological collections) and entomologists and encouraging networking; 'Scientific synthesis' relates to the discussion and dissemmination of information through publications and conferences; 'Scientific projects' is the role of co-ordinating scientific effort in areas of scientific importance such as Arctic invertebrate biology, studies of the aquatic insects of Newfoundland and arthropods of the Yukon. The CBS's other major roles ('Other interests') include raising consciousness through lobbying activities, and providing information. It is interesting to note that the Survey has proposed the concept of 'Regional Centres' to promote recording, a suggestion which parallels the situation in the UK (Danks, 1983).

Other Canadian museums (e.g. Royal Ontario Museum) are actively involved in environmental recording, but the overall picture of the various recording schemes decribed by Birker (1991) is somewhat confusing, with a variety of institutions in addition to museums promoting and directing surveys, but with no national system of co-operation. The CBS, acting as a catalyst for action, seems to point the way forward; there is also the potential of the Canadian Heritage Information Network (CHIN) to assist in the co-ordination of recording efforts, perhaps acting as an equivalent of the NFBR in Britain. The need for progress, and ultimately the identification of the status of the Canadian fauna and flora, is becoming increasingly urgent.

It is encouraging to see museums in China carrying out environmental surveys in order to determine the status of species in their locale. The Shanghai Museum of Natural History (SMNH) was instrumental in surveying the remaining wildlife habitats in Jinshan County from 1974, and in monitoring the area following the opening of a major petrochemical works in 1979 (Hua, Zhou *et al.*, 1992). The decline in plant and animal species was dramatic; despite this decline areas (Jinshan Island, Chongming East Beach) were identified which still had considerable diversity, and after considerable negotiation these two sites were declared as nature reserves in 1989. Museum scientists continue to monitor the area.

Museums are not always central to recording efforts, but are invariably linked to collections. In Mexico for example, the major thrust of biodiversity research is based in the Instituto de Biologica of the National University (IBUNAM), which has acquired the collections of former academic institutions, including those of the National Museum of Natural History which had been founded in 1820 (Sarukhan, 1993).

Providing information for environmental decision-makers – expert opinion

Most professionals (for example, planners, architects, landscape consultants, foresters) involved in making decisions about the future of natural or semi-natural sites cannot use raw biological or geological data. They have to rely upon experts to interpret such information, and to be available when needed to justify their interpretation, as for example at a planning enquiry. This need for expert help places museum biologists in a strong, yet potentially compromising position. They hold the data (which has usually been provided by a variety of individuals and organizations), and because of their working knowledge of their own region are well suited to provide value judgements about the status of species or sites; they can provide objective comment from a position of real authority. A number of museum-based record centres in Britain – Sheffield for example – work closely within the local authority framework, and are able to provide a major conservation input into the planning process. In most cases such expertise is valued by a local governing body, but the potential for compromise or conflict is always present, with differing ethical standpoints, allied to differing objectives, being taken by the various factions towards development. This is particularly evident when the authority itself is the developer, but can also occur even if the development is proposed by an outside agency. For this reason it has been argued that museum-based record centres should not interpret data, but simply supply other organizations, including environmental consultants, with the data that they hold. This may have been a reasonable stance to take in the early 1970s, but as record centres are now increasingly expected to be financially self-supporting, they are being forced into a consultant role, seeking contracts which not only ask for site/species surveys, but also for expert opinion. Ely (1994) notes that 'The role that the ... [record centre] ... sees for itself helps to determine what type of information it wishes to hold and therefore, which systems it needs (and which it feels are more appropriate for others to deal with)'; whatever position the centre takes Ely suggests that users of information adhere to a draft Code of Conduct (see Figure 9.1) which addresses issues of restrictions on data, including confidentiality.

Whether a museum acts as an environmental consultant in the widest sense, or simply provides information to another organization, there are a number of practical and ethical considerations to be faced, if conservation is to be best served. From a practical point of view, data must be accessible and include inventories of research studies, management plans and other published accounts if accurate and meaningful judgements are to be made. From an ethical standpoint, museum biologists have to consider their position regarding the provision of data to other organizations, and ecological consultants in particular. Although guidelines have been suggested for the way in which consultants might work (Knightbridge, 1988) these have never been formally adopted, and no list of recognized or approved, independent environmental consultants exists. The ways in which ecological and nature conservation considerations impact on development and how they are quantified are always open to criticism. Consequently it is

A Code of Conduct for Users of Biological Record Centres

- Each user will accurately inform the manager of the BRC of the purpose of the enquiry and the use to which the information will be put.
- Data provided for a specific purpose will not be retained, disseminated or otherwise used for a different purpose without the written permission of the BRC manager.
- While the BRC will not withhold information without reasonable cause, users will respect its right to withhold information either because of its sensitivity, its confidentiality or the requirements of the Data Protection Act.
- The BRC will respect a request for an enquiry to be treated in confidence and users will not, in turn, pass on information in an irresponsible or unethical manner.
- Although every effort is made to verify information within a BRC, it is accepted that such information is not definitive.
- Due acknowledgement will be given of the use of information from the BRC in any publication or report that has made use of that information.
- Users will abide by any copyright restrictions that may apply to information in the BRC and will ensure that the copyright implications of passing information on to a third party are fully understood by that party.
- Users will inform the BRC of new sites or information that come to light as a result of their research. Copies of, or access to, the completed research will be made available within a time-scale agreed with the manager of the BRC.

Figure 9.1 A Code of Conduct for Users of Biological Record Centres (after Ely, 1994).

important that record centres recognize the role of the consultant as a means of providing meaningful conservation input to planned developments, and especially as a means of reducing the impact of change. The record centre is therefore placed in an ethical dilemma – should it make data available to all consultants, selling its information to anyone who demands it (which might be a tactic critical to its survival), or should it only provide information to experienced and creditable consultants who are known for giving objective advice to their clients? The latter course is undeniably the correct ethical stance for any organization which has environmental conservation as a main objective; whether it is always a practical solution is another matter.

Recording activities as an aid to promoting natural history for the amateur

The amateur naturalist is the mainstay of biological and geological recording in most countries, and most major recording schemes need to enlist the dedicated amateur if reasonable geographic coverage is to be made. Talented and largely unpaid labour has been critical to the success of schemes to map the status and distribution of, for example, butterflies (Warren, 1988), lichens (Seaward, 1992) and moths (Woiwod and Riley, 1992) in Britain.

Another recent example which has more immediate value for conservation is the survey of freshwater mussels throughout Europe, co-ordinated by Fred Woodward from Glasgow Museum. The freshwater pearl mussel, *Margaritifera margaritifera* was given legal protection in Britain in January 1991 when it was added to Schedule 5 of the Wildlife and Countryside Act – the information which led to this new legislation was supplied by a dedicated group of active local naturalists, who mapped the distribution of the species and uncovered the serious extent of illegal 'pearl-fishing'. A simple identification key and recording card was provided by the museum as an aid to collecting relevant data. The species was earlier protected under the Berne Convention thanks to the survey work done by amateur naturalists; mussel recording has been extended into Eastern Europe and encompasses other mussel species which are endangered by water pollution (Jay, 1991).

Community recording projects

Involving the public in biological and geological recording poses a real challenge to museums. It is an expensive road to travel, requiring good publicity, the production of free supporting materials and effective and rapid publication of results, which is why most 'public participation' schemes have been carried out by major wildlife conservation organizations (e.g. Watch, Marine Conservation Society) with the aid of significant sponsorship. Despite the obvious disadvantages of finding financial support, there are many benefits to be gained by any museum prepared to involve the public; useful information can be gathered, provided the target organism (or habitat) is selected with care. Scientific results will certainly be limited, and the parameters of the project need to reflect this at the outset. The principal benefit will be good publicity for the museum and the records centre, an accompanying increase in public profile, and data which can be used to support exhibitions. The media take a delight in promoting schemes which (to them) seem somewhat bizarre, especially if accompanied by a suitably trite catch phrase ('Spot-the-Frog') or reward ('Free Hedgehog poster with every record. . . . ').

Examples of such surveys are many and varied, and interestingly the results sometimes provide the unexpected. Doncaster Museum's experience in using the kills of domestic cats to determine the distribution patterns of small mammals, a scheme cleverly marketed as 'What the cat brought in' (Howes, C. *pers. comm.*) has never been repeated, despite its tremendous public appeal and remarkable success. Museum biologists gained the distribution data they needed, but as a spin-off could also provide a list of the most-used cat names in Yorkshire, and identify the most popular tinned foods for felines. Attempts to plot the distribution of hedgehogs in northern England proved very attractive to the public, resulting in a huge response; unfortunately, but perhaps predictably, the end result resembled a perfect road map, as most 'sightings' were of road kills (Bainbridge, J. *pers. comm.*). Other surveys in northern England had more useful scientific results; a squirrel survey in the 1970s (Davis, 1979) charted the increasing spread of the Grey Squirrel (*Sciurus carolinensis*) in a region then a stronghold of the native Red Squirrel (*Sciurus vulgaris*). These data were especially useful during a more recent survey to determine the range of these two species, and their overlap, in northern England, carried out by a consortium of Wildlife Trusts and museums under the title of 'Red Alert' (see Plate 26). Other surveys in the region have provided significant information for biological conservation – amphibian and pond surveys enabled new sites to be found for warty newt, (*Triturus cristatus*), a protected species under the Wildlife and Countryside Act 1981. Way (1988) summarized the experiences which many record centres have had with such schemes, and makes particular reference to the importance of giving clear messages to participants, choosing the correct target species and providing feedback.

Environmental records and public education

A museum-based records centre is able to provide much more than information which enables nature conservation within the planning framework. For any museum planning a new exhibition, its databases provide a means of giving accurate and up-to-date information, ensuring academic respectability and alerting the visiting public to significant changes in their environment. This link between museum-based research and education/interpretation is one which cannot be over-emphasized, though it is seldom apparent to the general public. One of the most successful local natural history museum exhibits in the UK is that in Devizes Museum, Wiltshire, where through an interactive display it is possible to call up distribution maps of animal and plant species from the computerized database and superimpose upon them details of topography, geology and climate (G. Stansfield, *pers. comm.*). Such a display encourages the museum visitor to understand the reasons responsible for the distribution of species, providing evidence that some species are limited to riverside environments, moorland or limestone regions. Environmental data can also be used in publications aimed at the public and the amateur naturalist, and suggest suitable, robust sites for educational visits.

One of the major roles of a museum is responding to inquiries from the public, particularly with regard to the identification and status of species. The ability of biologists and geologists to provide rapid and correct answers to such queries (what is it? is it rare? does it bite? are they usually found in chocolate?) usually makes a considerable impression on visitors; for the

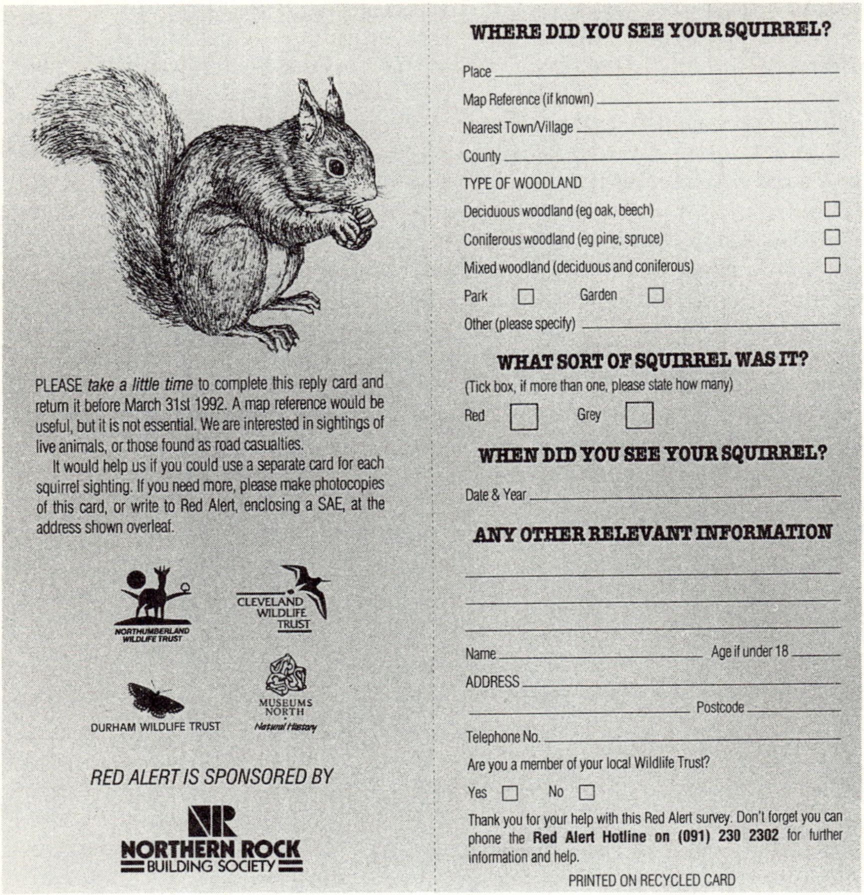

PLEASE *take a little time* to complete this reply card and return it before March 31st 1992. A map reference would be useful, but it is not essential. We are interested in sightings of live animals, or those found as road casualties.

It would help us if you could use a separate card for each squirrel sighting. If you need more, please make photocopies of this card, or write to Red Alert, enclosing a SAE, at the address shown overleaf.

NORTHUMBERLAND WILDLIFE TRUST
CLEVELAND WILDLIFE TRUST
DURHAM WILDLIFE TRUST
MUSEUMS NORTH *Natural History*

RED ALERT IS SPONSORED BY

NORTHERN ROCK BUILDING SOCIETY

WHERE DID YOU SEE YOUR SQUIRREL?

Place ______

Map Reference (if known) ______

Nearest Town/Village ______

County ______

TYPE OF WOODLAND

Deciduous woodland (eg oak, beech) ☐

Coniferous woodland (eg pine, spruce) ☐

Mixed woodland (deciduous and coniferous) ☐

Park ☐ Garden ☐

Other (please specify) ______

WHAT SORT OF SQUIRREL WAS IT?

(Tick box, if more than one, please state how many)

Red ☐ Grey ☐

WHEN DID YOU SEE YOUR SQUIRREL?

Date & Year ______

ANY OTHER RELEVANT INFORMATION

______

Name ______ Age if under 18 ______

ADDRESS ______

______ Postcode ______

Telephone No. ______

Are you a member of your local Wildlife Trust?

Yes ☐ No ☐

Thank you for your help with this Red Alert survey. Don't forget you can phone the **Red Alert Hotline on (091) 230 2302** for further information and help.

PRINTED ON RECYCLED CARD

Plate 26 'Red Alert' reply card – mapping the distribution of squirrel species using a public participation survey. (Reproduced by permission of Northumberland Wildlife Trust.)

curator it can also be entertaining and rewarding. Often aided by the production of standard reply sheets or information packs for common inquiries, the inquiry service is both served by the records centre and contributes to it – with a records centre to hand the 'status/rarity' questions are easily dealt with, and the documentation of inquiries can provide data which can be analysed for type of inquiry, seasonal variation, and even type of inquirer. Bristol Museum's experience with some 2,000 public inquiries each year (Burrell, 1988) has indicated the benefit of documenting natural history inquiries, not simply in the acquisition of new and useful distribution data, but in identifying their public and allowing effective development of the inquiry service.

Contracts for environmental survey – a museum role

Most local records centres in Britain have undergone a marked shift in emphasis following their origin in the 1970s. From the simple objective of species mapping they have become multi-purpose databases serving a wide community, and often providing complex data. This shift is matched by a change in museum activities, some embracing recording in a grand manner because it is seen as a positive environmental function, others using recording as a practical means of justifying (and funding) threatened posts.The Biological Records Centre has experienced a similar change (Harding, 1988), responding to commercial pressures as financial support for pure biogeographical work has dwindled. What money is available from the private sector for pure distribution studies has been used for publication of results (Mendel and Piotrowski, 1986), or for surveys of species with wide public appeal. In contrast to the limited resources for distribution studies, increasing requirements for wildlife appraisals by planning and conservation organizations mean that funds are now more widely available to fund contracts for which record centres, and therefore museums, can bid. If museums are to take an effective role in monitoring their local environment core funding is essential, yet the financial position of most museum governing bodies makes this difficult, if not impossible. In the longer term, museums in Britain may expect support from the Co-ordinating Commission for Biological Recording to assist them with the task, but in the meantime it is essential that museum record centres sell their identification and survey skills, and the information they house, if they plan to contribute to the conservation of their local environment. Museums must however face up to the questions of accuracy, objectivity, and legality. How ethical is it for museums to compete for contracts when they benefit from hidden subsidies, through premises and computing equipment for example? Other ethical issues revolve around access to data; why should data held in a public institution not be freely available?; when protection of rare species is concerned, what right do museums have to release or withhold information?

The idea of a museum actively seeking contracts for faunal or other surveys is a relatively new one in the UK, and usually makes only a very modest income. There are exceptions to this. Leicestershire Museums have generated a large proportion of survey income from outside sources, (G. Stansfield, *pers. comm.*) and Tyne and Wear Museums has created some additional income through survey work. In contrast, some museums in the USA generate a large part of their operating budget through such work, which can often help support other activities, including systematic research and more public-oriented museum programmes. The Academy of Natural Sciences in Philadelphia provides one of the best examples, with one of its three main administrative sections being designated the 'Division of Environmental Research' (DER) which has a particular reputation for its research into aquatic ecology, and being ' . . . uniquely qualified to provide a variety of organisations and individuals with the information necessary to make intelligent decisions about water resources – from small streams and creeks to large estuaries' (Thomson, 1992). From its three research laboratories in North America and its Field Station in Costa Rica (established in 1990) Academy scientists investigate a number of changes in land use which affect water quality, including the impact of farming, forestry and industry; in 1991/2 total income to the division in terms of research grants and contracts exceeded $5 million.

The DER in its promotional leaflet describes itself as 'A national leader in helping industry, government agencies, municipalities, civic groups and environmental consultants protect and manage our precious water resources.' As a self-financing research body, the DER could certainly survive alone; but owes a good deal of its success to the Academy association. Undeniably this gives it special status, and an advantage in the contracts market, using the Academy's historical associations and the world-wide reputation for systematic research and education to sell a particular image. The relationship is a symbiotic one, with the Academy gaining in its scientific reputation with the diverse interests and projects carried out by the DER. It was a brave move by the Academy to establish its Division of Environmental Research in 1948; by moving into the field of applied science it was departing from the norm for most museums at that time. The consequence of that change meant that the Academy was prepared for the new demands that environmentalism, and the associated new legislation, brought about. It has been able to capitalize as a result, it has surveyed most of the major rivers, lakes and estuaries in the continental USA over a number of years and has a database of considerable significance which can assist in predicting changes in water quality. Recent research has focused on the San Antonio Bay (1987), South Fork Holston River (1990) and White River (1991), in association with a variety of clients. The DER lists amongst its 'sponsors' the majority of major chemical and pharmaceutical companies, petroleum and gas manufacturers and government agencies; for example the Holston River survey is part of a long-term monitoring programme to assess water quality in the vicinity of an Eastman Chemical company plant.

The Academy's Systematics Division is also involved in field survey, but usually devoted to providing specialist advice about species or groups of species rather than looking at environmental quality. Research on the distribution of Unionidae (freshwater mussels) in Western Pennsylvania (Davis, 1993) was carried out under contract to the Pennsylvania Fish Commission and the Wild Resources Conservation Fund, and led to the creation of an important database from which the status of individual species and sites could be determined. The field survey was followed by the examination of all the major unionid collections for the state (which included the discovery of two undescribed species). It was concluded that of the 68 species recorded historically, 22 are now extinct and 23 are rare or endangered; in two major rivers (the Mononghala and Ohio rivers) all unionids are extinct. The report proposed a number of conservation measures, including a commitment to monitoring.

Much of the upsurge in contract work is due to the increase in environmental legislation and the demand for environmental impact statements.

These are now widely used throughout the developed world, and the response by museums to this trend has been significant in many countries, and has recently been highlighted in South Africa (de Jong, 1993). Environmentalism has been as significant in South Africa as in Europe and North America, with a 'new wave' of societies, including Earthlife Africa (ELA) and the Eco-Programme, and new approaches being adopted by state conservation agencies such as the National Parks Board, and newly formed environmental coalitions (Khan, 1993). The strength of the conservation organizations placed demands on the government for new legislation to protect the environment, which culminated in the Environment Conservation Act 1988, amended in 1992 (Coetzee, 1993). Environmental Impact Assessment (EIA) became a significant feature of this new legislation, following the lead taken by the USA in the National Environment Protection Act of 1969 and subsequently by most industrialized countries including the European Community (Wathern, 1988). Museums in South Africa are beginning to make a significant contribution to EIAs using their specialist survey and identification skills, for example the Natal Museum's contract to investigate the archaeology and history of the St Lucia Dune System which was threatened by mining operations (Dominy, 1993).

One particularly interesting contract is that carried out by the National Cultural History Museum for the Department of Environment Affairs to produce guidelines for the management of cultural resources – 'the intangible properties and material products, both natural and man-made, which are associated with human activity and history' – within nature reserves or forestry areas (de Jong, 1993a). This contract – the CANIS project – recognized that to conserve cultural resources within such a framework would necessitate the skills of archaeologists, anthropologists, historians and museologists, in addition to biologists, and would require careful survey, planning, maintenance and financing to be effective. The conclusion was reached that cultural resources, which until recently had largely been ignored in areas of high biological significance, should themselves be recognized as part of the environment, and carefully managed if their potential for education, recreation and research is to be fulfilled, through an integrated approach to management. This change is significant, and the fact that the policy is aimed at conserving cultural resources of native South Africans as well as the evidence of the impact of white settlers, it is a reflection of the need to create a new identity for a new South Africa, one which takes into account the rich cultural traditions of all its peoples.

References

Adams, K. J. (1990) 'Proposals for a 5-km^2 mapping scheme for eastern England', *Bulletin of the British Bryological Society*, 55, 14–17.

Adams, K. J. and Preston, C. D. (1992) 'Evidence for the effects of atmospheric pollution on bryophytes from national and local recording'. In Harding, P. T. (ed.) *Biological Recording of Changes in British Wildlife*, HMSO, London, 31–43.

Birker, I. U. (1991) 'Biological Recording – an environmentally friendly practice for natural history museums', *Muse*, VIII (4), 27–33.

Burrell, P. (1988) 'Answering wildlife enquiries – the Bristol Museum Enquiry Project'. In Stansfield, G. and Harding, P. T. (eds) *Biological Recording: The Products,* National Federation for Biological Recording, Cambridge, 51–56.

Coetzee, I. (1993) 'The what, who and why of Environmental Impact Assessment'. In de Jong, R. C. (ed.) *Museums and the Environment,* Southern Africa Museums Association, Pretoria, 185–200.

Cumbaa, S. L. (1991) 'Environmental programs of the Canadian Museum of Nature: keeping the bandwagon tuned up and on track', *Muse,* VIII (4), 38–40.

Danks, H. V. (ed.) (1979) 'Canada and its insect fauna', *Mem. ent. Soc. Can.* 108, 573 pp.

Danks, H. V. (1983) 'Regional collections and the concept of regional centres'. In Faber, D. J. (ed.) 'Proceedings of the 1981 workshop on the care and maintence of natural history collections', *Syllogeus* 44, (National Museum of Canada), 151–160.

Danks, H. V. (1993) 'The Biodiversity Crisis, A National Initiative: The Biological Survey of Canada (Terrestrial Arthropods)', *ASC Newsletter,* 21 (2), 17–22.

Davis, G. (1993) *The Unionidae of Western Pennsylvania: Rare, Endangered, Extinct,* Concluding report (SP17139) to the Pennsylvania Fish Commission and Wildlife Resources Fund, Academy of Natural Sciences, Philadelphia.

Davis, P. (1979) 'The results of a squirrel distribution survey in north-east England', *Vasculum,* 64 (3), 17–26.

Dawson, D. and Game, M. (1988) 'Biological Survey for Nature Conservation Planning'. In Stansfield, G. and Harding, P. T. (eds) *Biological Recording: The Products,* National Federation for Biological Recording, Cambridge, 16–26.

de Jong, R. C. (ed.) (1993) *Museums and the Environment,* Southern Africa Museums Association, Pretoria.

de Jong, R. C. (1993a) 'The Canis Project'. In de Jong, R. C. (ed.) *Museums and the Environment,* Southern Africa Museums Association, Pretoria, 151–163.

Dominy, G. A. (1993) 'Contested terrain: the politics of environmental impact assessment in a "landscape of heritage": the debate over the eastern shores of Lake St. Lucia'. In de Jong, R. C. (ed.) *Museums and the Environment,* Southern Africa Museums Association, Pretoria, 201–205.

Ely, B. (1994) 'Natural History Museums and biological recording'. In Stansfield, G., Mathias, J. and Reid, G. (eds) *Manual of Natural History Curatorship,* HMSO, London, 167–189.

Everett, W. T. (1990) 'Protecting the Treasures of Islas Los Coronados; museum research and collections play a vital role', *Environment West,* 1 (3), 14–17.

Foster, G. N. (1992) 'The effect of changes in land use on water beetles'. In Harding, P. T. (ed) *Biological Recording of Changes in British Wildlife,* HMSO, London, 27–30.

Genoways, H. H., Choate, J. R., Pembleton, E. F., Greenbaum, I. F. and Bickham, J. W. (1979) 'Systematics, other users, and uses of north American collections of recent mammals'. In Banks, R. C. (ed.) *Museum Studies and Wildlife Management,* Smithsonian Institution, Washington DC, 20–48.

Gray, D. R. (1979) 'Nature conservation in the north: the contribution of the National Museums of Canada', *Gazette,* 12 (1), 35–42.

Harding, P. T. (1988) 'Funding biogeographic recording'. In Stansfield, G. and Harding, P. T. (eds), *Biological Recording: The Products,* National Federation for Biological Recording, Cambridge.

Hua, H., Zhou, M. and Dooley, P. (1992) 'Natural history museums and their role in conservation: a Chinese perspective', *Curator,* 35 (3), 206–214.

Jay, S. (1991) 'Good news for mussels', *BRISC Recorder,* Issue 12, 2.

Johnels, A. G. (1973) 'Role of Natural History Museums', *Museum*, 25, (1/2), 54–58.

Kelly, T. M., Jones, J. D. and Smith, G. R. (1979) 'Historical changes in mercury contamination in Michigan Walleyes (*Stizostedion vitreum vitreum*)'. In Banks, R. C. (ed.) *Museum Studies and Wildlife Management*, Smithsonian Institution, Washington, 262–271.

Khan, F. (1993) 'Environmentalism in South Africa and the role of museums'. In de Jong, R. C. (ed.) *Museums and the Environment*, Southern Africa Museums Association, Pretoria, 164–184.

Klaas, E. E., Ohlendorf, H. M. and Heath, R. G. (1979) 'Avian eggshell thickness: variability and sampling'. In Banks, R. C. (ed.) *Museum Studies and wildlife management*, Smithsonian Institution, Washington, 253–261.

Knightbridge, R. (1988) 'The involvement of ecological/nature conservation consultants in development and planning issues'. In Stansfield, G. and Harding, P. T. (eds) *Biological Recording: The Products*, National Federation for Biological Recording, Cambridge, 27–34.

Mendel, H. and Piotrowski, S. H. (1986) *The Butterflies of Suffolk: An Atlas and History*, Suffolk Naturalists' Society, Ipswich.

National Museum of Natural History (1976) *The Use of Museum Specimens for Past Pollutant Level Documentation – An Evaluation*, Smithsonian Institution, Washington DC.

Pettitt, C. (1994) 'Using natural history collections'. In Stansfield, G., Mathias, J. and Reid, G. (eds) *Manual of Natural History Curatorship*, HMSO, London, 144–166.

Ratcliffe, D. A. (1970) 'Changes attributable to pesticides in egg breakage frequency and eggshell thickness in some British Birds', *Journal of Applied Ecology*, 7, 67–115.

Rose, C. I. and Hawksworth, D. L. (1981) 'Lichen re-colonization in London's cleaner air', *Nature*, 289, 289–292.

Sarukhan, J. (1993) 'National efforts to study the Biodiversity of Mexico, a megadiverse country', *ASC Newsletter*, 21 (4), 41–44.

Sears, S. R. (1990) 'Guarding an uncommon natural legacy', *Environment West*, 1 (3), 6–9, 21.

Seaward, M. R. D. (1992) 'Twenty-five years of lichen mapping in Britain and Ireland', in Harding, P. T. (ed.) *Biological Recording of Changes in British Wildlife*, HMSO, London, 76.

Thomson, K. S. (1992) *The Academy of Natural Sciences, Annual Report 1991–1992*, The Academy of Natural Sciences, Philadelphia.

Warren, M. S. (1988) 'Supporting butterfly conservation research'. In Stansfield, G. and Harding, P. T. (eds) *Biological Recording: The Products*, National Federation for Biological Recording, Cambridge, 40–50.

Wathern, P. (ed.) (1988) *Environmental Impact Assessment: Theory and Practice*, Unwin Hyman, London.

Way, L. S. (1988) 'Surveys of wildlife by the public: funding and resources'. In Stansfield, G. and Harding, P. T. (eds) *Biological Recording: The Products*, National Federation for Biological Recording, Cambridge, 57–60.

Woiwod, I. P. and Riley, A. M. (1992) 'The Rothamstead Insect Survey light trap network'. In Harding, P. T. (ed.) *Biological Recording of Changes in British Wildlife*, HMSO, London, 76–77.

10. *Live issues I – zoological gardens and aquaria*

'But if I ran the zoo'
Said young Gerald McGrew
'I'd make a few changes,
That's just what I'd do . . . '
(Dr Seuss, *If I Ran the Zoo*)

Natural history museums and institutions housing living collections share an ultimate goal, to foster a conservation ethic. Museums in general have many common features with zoos, aquaria and botanic gardens – all are institutions housing collections and performing scientific, educational and recreational roles. Indeed, the American Association of Museums recognizes such institutions as museums in its definition 'an organised and permanent non-profit institution, essentially educational or aesthetic in purpose, with professional staff, which owns and utilises tangible objects, cares for them, and exhibits them to the public on some regular schedule'. The broad definition of a museum adopted by the International Council of Museums also accommodates institutions with living collections. Consequently, should we consider zoos, botanic gardens and aquaria as specialized natural history museums, where collections are housed in enclosures rather than cabinet drawers or display cases? Undoubtedly the answer from such institutions would be a firm 'no', they do not see themselves as museums, they have their own objectives and philosophies, yet they are driven by the same desire to understand the natural world and communicate that knowledge to a wider public. However, like museums, they have had to react to environmentalism; zoos in particular have had to re-examine their role in the face of determined criticism about the display of live animals. As a result, many modern zoological gardens, (and aquaria and botanical gardens) have pioneered successful techniques and approaches that museums might adopt to their benefit. The underlying philosophy that is common to natural history museums and collections of living organisms makes it imperative that the most effective means of delivering conservation messages are employed, and it might be argued that the key to success is to combine successful elements from all such institutions within a biopark concept. Possibilities of co-operative ventures between organizations also needs to be explored in order to deliver a more forceful conservation message.

Zoological gardens

During the late Middle Ages and the Renaissance the practice of keeping collections of wild animals became a symbol of status in Europe. Hunting animals, especially dogs and falcons, had always been valued by kings and noblemen, but by the fourteenth century exotic birds and animals were favoured as a mark of rank and influence. Philip VI of France (reg. 1328–1350) is known to have kept lions and leopards at the Louvre; in the sixteenth century, under the Medici Pope Leo X, the Vatican menagerie imported bears from Hungary, lions and leopards were sent from Florence, and gifts including an elephant and a snow leopard were accepted from King Manuel of Portugal (Loisel, 1912). Exotic collections were not confined to Europe, and the Aztec emperor of Mexico, Montezuma II, had a magnificent collection of birds of prey, waterfowl, snakes, and big cats, housed in special enclosures and cared for by hundreds of keepers. Hernan Cortes, conqueror of Mexico in 1517–21, and his lieutenant Bernal Diaz del Castillo, provided detailed accounts of Montezuma's palace and his zoological collection, which in addition to wild animals included human albinos and individuals with deformities (Cortes,1972; Castillo, 1928).

Although most animal collections remained in aristocratic hands, some European menageries, notably that founded by the citizens of The Hague in 1590, were opened to the public, providing them with their first view of strange animals from newly discovered parts of the world (Dembeck, 1966). Louis XIV's menagerie at Versailles, founded in 1664, was opened to public and scientists; dissections carried out on animal casualties by members of the Academy of Sciences helped to promote the study of comparative anatomy (Loisel, 1912). From the eighteenth century European menageries increased in number and accessibility, and travelling collections fulfilled a useful educational role, as well as being entertaining (Middlemiss, 1987).

The first zoological collection to be specifically founded for scientific purposes was that of the Zoological Society of London in 1826 (see Plate 27). It was also this Society that gave the word 'zoo' to the English language as the Great Vance sang 'The O.K. thing on Sundays is walking in the zoo' in 1867 (Cherfas, 1984). The original scientific purposes for the animal collections stated by Sir Stamford Raffles was domestication and acclimatization – the introduction and release into the wild of non-endemic animals – a concept in vogue at the time, but one which seems incomprehensible now. Not surprisingly, these goals were never achieved. Perhaps it was inevitable that the poor conditions of housing and hygiene at the zoo, coupled with ignorance of the dietary and other needs of animals, resulted in the main scientific thrust being directed towards dissection and anatomical studies of animal corpses.

Zoos used to play a major role in taxonomic studies in the nineteenth century, often forming museums in addition to live animal collections. The founders of the Zoological Society of London had proposed in 1825 that the

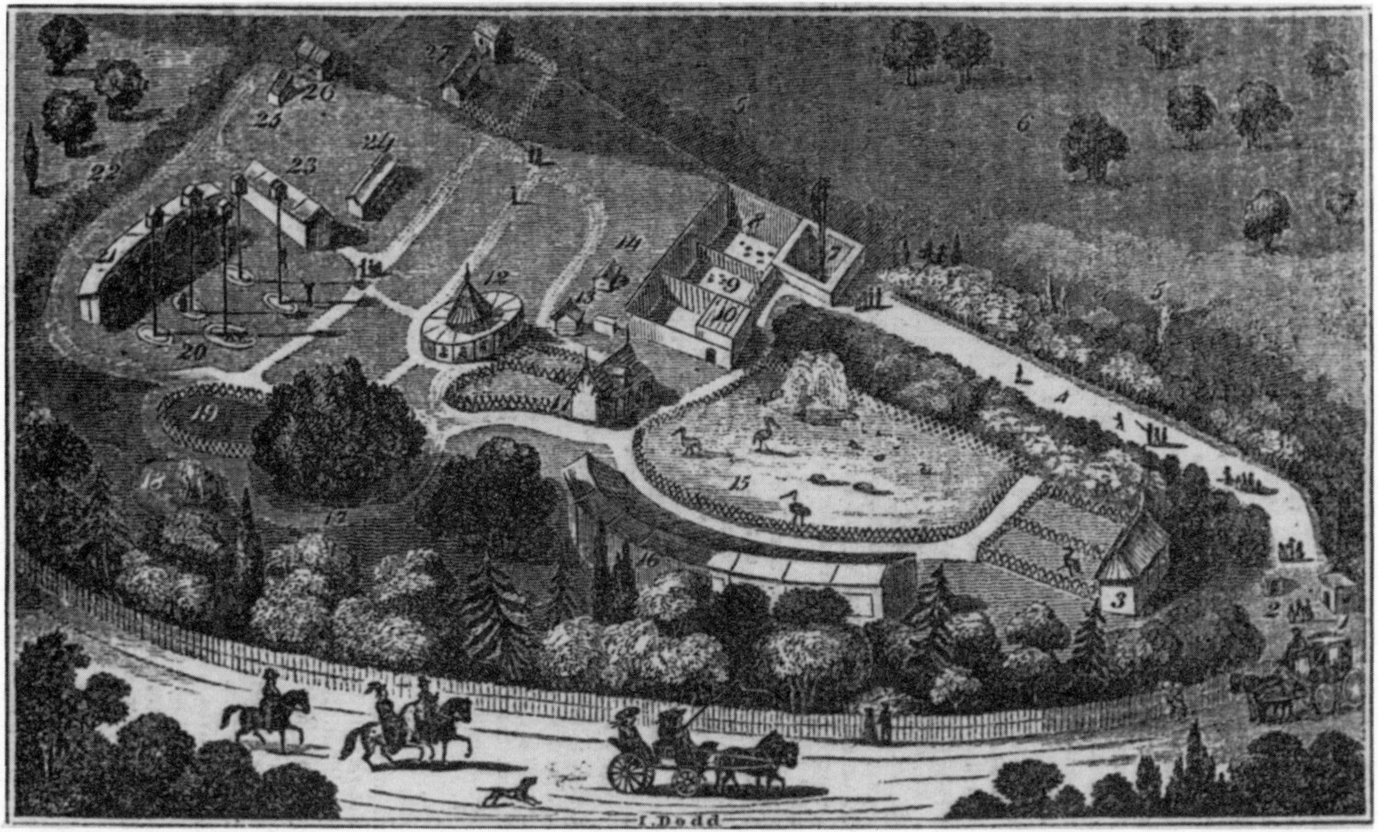

Plate 27 The Zoological Gardens, Regent's Park, 1828. (Original engraving published in *The Mirror of Literature, Amusement and Instruction*, 6 September 1828.)

Society should have 'A Museum of prepared specimens', (Blunt, 1976), as well as a library, and £1,000 was put aside for development. In the first ten years the growth of the museum was dramatic, and by 1835 the Society had to rent rooms at 28 Leicester Square to accommodate their collections. It was not until 1850 that the Council, recognizing the improvement in the storage situation at the British Museum noted that ' . . . this renders the maintenance of our own museum as a separate collection no longer an object of the importance which it had formerly possessed', and in 1857 the museum collections were dispersed. The mammal specimens were arguably the most important historical collection ever received by the British Museum, and included, for example, important Australian material from the naturalist and artist John Gould (1804–1881).

It is worth emphasizing that specimens were collected specifically for the museum – it was not simply a repository for the remains of casualties at the zoo. However, there is no doubt of the significance of the link between the live animal displays and the museum, and it is nowhere better demonstrated than in the daily reports sent from the menagerie in Regent's Park to the Society's offices in central London. These forms, headed 'Occurrences at the Garden', (now held in the Society's Library) give a fascinating insight into the day-to-day happenings at the zoo, the arrival of new exhibits, the health of the animals, and the fate of casualties. Consider the following entries extracted from the 'occurrences' of 1828:

27 February	Emu laid her fifth egg
28 February	Received the bones of a kangaroo for maceration
2 March	Wolf still slightly unwell
11 March	Emu laid her sixth egg
13 March	Wolf slightly unwell
14 March	Wolf very unwell. Emu laid her seventh egg
15 March	Wolf died. All other animals well
18 March	Body of wolf put into maceration

The provision of anatomical specimens is no longer a core activity for zoos. As Bostock (1993) points out, 'this is a function that zoos perform less the better they become, but it is responsible of them to ensure that the best use is made of their animals after death'. Science in zoos is now mainly devoted to reproductive technology, veterinary studies, behaviour and genetics, and (sadly) the visiting public is largely unaware of this activity. For them, zoos are recreation and entertainment, education and leisure. Looking at exotic animals is fun.

A new role for the zoological garden

Displaying wild animals for the amusement of the public (see Plate 28) has always been a key role for the zoo. The ethics of keeping live animals has been discussed by Bostock (1993), who provides a thought-provoking account of the philosophy of zoological gardens which argues that their role in conservation justifies their existence. Reid (1994) also refers to ethical and conservation issues relating to keeping captive animals, both in zoos and within a museum context.

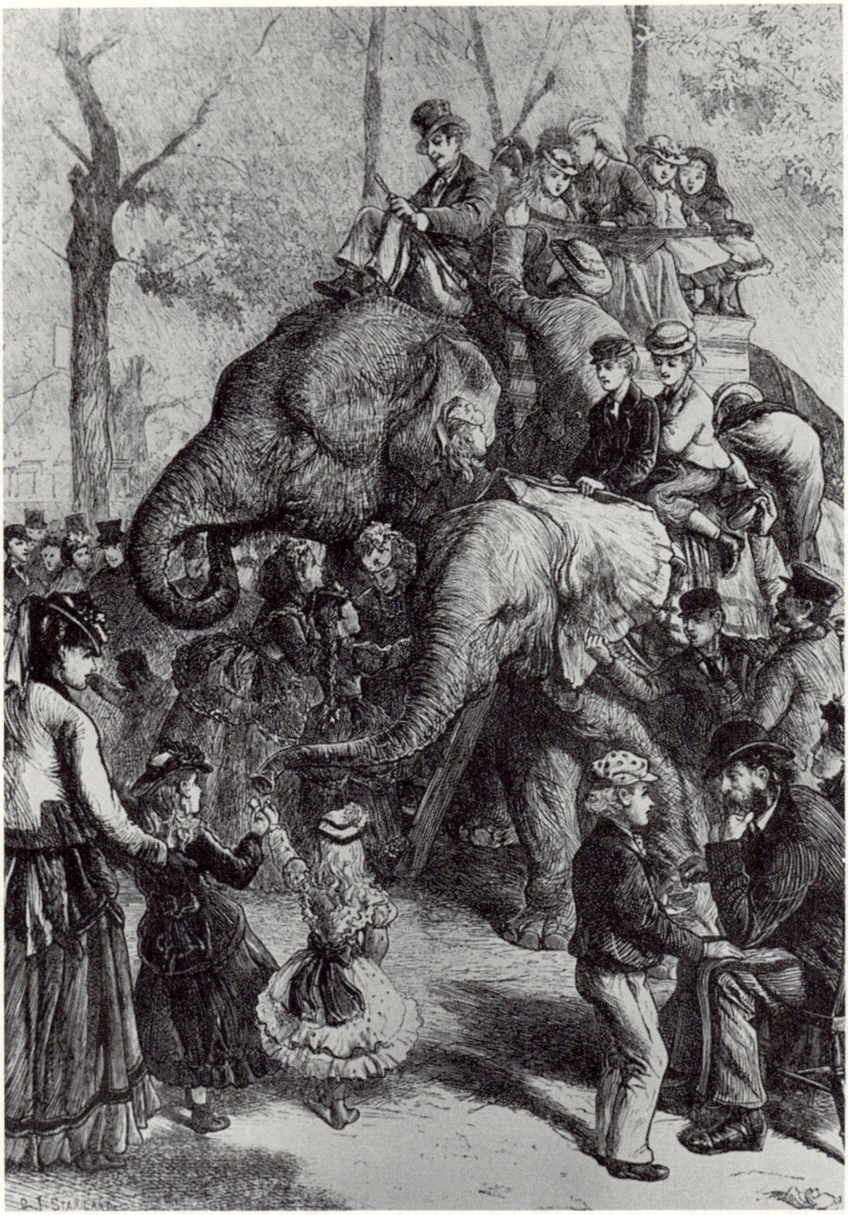

Plate 28 'Monday afternoon at the Zoological Society's Gardens' by C. J. Stanisland. (Reproduced by permission, Mary Evans Picture Library.)

In recent years zoos have had to reconsider their role, often due to dwindling attendances and budget shortfalls. However, there has also been change due to pressure from the environmental lobby, as concern grows for the preservation of wild stocks of animals, and the impact of trade in wildlife, which in part serves the zoo community. The animal rights movement, its concerns about cruelty to animals and the way in which zoos can be seen as a demonstration of domination and power, has also had an undeniable impact on the welfare of captive animals, leading to a re-evaluation of the mission of zoological gardens.

People are passionate about zoos – the recent (1991) financial problems faced by London Zoo led to a wide-ranging debate which featured philosophers (Cooper, 1991) and politicians (Hattersley, 1991), as well as zoo directors (Anon, 1991). The argument extended well beyond the fate of the world's first zoological garden, examining the role of zoos in general and evoking emotive statements that '... zoos exploit, not save lives by keeping animals in a cruel captivity to act out a miserable imitation of their natural existence ... the zoo community's educational, conservation and research claims are overstated if not specious.' (Clifton, 1991). The debate became a struggle between the animal rights movement and the zoological/natural sciences community. There was no consensus view delivered by the zoologists, and even within the close-knit world of zoo directors, there was healthy argument and differences in opinion about the way forward for London Zoo. David Hancocks, Director of the Arizona-Sonora Desert Museum suggested in an editorial in the *Washington Post* that it may not be such a bad thing for closure to take place if it encouraged other large urban zoos to renovate their facilities and alter their missions away from the nineteenth-century tradition of leisure-time displays of exotic beasts. Hancocks argued that:

> zoos as they are currently constituted are handicapped in fulfilling their potential to effect environmental change by developing exhibits that are object-driven, not story driven .... The new demands on zoos mean that a great broadening of their scope is needed. They must find ways to demonstrate the connections and interdependence between all components of nature. They must create exhibits that explain the relationships between soils and flowers and frogs and snails and trees and elephants. And evolution. And extinction. (quoted in Anon, 1991.)

His hope, he concludes, is that the current controversy means the end of the traditional zoo and the development of a new one in which the full story of nature, not just zoology, is told.

The case for captive breeding

Alexander (1979) states 'Today's well-run zoos with emphasis on animal health and protection have truly become the safest havens for wild beasts, the only sure way to avoid their total extinction'. Captive breeding for conservation purposes is promoted as a major role of zoos, yet can only be successful for a very small number of threatened species. Alexander's statement is naïve, though well intentioned, yet indicates a lack of understanding of the diversity of natural ecosystems, and of the poor success rate of the majority of zoo breeding programmes. Despite a real commitment to saving endangered species which is inherent in *The World Zoo Conservation Strategy* (IUDZG/CBSG, 1993), and a long-held view that captive breeding is an essential element in conservation (e.g. Woodroffe, 1981; Brambell, 1993) there is considerable scepticism about such programmes, especially because of the poor standards of husbandry seen in many of the world's estimated 10,000 zoos (Ormrod, 1994) – those in Romania have been identified as some of the most shocking examples (Bristow, 1994).

However, some zoos have had remarkable successes in breeding rare species, often going to extremes to ensure that captive breeding programmes are well managed. For example, Turner (1984) relates:

> At the Smithsonian's National Zoological Park in Washington D.C., the golden lion tamarin has been carefully and successfully raised. This is not merely a matter of placing a male and female together but involves extensive research into the complex behaviour of these tiny primates, and sometimes the patient and intricate job of hand-raising.

The Bronx Zoo of the New York Zoological Society has focused its attention on Andean condors. To speed up the reproductive rate of these huge birds, the female's first egg is removed, stimulating the production of a second. The condor chicks are then hand-raised, but, as they are destined to be returned to a sanctuary in the Andes Mountains of South America, zookeepers ensure condor identity by using adult condor glove-puppets when moving, cleaning and feeding the chicks. Co-ordination of captive breeding programmes in American zoos is achieved by the American Association of Zoological Parks and Aquaria (AAZPA); this body has also developed and implemented a Species Survival Plan (American Association of Museums, 1984), linking zoos throughout the world, and supports the International Species Inventory System (ISIS). This computer database stores information on the genetic backgrounds and general histories of thousands of individual zoo animals; 51,000 had been inventoried by 1984. This means that smaller zoos, against which much criticism has been levelled (e.g. Midgely, 1987) can become part of an integrated management system for captive animals, and play a real conservation role.

Exactly what proportion of the estimated number of species in the world can zoos attempt to save? – and where should zoos be putting their efforts? Zoos by definition have to concentrate on those species which the public wish to see, mainly mammals and birds; consequently the zoo can only be a very restricted Ark, it cannot conserve tropical reef invertebrates, or the fauna of leaf litter, or marine plankton, all arguably far more important in terms of global ecology than the giant panda or white-naped crane. Even the most ardent of zoo supporters recognizes that the preservation of habitats is the key to species survival, not captive breeding programmes (Tudge, 1991).

Yet habitats themselves are at risk, primarily from the continued rise in the human population; people need the space, need the resources. Even in the USA, wilderness areas such as the Great Smoky Mountains are under threat (Nevin, 1993). When any natural habitat is in decline, it is the larger species that are lost first (Foose *et al.*, 1985). Because zoos concentrate their captive breeding efforts on these larger, more attractive animals and birds they do play a vital role in species conservation. There must however be some rationalization of breeding programmes – should all zoos be trying to save species in this way? It might be argued that smaller zoos should concentrate their resources on better facilities for fewer captive animals, or on providing more educational activities for the zoo visitor, rather than become involved in costly scientific programmes. Even the best funded of zoos cannot afford to be optimistic about the impact of captive breeding – Benjamin Beck of the Washington Zoo, despite his success with the re-introduction of the Golden Lion Tamarin into the wild, has commented that 'Anyone who tries to claim success with conservation in his own lifetime is simply not being realistic' (Greene, 1989).

Many species which zoos rear are under threat for reasons other than habitat loss, including persecution or trade (Stuart, 1991). As Bostock (1993) notes:

> It is now recognised by the International Union for the Conservation of Nature (IUCN), and so by many conservationists who are not zoo-based, that captive breeding with a possibility of reintroduction is a significant supplement to the direct protection of wild habitats and their occupants.

However, there is another and equally important reason to support zoo research into the reproductive physiology of rare species, the promotion of breeding in captivity, and the return of individuals to the wild. More than anything this huge effort is a symbol of a caring attitude towards the environment, an indication that there are individuals who are deeply concerned that species are endangered, and want to prevent the extinction of a beautiful animal. That breeding programmes concentrate on mammals and birds is in some ways essential – the public can understand the need to maintain wonderful animals such as gorilla, golden lion tamarin or Californian condor. But the public need to be informed of the zoos' activities, of their missions in conservation which are too often hidden from view. Zoos need to shout about the scientific work they undertake if they are to convince the public about their new role; they need better and more interpretation to excite and even worry the visitor a little. As William Conway, Director of the Bronx Zoo has remarked

> I couldn't care less if a zoo visitor never learned the difference between an African and an Asian elephant. I do care about making the visitor say 'What a magnificent creature! We've got to save this animal!' And 'Gosh, what a place! What nature is! What are we cutting down? What are we destroying?' (quoted by Greene, 1988b).

Even this is not enough – the zoo needs to add 'This is what we are doing about it.'

Fortunately some zoos are beginning to be more explicit in their conservation messages, for example, Fort Worth Zoological Park's multimedia production, *A Place for Life*, told visitors of the scientific goals of modern zoos, and a more recent programme, *So They May Live*, also addressed issues of habitat loss and species extinction, and explained the zoo's role in the world-wide conservation effort (Turner, 1984), demonstrating the commitment of its staff to an environmental goal.

Displaying live animals

Living animals generate great excitement and enthusiasm (Birney, 1988), and zoos have traded on this phenomenon since their inception. Environmentalism has encouraged zoos, like museums, to go beyond the immediate fascination with animals, to begin to find ways to change the way that people think and behave. As William Conway of the Bronx Zoo has commented 'Our intent is to inspire interest and concern, for only then are people receptive to education and modifying the way they live' (Craig, 1988). Michael Robinson, Director of the National Zoo in Washington is equally committed to the role of zoos in promoting environmental literacy:

> Zoos can help to create effective pressure for environmental aid programmes, but if such pressure is to develop, then ordinary people need to campaign for the protection of habitats in foreign lands as effectively as they campaigned for the protection of their own environmental treasures at the height of the environmental movement in the United States. Zoos must educate us into attitudes to the living world that lead us to try and preserve it, lead us to treat it as an irreplaceable heritage, and react to it with compassion and humanity (quoted in Craig, 1988).

The need for zoos to attempt to change lifestyles and attitudes has been echoed by DeLapa (1994). To do so they have developed many innovative displays and educational programmes. Often new displays have meant the rethinking of zoo architecture and spatial design, leading to more pleasing, exciting environments, both for animals and zoo visitors (see e.g. Winemann and Kyung Choi, 1991; Greene, 1988a). Carl Hagenbeck's Hamburg Zoo pioneered this approach in the early twentieth century, and now the concept of the landscape-immersion zoo, or immersion habitats further remove the familiar zoo security, they confront the visitors, giving many choices from which to construct pathways and educational opportunities; controlled views and intimate viewing areas enhance personalized interpretation.

Grant Jones, zoo designer, illustrates this new thinking in his comments on his gorilla exhibit at the Woodland Park Zoo in Seattle:

> When they asked me to design a gorilla exhibit, I naturally re-phrased the problem in my own mind as designing a landscape with gorillas in it. In what sort of landscape would I want to behold gorillas? I would want to include mystery and discovery. I'd like to see the gorillas from a distance first, and then up close. I'd like to be able to intrude on them and see what's going on without their knowing I'm there. I'd want to give them flight distance, a place to back off and feel secure. And I would want an experience that would take me back to a primordial depth. How did I spend my day millions of years ago living in proximity to this animal? (Greene, 1988a).

John Coe, another expert in zoo design, is scathing about traditional zoos. Greene (1988a) writes:

> When walking through a client zoo for the first time he offers a few suggestions: Get rid of the tire in the chimp exhibit. Get rid of the signs saying Nimba the Elephant and JoJo the Cheetah. Stop the publicised feeding of the animals, the elephant's birthday party, or any other elements contributing to either an anthropomophised view, or to a view of animals as tame pets.

Similarly unnatural enclosures, garden-like settings and moated displays must be abandoned, simply because design affects the way in which the visitors perceive the animals.

The Bronx Zoo's *Jungle World* is an indoor re-creation of a rainforest, and widely acknowledged as one of the best immersion habitats; barriers between animals and people are so well camouflaged that zoo-goers may be uncertain whether an animal has access to them or not. It has been followed by Brookfield (Chicago) Zoo's $10.5 million *Tropic World*, where a huge exhibit centres around different rainforest ecosystems of Africa, Asia and South America. The emphasis is on primates, although the habitat includes birds, mammals and a variety of plants, with a 58-foot tall waterfall as a dramatic centrepiece. Other 'rainforests' have been created at North Carolina Zoo in Ashboro, Topeka Zoo and Cleveland Zoo. Environmental messages are more conspicuous in these habitats. Visitors are certainly affected by the two clocks at the end of the *Jungle World* exhibit; one records the 125 new human lives per minute as the other ticks off the 50 acres of rainforest that have been lost in that time. Messages are reinforced in accompanying guidebooks, that for *Tropic World* suggesting visitors contact and support the Rainforest Action Network, and act as wise consumers regarding rainforest products.

Aquaria and marine life centres

Aquaria in Britain used to be part of the zoo experience, a cool refuge from the summer heat, often hidden away in a forgotten corner. They were the Cinderella at the zoological ball, ignored by the zoo visitor who wanted to see more spectacular animals and birds, the gorillas, tigers, giraffes or penguins. Furry animals and anthropomorphic birds were so much more fun. But not anymore. The environmental revolution of the past 30 years has resulted in growing public awareness, new conservation groups have been formed dedicated to conserving particular species, groups of animals and plants or individual habitats. Foremost in these new organizations have been those devoted to marine life. Greenpeace took direct, non-violent action to prevent hunting of cetaceans, and drew the world's attention to the impact of drift-net tuna fishing on dolphin stocks. As a result whales and dolphins became potent symbols for the conservation movement. On a national scale, in Britain the Marine Conservation Society (founded in 1983) began a series of research programmes to document the marine resources of the British Isles and promote the development of Voluntary

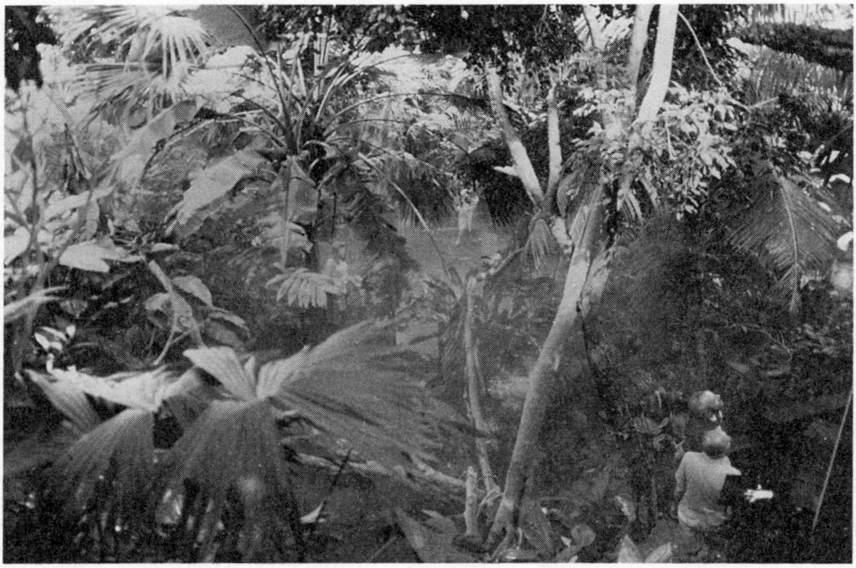

Plate 29 The Rain Forest exhibit, The National Aquarium, Baltimore – an example of an immersion habitat.

Marine Nature Reserves; successful campaigns to clean up Britain's beaches and act against the trade in marine curios were launched. A series of new underwater spectaculars hit the television screens, featuring the colourful world of coral reefs, the marine life of deep-sea vents, the charm of sea otters and the majesty of the humpback whale. New technology enabled live broadcasts to be made from exciting underwater locations around the British Isles. Marine life had a new image, and the public wanted to see it.

As a consequence there has been a growth industry in aquaria, often given other titles such as 'Sea Life Centres' or 'Marine Life Centres'. These institutions are far removed from the 'Dolphinaria' of previous years, with a totally different mission. It is no longer necessary to have performing mammals as a star attraction – if anything the opposite is true. In Britain, localities as diverse as Oban, Blackpool, Scarborough, Weymouth and Edinburgh now all have excellent aquaria; across the Channel Boulogne boasts a major new facility, *Nausicaa* (Anon, 1992a); more specialized is the *Maison du Saumon* at Brioude (Anon, 1992b). In the USA a new National Aquarium has been developed on the waterfront in Baltimore (see Plate 29), and the State of New Jersey provided funds of $40 million to build the new aquarium at Camden in 1992; these exciting developments are paralleled on the west coast, with Monterey Bay Aquarium having the reputation of being the best anywhere in the world, and in one of the most scenic locations. In her description of the education programme *Live from Monterey Canyon* (the use of a roving submersible to send images of life in the 10,000-feet deep Monterey Canyon back to the aquarium), Judith Connor refers to the best aquariums as 'wet museums ... a pleasant setting where people see living organisms and learn about their adaptations and natural history' (Connor, 1991). The philosophy of this aquarium is further evident in its mission statement: 'The purpose of the Monterey Bay Aquarium is to stimulate interest, increase knowledge and promote stewardship of Monterey Bay and the world's ocean environment through innovative exhibits, public education and scientific research' (Monterey Bay Aquarium, 1990).

This new breed of aquaria provide a spectacle equal to any provided by zoos or museums. Advances in materials, technology and design have resulted in aquarium tanks of immense size and ingenuity. The Whitechapel surgeon, Nathaniel Bagshaw Ward could never have imagined where his simple invention of the early 1830s would lead. Visitors to today's aquaria gaze spellbound as sharks, rays and barracuda glide silently by, and marvel at the underwater acrobatics of harbour seals. Yet modern marine life centres also provide much more than displays of live fishes, invertebrates or mammals, including demonstrations, lectures, hands-on displays, touch-tanks and a variety of educational programmes. Like museums and zoos, they can provide an atmosphere which inspires and excites, and consequently have tremendous potential for putting across environmental messages. However, few achieve their potential in the exhibits themselves, a fault frequently rectified by the staff in lectures and demonstrations. For example, at talks about seals at the Oban Marine Life Centre, stress is laid on the fact that the animals are only in captivity because they have been brought to the centre as sick or injured animals – once well they are returned to the wild. The conservation of seals forms the centre-piece to the talk. At a similar demonstration at Baltimore Aquarium in 1990 demonstrators discussed the biology of the two white whales (beluga) that they held captive, and constantly emphasized that the animals were not about to perform tricks, but deserved our respect and admiration. Again, conservation measures necessary for the survival of the species were highlighted.

The most recent example of a new aquarium development in the USA is at Camden, New Jersey (see Plate 30), opened in February 1992, at a cost of $52 million. It has proved remarkably popular, with 1.2 million visitors in its first year of operation. The main aquarium tank is huge, holding 760,000 gallons of sea water, the design allowing visitors to see into the tank from a variety of angles – all the fish have been caught locally, a reflection of the aquarium's original mission to concentrate on the marine life of the western Atlantic. An immense clear plastic wall (no less than 13 inches thick) provides an impressive viewing platform. At certain times of day divers enter the tank to feed and swim with the fish, and using sophisticated helmets and microphones they are able to talk to the public about the fish they can see, their adaptations, diet or distribution. Smaller aquaria house fish which are used to demonstrate other aspects of their biology – camouflage and coloration, shoaling behaviour and venom for example. Invertebrates, as always, get less attention, though they are present in most tanks – anemones, bivalves and starfish; jellyfish get a tank of their own, pulsating bells swirl in a tank of moving sea water, one of the most attractive of all the smaller exhibits.

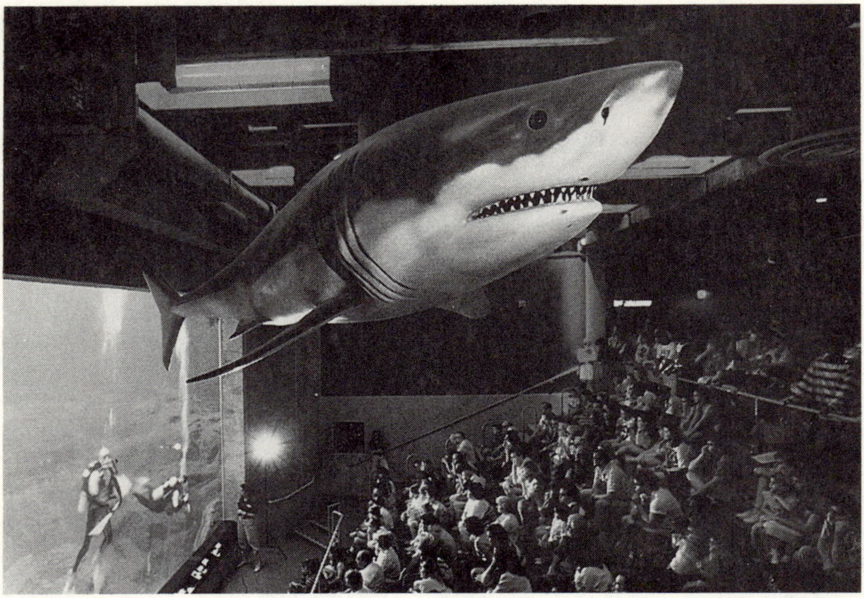

Plate 30 The spectacular viewing area at the Camden aquarium, New Jersey. (Courtesy of the Thomas H. Kean Aquarium, Camden, New Jersey.)

Although the mission statement of the New Jersey Academy for Aquatic Sciences stresses the role of the aquarium in environmental conservation, ('The primary mission ... is to educate the public about aquatic environments ... to instil in the public an appreciation of the beauty, value and fragility of these environments' (Anon, 1993a)), this is not evident in the labelling of the living exhibits. Only two small tanks containing sea turtles have labels which mention conservation – the fact that they are now rare and endangered. Wouldn't it be better to have an empty tank labelled 'We'd like to show you turtles but aren't going to because. . . . '? There is a real need to upgrade aquarium labelling – like labels in botanical gardens it is too often assumed that all the public want is a name. However, the spectacle of the new aquarium cannot be denied – it is exciting, dramatic, a wonderful experience which lifts the spirit.

Conservation messages come elsewhere, and mainly through the team of docents (the aquarium has about 200 trained volunteers) which patrols the exhibits, talking to visitors. Other activities, including puppet shows, drama, slide shows and a good selection of films continue to stress conservation. Demonstrations with live fishes in handling tanks also mention conservation of fish stocks, and teach the public that fish need to be given care and respect, that they are sensitive living creatures. (It is pleasing to note that fish are only held for a short time in these handling areas before being returned to the main tank or the open sea.) The aquarium takes elements from all these activities for its outreach programmes to schools and adult groups, two vans are fully occupied taking the 13 Troutreach (!) programmes to a variety of venues in New Jersey, Pennsylvania, Delaware and Maryland. Like museums, the aquarium runs special activities for children during school holidays, the Summer Sea Camps enabling children to participate in field trips, live animal interactions, science experiments, art, craft and drama, often with an environmental theme.

There is also a temporary exhibit space, in 1993 devoted to an excellent exhibition on whales and dolphins, and an interactive exhibit area which seeks to explain how fish function – sight, movement, coloration and body shape. It is interesting that this latter exhibit is being abandoned, not because it is ineffective, but simply because of the need to increase revenue at the aquarium. This enforced change provides an example of a very real dilemma that many such institutions face. On the one hand they have a primary objective of environmental conservation, on the other the need to raise money to ensure their survival. Camden's start-up funds were provided by the State of New Jersey, but it is now totally reliant on admission income and shop sales for its survival. It has to get more people through the door, and it has to provide what people want. Unfortunately visitor surveys have indicated that the (mainly) grey and brown fish of the local coastline fail to excite the public, that what they expect to see is coral reefs and brightly coloured fishes, and this is what is needed to encourage repeat visits (Jane Herman (Horowitz), *pers. comm.*). Consequently the interactive area is to be swept aside to make way for a large coral reef aquarium and many smaller warm-water marine exhibits – the original intention of devoting the aquarium to local marine life has been abandoned. The consequences of this change are considerable, and not least for the tropical marine environment. Creating a coral reef on the scale envisaged will inevitably mean taking large quantities of corals, other invertebrates and fish from the wild; coral reefs are a habitat threatened by many forms of exploitation, already many coral reefs throughout the world have been devastated by harvesting animals for the aquarium and curio trades (Wood and Wells, 1988; Wells and Wood, 1991; Wells and Barzdo, 1991). Unlike freshwater tropical fishes, it is still not possible commercially to breed most coral reef fishes, as a consequence deaths in the tanks will mean more cropping of wild stocks. If it is difficult to quantify the impact of such a large-scale development on the marine environment itself, it is even harder to assess the benefits of such an exhibit in terms of biological conservation. It can only be hoped that the new displays will be used to relay strong conservation messages about coral reefs and their significance. A large label which stated 'Don't try this at home' might be a good starting-point. Alternatively, part of the entrance fee to the new exhibit could be channelled into a major coral reef conservation project.

The Monterey Aquarium's programme *Live from Monterey Canyon* (Connor, 1991; Anon, 1991a) is a marvellous example of the high technology used in the public domain. As well as being an exciting experience, perhaps the ultimate interactive exhibit, it also draws the visitor's attention to the research being undertaken at the Aquarium in the fields of deep-sea biology, chemistry and geology, and the testing of new technologies. The

JASON project (Kool and Trayner, 1990) provided museum and aquarium visitors with a similar experience. Research is an important function in many of the larger aquaria, including the development of husbandry techniques for new exhibitions, but often more fundamental in nature – Monterey has completed long-term field studies on the ecology of kelp forests and the biology of the sea otter for example. Staff of the Baltimore aquarium have forged links with various scientific projects, including the Project Puffin reintroduction programme on the coast of Maine, and the Maryland Department of Natural Resources to conduct surveys of the threatened bog turtle in northern Maryland.

What can museums learn from aquaria? They have already learned the need for the spectacular exhibit to attract the public. Museums have learned that fish are attractive – many of the temporary 'blockbuster' exhibitions now include aquaria as a feature – The Yorkshire Museum's Monsters of the Deep is a good example. In France, a strong case has also been made to promote the use of live collections in museums to encourage conservation of the natural heritage (Maigret, 1989); the museum of Besançon with its aquarium and 'insectarium' (Robert and Galliot, 1994) is an excellent example. Perhaps more remarkable is the development of an aquarium in association with an art museum, at the Musée National des Arts d'Afrique et d'Océanie in Paris (Hignette, 1994). The NMNH, Washington has created not just an aquarium, but a self-supporting coral reef ecosystem (Kelly, 1983; Cherfas, 1988). The Horniman Museum's *Living Waters* exhibit (see Plate 31) not only promotes conservation (Reid, 1993), but is also linked to a breeding programme for endangered species of cichlid fishes (Reid 1990a, 1990b). This is a rare example of a museum (as distinct from a zoo or aquarium) actively engaged in such activity.

A real concern for the welfare of fish is demonstrated by the modern aquarium, and this is something that museums need to note. Living displays deserve constant attention – but on too many occasions neglected and run-down aquaria or vivaria feature in museums. If they are to form a major feature then sufficient funds and manpower need to be found to care for them. Care also needs to be taken with the choice of exhibits, captive-bred species being mandatory, and ease of maintenance considered.

Docent programmes are a big feature in museums, zoos and aquaria in the USA. For example, the Camden Aquarium recorded 2,495 hours of freely given time in May 1993, the Academy of Natural Sciences of Philadelphia 25,590 hours in 1992. It is this volunteer force which enables those institutions to deliver more at every level – in collections care, but especially in educational activities. Demonstrations and informal talks with museum visitors are the realm of the docent, and it is often through these that informed statements about conservation are made; it is a feature which all museums need to emulate.

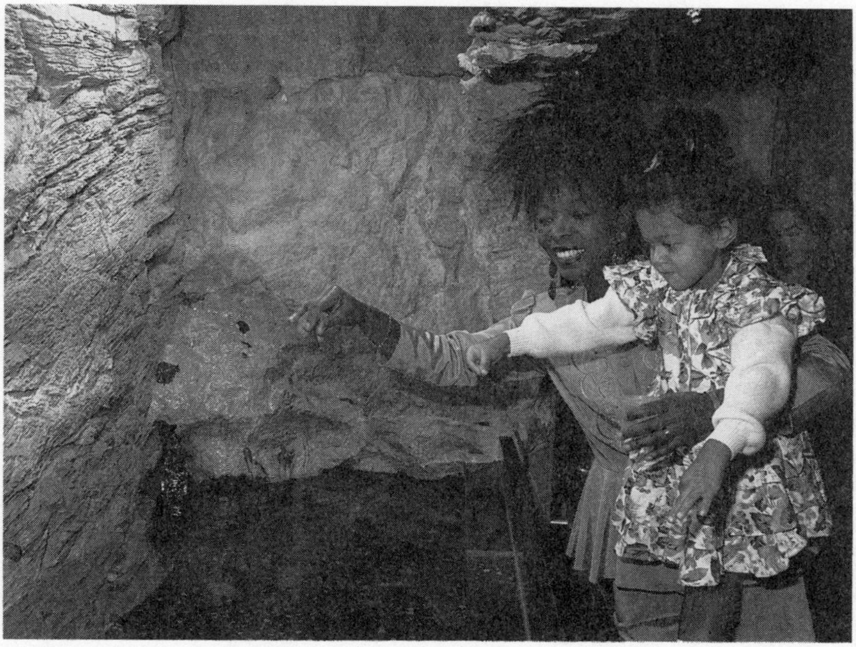

Plate 31 TV presenter Floella Benjamin and her daughter explore a touch pool in the display 'Living Waters' at the Horniman Museum, Forest Hill, London. (Courtesy of the Horniman Museum.)

A new approach – The case for bioparks

The idea of the biopark has already found a place in Europe through the award-winning Noorder Dierenpark at Emmen in the north east corner of The Netherlands. The dedicated museum-goer Kenneth Hudson commented that 'In my judgement, this is the best-planned, the best run and the most cost-effective museum in Europe' (Hudson, 1993). Some 1.7 million paying visitors in 1992 give weight to his statement. Created around a small zoo established in the 1930s, the Noorder Dierenpark (Hudson translates this as 'park for wild creatures in the north of Holland') breaks down traditional barriers between zoo, botanic garden and museum, as well as taxonomic barriers.

Hudson (1987) notes:

> It is curious that what would seem to be a natural and inevitable development, the hybridisation of a zoo and a natural history museum, should have been so slow in arriving. The reasons for the delay were partly physical, partly philosophical and partly a matter of empire-building. The major natural history museums were in the centre of cities, where there was no adjacent land on which to create a zoo, running a zoo suggested a degree of showmanship from which the more academic and influential zoologists shuddered, and the blending of a museum with a zoo involved merging two empires into one, a process which always, and in any field, leads to fierce resistance.

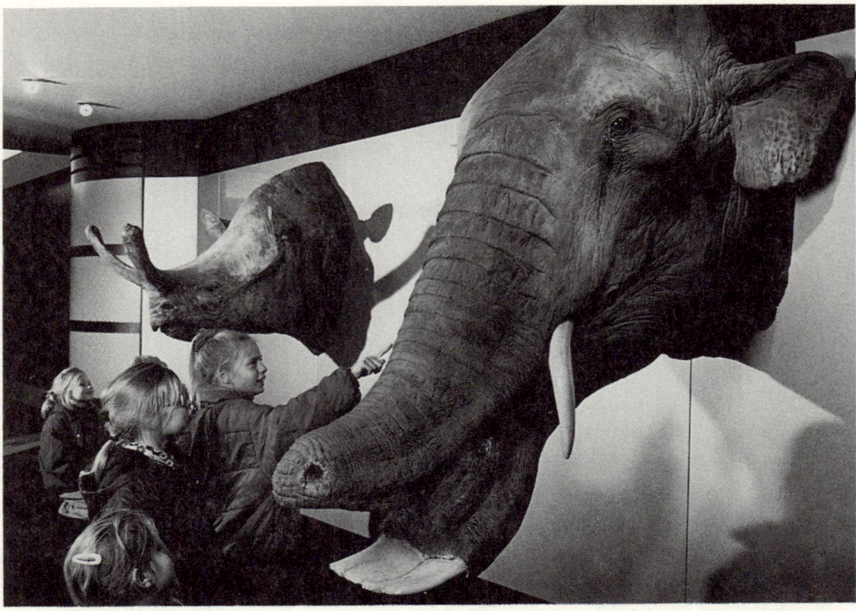

Plate 32 Young visitors meet extinct pachyderms in The Biochron, Emmen Zoo, The Netherlands. This unconventional reception area takes the public through geological time before entering the world of present-day animals. (Courtesy Noorder Dierenpark Emmen.)

Since its adoption of a new policy towards the presentation of its exhibits in 1970, Emmen (see Plate 32) has successfully combined elements of a museum (the 'Biochron'), with botanic garden and butterfly house, and an immersion-habitat zoo. The Biochron is an attempt to put the live exhibits into a geological time-scale, with visitors walking through exhibits which create dioramas of landscapes inhabited by models, and (where appropriate) live animals. The philosophy of the Biochron is boldly stated:

> Anyone visiting a zoo and seeing all those animals from every corner of the world will be amazed at the immense variety of forms. The question will inevitably arise as to why there is this incredible wealth of species. In order to answer this question, it is necessary to go way back in time to the moment when life began and then to follow the trail up to the present day (in Hudson, 1987).

The visitor emerges from the Biochron into the humidity of a botanical glasshouse, where a variety of tropical butterflies fly amongst dense foliage, then proceeds to the zoo itself. A deliberate policy has been taken at Emmen to keep only those species which are known to survive well in captivity – there are no pretensions to be involved in captive breeding programmes here – and to keep animals in an illusion of natural habitats, (African savannah for example), and within discrete geographical zones. The animal exhibits are complemented by a changing programme of exhibitions, games, publications and even puppet shows which seek to promote conservation themes.

Emmen is a remarkable success. It has discarded the traditional approach of the zoological garden, set clear educational and recreational objectives (above all it is fun to visit!), and through a combination of good research, considerable design flair, novel educational programmes and much hard work shown that it is possible to combine the best elements of live and static exhibits – a biopark before its time.

The word 'Biopark' was first promoted by Michael Robinson of the National Zoo in Washington (Robinson, 1988, 1989). It is his mission to introduce plants into a zoo setting to enable important conservation messages to be told, not simply as a decorative backdrop to an animal exhibit. He foresees a time when zoos will take a holistic approach to habitat creation, and address other aspects of biological sciences including functional anatomy, comparative nutrition, physiology and veterinary medicine. It is difficult to imagine how this could be done other than by combining the techniques and strategies used by botanic gardens, zoos and museums. *Amazonia*, the first major display of the biopark concept in Washington, will do just that, combining a Rainforest Habitat re-creation with an Amazonian Science Gallery (Anon, 1993). Although the Biodôme de Montréal has been labelled 'the first garden-museum dedicated to planet Earth' (Bourque, 1992) it is essentially a biopark; it is a 10,000 square metre garden which re-creates four major ecosystems – the polar world, tropical rainforest, mixed forest and the marine environment. Each habitat not only has characteristic plants of the environments, but live animals – the marine environment, focusing on the estuary of the St Lawrence river, not only has representative fish (cod, shark, rock bass) but also beluga (white whale) and seabirds.

The Arizona Sonora Desert Museum, in its new long-range plan (Hancocks, 1992) has set down fundamental principles for its future development which concentrate on demonstrating the interrelationships between geology, climate, plants and animals in a desert ecosystem. New live animal exhibits will concentrate on invertebrates ('It is not surprising that we fail to see the ways in which we destroy our environment when our views of nature are so myopic and so concentrated on the charismatic, the cute and the cuddly' (Hancocks, 1992)), and major exhibits will replicate biological communities found in the Sonora Desert region. The primary experience for the visitor is seen as immersion in the Desert, and exhibit guidelines dictate careful planning to ensure this experience. This progressive plan must be admired, and demonstrates the biopark concept admirably.

Whatever the final outcome of interaction between zoo, aquarium and museum, it is interesting to consider that the millions of people who gain an experience of the natural world through visiting them could never venture into wilderness, or experience 'real' wildlife, without causing extensive environmental harm.

References

Alexander, E. P. (1979) *Museums in Motion,* American Association for State and Local History, Nashville.

American Association of Museums (1984) *Museums for a New Century,* AAM, Washington DC.

Anon. (1991) 'Threatened closing of London Zoo evokes a heated reaction', *Museum News,* November/December, 10–11.

Anon. (1991a) 'Technology brings sealife up from the ocean and into the Aquarium', *Museum News,* March/April.

Anon. (1992a) 'A la découverte de Nausicaa, a l'écoute de la mer', *La lettre de l'OCIM,* 20, 15–18.

Anon. (1992b) 'A la découverte de Plampougnit au pays du saumon', *La lettre de l'OCIM,* 22, 27–29.

Anon. (1993) '*Amazonia* signals change at nation's zoos', *ASTC Newsletter,* 21 (2), 6.

Anon. (1993a) *First Year Summary,* Briefing paper from the Thomas H. Kean New Jersey State Aquarium at Camden.

Birney, B. A. (1988) 'Criteria for successful museum and zoo visits: children offer guidance', *Curator,* 31 (4), 292–316.

Blunt, W. (1976) *The Ark in the Park: The Zoo in the Nineteenth Century,* Hamish Hamilton, London.

Brambell, M. (1993) 'The evolution of the modern zoo', *International Zoo News,* 40 (248), 27–34.

Bostock S. St C. (1993) *Zoos and Animal Rights,* Routledge, London.

Bourque, P. (1992) 'The Biodôme de Montréal, the first garden-museum dedicated to Planet Earth'. In Saucier, C. (ed.) *Museums: Re-thinking the boundaries,* Proceedings of the 16th ICOM General Conference, Quebec, Canada, 1992, 59–61.

Bristow, M. (1994) 'The back of beyond', *BBC Wildlife* 12 (7), 45–47.

Castillo, B. Diaz del (1928) *The Discovery and Conquest of Mexico,* 1517–1521, translated Maudslay, A. P., Routledge, London.

Cherfas, J. (1984) *Zoo 2000–A Look Beyond the Bars,* BBC, London.

Cherfas, J. (1988) 'The ocean in a box: a museum exhibit houses the most productive ecosystem in the world, with applications far beyond its academic origins', *New Scientist,* 117 (1602), 46–50.

Clifton, M. (1991) Letter to the Editor, *The Times,* 16 June.

Connor, J. L. (1991) 'Promoting deeper interest in science', *Curator,* 34 (4), 245–260.

Cooper, D. E. (1991) 'How we burden our beasts', letter, *Guardian,* 11 April.

Cortes, Hernando (1972) *Letters from Mexico,* translated Pagden, A. R., Oxford University Press, Oxford.

Craig, T. L. (1988) 'Changing the way people think', *Museum News,* September/October, 52–54.

DeLapa, M. D. (1994) 'Interpreting Hope, Selling Conservation: Zoos, Aquariums and Environmental Education', *Museum News,* 73 (3), 48–49.

Dembeck, H. (1966) *Animals and Men,* Nelson, London.

Foose, T. J., Seal, U. S. and Flesness, N. R. (1985) 'Conserving animal genetic resources', *IUCN Bulletin* 16 (1–3), 20–21.

Greene, M. (1988a) 'Invisible architecture – new zoo design compels respect', *Museum News,* May/June, 40–43.

Greene, M. (1988b) 'Stalking the average North American zoogoer', *Museum News,* September/October, 50–51.

Greene, M. (1989) 'Breeding zoo stock', *Museum News,* January/February, 58–9.

Hancocks, D. (1992) 'Foundations of the 1992 Long Range Plan', *Sonorensis*, Summer, 3.

Hattersley, R. (1991) 'Today's zoo as Victorian caricature', *Guardian*, 13 April.

Hignette, M. (1994) 'L'Aquarium du Musée national des Arts d'Afrique et d'Océanie (MAAO)', *La Lettre de l'OCIM*, 35, 3–6.

Hudson, K. (1987) *Museums of Influence*, Cambridge University Press, Cambridge.

Hudson, K. (1993) Letter from Emmen, *Museums Journal*, 93 (9), 24.

IUDZG/CBSG (International Union of Directors of Zoological Gardens/Captive Breeding Specialist Group) (1993) *The World Zoo Conservation Strategy: A View of the Zoological Gardens and Aquariums of the World and their Roles in Global Conservation*, International Union of Directors of Zoological Gardens, Chicago.

Kelly, E. (1983) 'The coral reef: living system for exhibits and research', *Museum*, 35 (2), 119–120.

Kool, R. and Trayner, K. (1990) 'The JASON project: an evaluation', *Curator*, 33 (4), 261–271.

Loisel, G. (1912) *Histoire des ménageries de l'antiquité à nos jours*, 3 vols, Octave Doin et Fils et Henri Laurens, Paris.

Maigret, J. (1989) 'Un patrimoine negligé – les collections d'histoire naturelle vivantes', *Musées et Collections publiques de France*, 184–5, 35–37.

Middlemiss, J. L. (1987) *A Zoo on Wheels: Bostock and Wombwell's Menagerie*, Dalebrook, Burton on Trent.

Midgley. M. (1987) 'Keeping species on ice'. In McKenna, V., Travers, W. and Wray, J. (eds) *Beyond the Bars*, Thorsons, Wellingborough.

Nevin, D. (1993) 'Tranquillity, tourism and trouble: the Great Smokies have it all', *Smithsonian*, August, 20–31.

Ormrod, S. (1994) 'Showboat as Ark', *BBC Wildlife*, 12 (7), 40–44.

Reid, G. M. (1990a) 'Captive breeding for the conservation of cichlid fishes', *Journal of Fish Biology* 37(A), 157–165.

Reid, G. M. (1990b) 'A fish rescue and breeding centre at the Horniman Museum', *International Zoo News*, 37/4 (221), 8–10.

Reid, G. M. (1993) 'An aquarium for the future', *Journal of Biological Curation*, 1 (3/4), 61–66.

Reid, G. M. (1994) 'Live animals and plants in natural history museums'. In Stansfield, G., Mathias, J and Reid G. (eds) *Manual of Natural History Curatorship*, HMSO, London, 190–212.

Robert, J-Y, and Galliot, G. (1994) 'L'insectarium du Muséum de Besançon; Un zoo miniature à multiples facettes', *La Lettre de l'OCIM*, 36, 9–12.

Robinson, M. (1988) 'The Peaceable Kingdom Myth', *Museum News*, July/August, 80.

Robinson, M. (1989) 'No longer a zoo, but a biopark'. In *From Bison to Biopark – 100 Years of the National Zoo*, Friends of the National Zoo, Washington, DC.

Stuart, S. N. (1991) 'Re-introductions: to what extent are they needed?', *Symposia of the Zoological Society of London*, 62, 27–37.

Tudge, C. (1991) *Last Animals at the Zoo*, Hutchinson Radius, London.

Turner, E. (1984) 'Zoos aren't just for kids anymore', *Museum News*, December, 27–32.

Wells, S. M. and Barzdo, J. G. (1991) 'International trade in marine species – is CITES a useful control mechanism?', *Coastal Management*, 19, 135–154.

Wells, S. M. and Wood, E. M. (1991) *The Marine Curio Trade*, Marine Conservation Society, Ross-on-Wye, UK.

Winemann, J. and Yoon Kyung Choi (1991) 'Spatial/visual properties of zoo exhibition', *Curator*, 34 (4), 304–315.

Wood, E. M. and Wells, S. M. (1988) *The Marine Curio Trade: Conservation Issues*, Marine Conservation Society, Ross-on-Wye, UK.

Woodroffe, G. (1981) *Wildlife Conservation and the Modern Zoo*, Saiga Publishing, Surrey.

11. *Live issues II – botanical gardens*

When he has learnt that bottiney means a knowledge of plants, he goes and knows 'em.

Charles Dickens, *Nicholas Nickleby*

All gardeners know better than other gardeners

Chinese proverb

Plants provide us with food and medicines as well as being a source of delight and inspiration. As a consequence the maintenance of plant collections for scientific study and educational purposes has a long and distinguished history. According to Hill (1915) the Chinese should be credited as the founders of the botanical garden; the Han Emperor Wu Ti (140–86 BC) is known to have collected and grown a wide variety of food plants. The first model European garden may well have been that founded at Padua in 1545 (Alexander, 1979) which to this day remains largely unchanged on a formal geometrical plan which dominated the layout of gardens in Europe for 300 years. The garden at Pisa may have been established in the same year, (there is continued debate about which garden was established first – Pickering (1992) for example cites Pisa as the first European garden, and gives 1543 as the date of foundation) though it was moved to a new site in 1562. The botanist Luca Ghini (*c.*1490–1556) created the Pisa garden and is recognized as the plant collector responsible for first preserving pressed plants on card – the herbarium specimen; he was followed by the distinguished scholar Andrea Cesalpino (1519–1603), one of the first men to try to establish a classification system for plants. This pattern of building collections and classifying plants has been the core activity of botanical gardens ever since the sixteenth century, and many of the earliest gardens still survive today on their original site. Bologna (1547), Zurich (1560), Leiden (1577) and Montpellier (1598) are examples from continental Europe; in Britain the gardens of Oxford (1621), Chelsea (1673) and Edinburgh (1680) have unbroken lineage. The majority of early gardens were established by universities, and as a consequence botany quickly became a recognized science as plant taxonomy and the study of plant reproduction were added to the initial functions of growing medicinal plants and collecting new species.

Although the early botanical gardens had a research and education function, many were made available for public enjoyment – Alexander (1979) notes that the Leiden garden was open every day as early as 1740, though a small fee was charged and 'couples openly in love are on no account admitted'. This move to public access has been a feature common to all botanical gardens, irrespective of their origins or geographical location. Gardens have been created by private individuals such as John Tradescant (1570–1668), by royalty (Kew), as a result of the expansion of Empire and to profit from new economic resources (Calcutta); in nineteenth-century Britain many enlightened local government authorities (Sheffield, Glasgow) established botanical gardens as havens of green space in industrial cities. Thus gardens have long been in a position to promote the appreciation of the natural world, as well as carrying out research which has benefits for biological conservation and mankind.

Unlike museums and zoological gardens, the profile of botanical gardens (with some notable exceptions) had traditionally been low; as places of quiet enjoyment they have seen no need to respond to a rapidly changing world. However, from the 1960s economic pressures have changed the face of public botanical gardens; municipal gardens, once created with great pride, are now perceived as financial burdens. Botanic gardens maintained by universities have suffered as the teaching of taxonomic botany, once the keystone of all botanical study, has declined. The traditional botanical garden fails to attract public attention and support. There appears to be two great ironies as a result, first the reduction of taxonomic skills at a time when up to 100,000 plant species may be threatened with extinction by the end of the century, secondly, that the public interest in plants and horticulture has increased enormously in recent years at a time when the status of botanical gardens has waned. It might be argued that this growing demand for gardens and garden plants is in itself another impact of environmentalism; at least it must be perceived as a backlash against urban and suburban living, a demand for aesthetic beauty. Botanical gardens might have capitalized more on this interest, making horticulture part of their mission, yet many have remained purist in outlook and ignored the great upsurge of interest in gardens, gardening and hybridization.

If botanical gardens are going to survive, they need to adapt to new roles and pursue new objectives. At the same time, it is vital that they are recognized as an integral part of a nation's heritage and culture – their collections only differ from those in an art gallery by being alive. Collectively the plants in any garden reflect the history of the work of the collectors, scientists and horticulturists who have been associated with it. In addition, gardens need to change their role to be more relevant to modern-day society, as well as the demands of the conservation movement. The best gardens have moved with the times, have begun to provide more facilities and activities for the public, execute exhibitions on a variety of themes, and change garden design to provide both emotional and intellectual stimulation.

Botanic gardens – the new approach

Botanic gardens are a major public attraction. Kew alone has 1.3 million visitors each year; the British Tourist Board listed 153 UK gardens which had a minimum of 5,000 visitors, and a total of 8.7 million visitors in 1990; world-wide botanic gardens attract 150 million visitors annually according to the Botanic Gardens Conservation Secretariat. The potential for conservation education is therefore considerable. As Pickering (1992) notes 'If education through and by botanic gardens increases awareness and exerts political pressure to conserve habitats and species *in situ*, then they will play a valuable role.' It is surprising then that a recent survey seems to indicate that this conservation role has not yet been fulfilled in British botanical gardens, with only 38 per cent of respondents considering that the garden for which they had responsibility had any role in increasing environmental awareness (Dixon, 1991). That gardens vary in their educational aims is verified by Pickering's research, where 41.5 per cent of visitors to the Royal Botanic Garden Edinburgh found the experience educational, but only 25.8 per cent gave the same response for the Cambridge Garden (Pickering, 1992). The impression from this recent research is that botanic gardens in the UK have still have much to achieve in terms of environmental awareness.

However, the influence of environmentalism on botanical gardens has been more evident in other countries. Nowhere is this trend more notable than in the USA, where many gardens have taken on the mantle of conservation, frequently making environmental goals their primary mission. The Atlanta Botanical Garden began its 1990 Annual Report with the following statement:

> 1990 launched 'The Decade of the Environment'. In the Atlanta community, the Atlanta Botanical Garden led the way. Various staff members became actively involved in the public's interest in environmental education, each using his or her special talents. Recovery programs for rescuing endangered species native to Georgia were established. The Garden was the site of Atlanta's first environmental art exhibition, The Earth Affair. On April 22, Earth Day 1990, the Garden ... hosted photographer Robert Glenn Ketchum to speak on the disappearing Tongass National Forest in Alaska, attracting 400 people. Education programs offered conservation messages throughout the year. Merchandise in the Gift Shop focused on environmentally friendly products. And a significant commitment to maintain our plant collections with environmentally sound practices became a priority. Botanical gardens around the country are making conservation an integral part of their mission.

Botanical gardens have always played an important role in documenting the world's flora, but the impact of environmentalism has been so great that even the older and scientifically active gardens are re-evaluating their objectives, often taking a more conspicuous conservationist role. The Missouri Botanical Garden's mission for example, is 'To discover and share

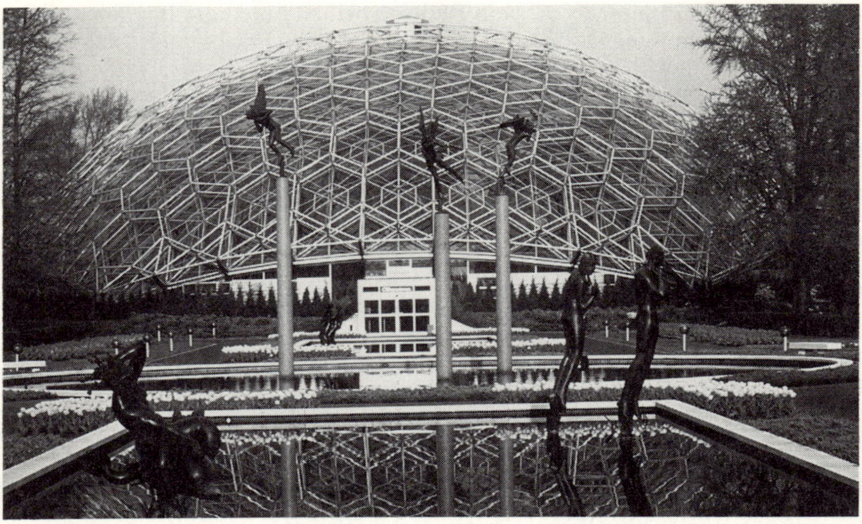

Plate 33 The Climatron geodesic domed greenhouse is the shining centrepiece of the Missouri Botanical Garden, St Louis. (Courtesy of Missouri Botanical Garden, St Louis.)

knowledge about plants and their environment, in order to preserve and enrich life.' Such statements of intent are significant, giving a new direction and impetus, even to long-established institutions. Raven (1992) indicates just how important he feels such a directive is for individual members of staff of the Missouri Garden:

> These are compelling words. Following through, making sure that we maintain our focus, requires each member of staff to take responsibility for his or her actions, and to take steps that further public understanding of the issues. We must use information about plants in ways that produce continuing benefits, and we must work to create a better understanding of the relationships between plants, people and their environment.

The Missouri Botanical Garden in St Louis is the oldest in the USA, established in 1859. The gardens occupy 79 acres with the 'Climatron' geodesic domed greenhouse as its centre-piece. This remarkable structure (see Plate 33) encloses more than half an acre, a lush tropical environment highlighting the diversity and ecology of tropical rainforests. A computerized climate control system maintains high temperature and humidity levels which the 1,200 species of plants require. The nature of the experience is evoked in the publicity leaflet about this area:

> Visitors enter the greenhouse and immediately experience the tropics: dense green foliage, a small native hut, sparkling waterfalls and rocky cliffs, a river aquarium with exotic fish, and a bridge from which the forest canopy and associated plants can be viewed. Several exhibits capture the imagination. 'The

> Talking Orchid' presents the story of epiphytes, and orchids in particular, in a creative and amusing way. 'The Talking Orchid' not only talks to visitors, but sips water from their hands. 'Products from Plants' is a series of exhibits about economic tropical plants. Another focuses on food from the tropical forests. Dispensers allow visitors to sample products from several tropical trees represented in the plantings.

Here is a very special example of the new approach of botanical gardens – an immersion habitat with live plants and animals, unafraid of a bold, even theatrical approach to deliver fascinating information. And that information is not simply about plants, but the ways in which we, as humans, depend on them.

Conservation messages are reinforced in the Brookings Interpretive Centre, the nucleus of the Climatron, and the only area not to house living plants. Its function is to explore the political, social, economic and natural factors which are responsible for rainforest destruction, with secondary themes devoted to habitat destruction world-wide. The Centre outlines human dependence on rainforests, but equally the rainforest's dependence on us, and seemingly remote and unrelated human activities come into sharp focus for their dramatic impacts in terms of species extinction and global warming. The visitor is finally confronted by the exhibit which implores them to ' ... translate this information into their activities as consumers, voters, and citizens of the world, through both personal and group action.'

The gardens themselves include a sculpture garden, a desert house, an English Woodland Garden and a Japanese garden. However, probably one of the most fascinating areas is the 'Centre for Home Gardening', a display of 20 distinct residential-scale gardens where adults and children can learn about vegetable gardening, growing flowers which attract butterflies or birds, planting ornamental species, landscaping and keeping indoor plants. A series of educational programmes is designed to deal with common gardening problems, offering advice on garden design, garden tools, soil development, and even providing a plant 'clinic', where consultations with an expert 'plant-doctor' are possible. This educational attraction for the keen gardener is indicative of the philosophical approach taken by American botanical gardens – they are prepared to step beyond the boundaries of the purist 'species-only' approach adopted by their European counterparts.

A less well-known, but equally progressive botanical garden is the Morris Arboretum of the University of Pennsylvania (see Plate 34). Founded in 1887 as the private estate of John Morris and his sister Lydia, the 166-acre garden on the outskirts of Philadelphia is internationally known as an outdoor museum that integrates art, science and the humanities, being far more than its collection of living trees. It is now recognized as the official arboretum of the Commonwealth of Pennsylvania, is listed on the National Register of Historic Places, and is fully accredited by the American Association of Museums. Research at the arboretum focuses on practical issues including the systematic study and mapping of the flora of Pennsylvania, the evaluation of trees and shrubs for introduction into urban landscapes, and

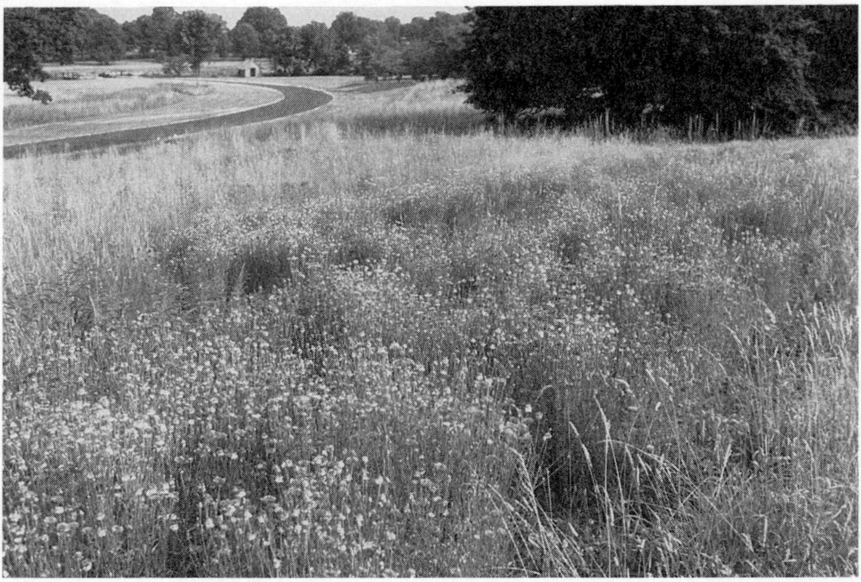

Plate 34 The Morris Arboretum, Philadelphia – the wild garden. (Courtesy of the Morris Arboretum, Philadelphia.)

investigations to find environmentally safe ways to control pests and diseases. The organization runs a comprehensive environmental education programme, including teacher training, professional courses in arboriculture, the production of publications and exhibitions, and delivers lectures and workshop programmes, including outreach activities. In essence, it does much that the traditional museum undertakes.

It is in the gardens themselves that new ideas and approaches are identifiable. There are a variety of garden styles, including a rolling 'English Landscape Park', formal Japanese gardens and tree-lined drives, all providing a magnificent setting for an outdoor sculpture collection. Edward F. Fry, serving in 1987 as the arboretum's consulting curator in fine arts, explained the philosophy behind the sculpture collection (Anon, n.d.):

> ... it is another aspect of the modern sculptural tradition that has proved to be particularly suitable for the Arboretum. These are works which not only benefit from the physical setting of the Arboretum but which by their very lyricism and complementarity to nature, also create situations of a heightened interactiveness with their natural environment. ... The selection of a work of sculpture and its carefully chosen placement have been made with a goal in mind of creating a mutually transforming interaction between art and the natural environment.

The use of sculpture was not new on the site – the Morrises had from the beginning introduced fine art into their Victorian estate landscape. However, the rekindling of the concept in the 1970s, with the emphasis on the interaction between art and nature, is an interesting phenomenon which must owe something to the impact of environmentalism. In Britain the

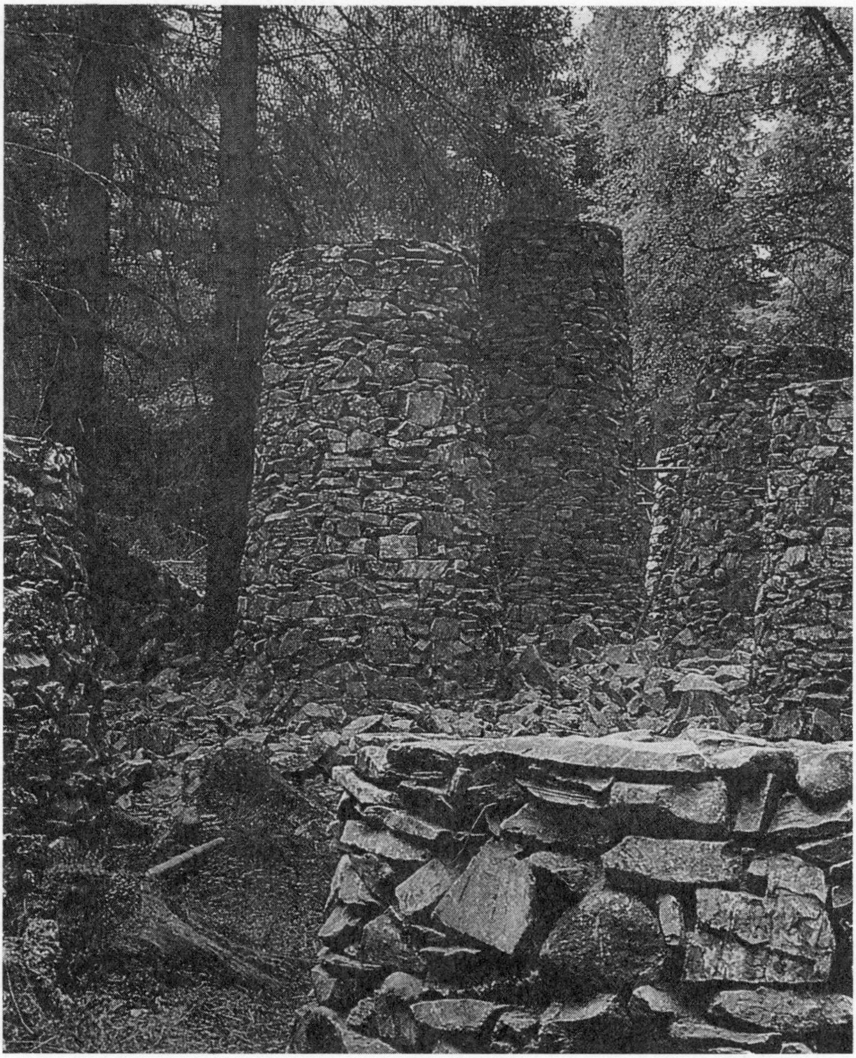

Plate 35 Sculpture in a natural setting – Kimio Tsuchiya's 'Stone Forest', Grizedale Forest, The Lake District, Cumbria. (Photograph by Val Corbett reproduced courtesy of the Grizedale Society.)

concept of the sculpture park, and the increasing number of sculptures placed in woodland, rural or urban fringe settings from the 1960s mark a change in attitude to the environment. The Grizedale Forest area of the Lake District was a pioneer venture, and is now internationally recognized (see Plate 35). Sculptures in Hamsterly Forest, County Durham, or on the riverside below Durham Cathedral, or those in botanic gardens such as the Royal Botanic Garden Edinburgh are set within frameworks of formal

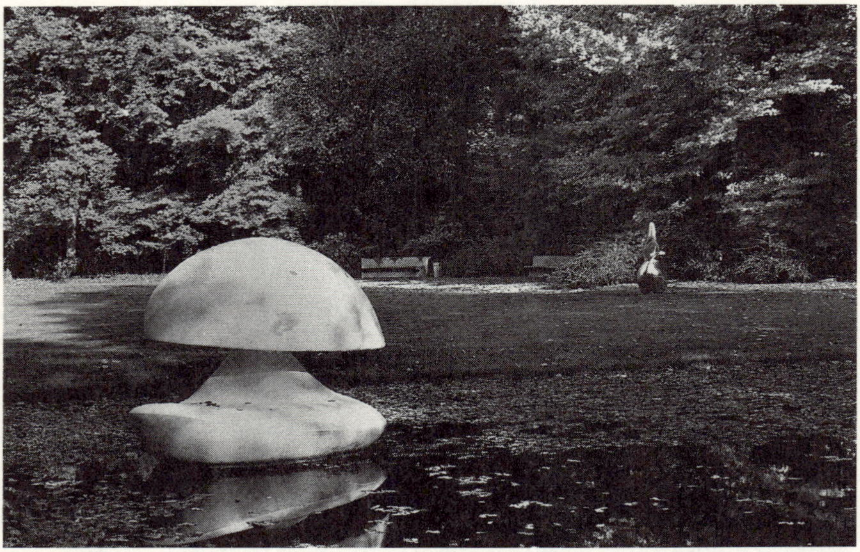

Plate 36 The Kröller-Müller Museum, The Netherlands includes a large-scale sculpture park set within the country's largest nature reserve. Marta Plan 'Floating Sculpture Otterlo', 1960–61 and Jean Arp 'Cloud Shepherd', 1953. (Courtesy of the Foundation Kröller-Müller Museum, Otterlo, The Netherlands.)

gardens or open woodland, often featuring sculptures taking their inspiration from nature, and using organic and natural materials to explore interactions between art and environment. The Kröller-Müller Museum in The Netherlands (see Plate 36), set within the country's largest nature reserve, seeks to do this – 'The fundamental idea of relating art and nature was . . . behind the . . . sculpture garden (1961), the Rietveld pavilion (1965), the sculpture park and sculpture wood (1988). The same relationship is an important feature of the new wing . . .' (Foundation Kröller-Müller Museum, 1994). It is puzzling that so many museums placed in rural or parkland settings fail to capitalize on the potential of these areas to explore such interactions – it is a field where botanical gardens have shown the way forward, with the Morris Arboretum a fine example.

Botanical gardens are frequently very formal, with expanses of well-mown lawn and manicured flower beds. Although many of them, especially those in city centres, act as a refuge for wildlife, they do so by default rather than by active management. This is another feature of the Morris Arboretum, where more than one-third of the park is managed as 'wilderness', with extensive semi-natural woodlands bordering the Wissahickon Creek, providing food and cover for a variety of breeding birds as well as mammals such as woodchuck and white-tailed deer.

Delightful to the eye are the acres of wet meadowland which greet the visitor on the drive through to the formal gardens, a blaze of colours provided by a distinctive flora. The golden meadows are alive with insects, bees collecting nectar, dashing dragonflies, huge tiger swallowtail butterflies dancing in the sunlight. This is good creative management, achieving greater species diversity and saving money in maintenance costs – more flowers for less cost, what could be better for a botanic garden? The area feels wild. It is remote from the main visitor area, and an effort needs to be made to get to it. There is no interpretation on-site, no information boards naming the flowers or large insects, nothing to say how the area is managed or why it is being left to meadowland. In essence there is nothing to detract from the experience, nothing intruding into the feeling of being somewhere that is different; there is only a brief reference to the wild areas in the visitor's guide. Consequently few individuals experience its beauty; it remains largely undisturbed, an oasis for wildlife in an intensively managed landscape.

Because of its location and lack of interpretation the meadows fail to achieve their potential in conveying principles of, for example, diversity, plant-insect relationships, or wildlife management to a wider public. Yet for those making the effort to reach them, they speak for themselves. Scientific names, ecological concepts, deeper undertandings may not be necessary in a place like this, simply an appreciation of beauty, a feeling of wonderment. Perhaps there are parallels here to the appreciation of works of art – curators of art galleries frequently argue that paintings need nothing more than a name or title, that a painting speaks for itself. If this is a manifestation of élitism in art, can biologists perhaps sometimes be accused of environmental élitism when appreciating wilderness or a special habitat? Interestingly Alexander (1979) makes the comment that 'Botanical gardens in the past have been much like art museums, in that the beauty of their collections has attracted an audience sensitive to aesthetic appeals – in this case, color, texture, design, perfume and natural growth.'

The scale of the Morris Arboretum allows it the luxury of dedicating space to more natural habitats; few botanic gardens follow its example, largely due to the more urgent demands of their living collections. Interpreting those living collections poses a real challenge to garden directors – do they continue to adopt a discreet approach which leads to aesthetic appreciation (and I have cited just one example where this is appropriate), or do they begin to take a more radical approach which enables them to discuss environmental issues – or at least explain why plants are important, the relevance they have in today's world?

Labelling in botanic gardens has until recently been minimalist, and indeed most gardens still display fine collections of very small labels, strategically hidden to confuse the visitor. Neilson (1985) has identified labelling as a major source of complaint from garden visitors. Why? – the effort of finding such labels is generally rewarded by discovering scientific name, collection inventory number, and country of origin – '*Ginkgo biloba*, 423/93/2, E. China'. No information about the tree's distinguished fossil record, or that the leaves contain a chemical (ginkgolide B) which is effective in preventing heart arrhythmia, or that in Korea the steam of boiled *Ginkgo* leaves is used as a cure for asthma. What a lost opportunity!

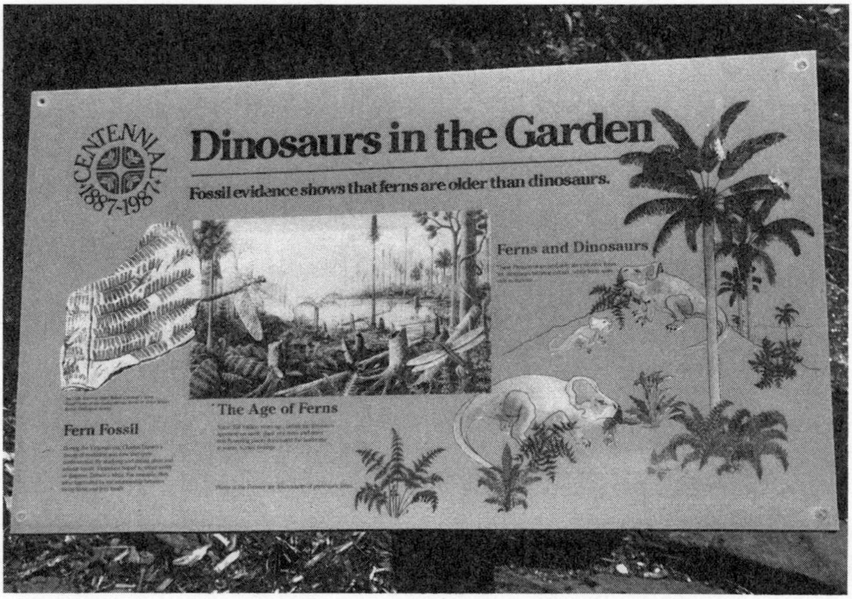

Plate 37 Labelling in botanic gardens – cycads linked to the age of dinosaurs at the Morris Arboretum, Philadelphia.

The growing interest in herbal medicine, and in ethnobotany generally, provides an important way of relating plants to people. For example, how better to increase understanding of the importance of saving rainforest than to refer to the medicinal plants which have been discovered in that habitat, and the potential for finding others? Or which plants of the rainforest are used as staples by the native peoples? The significance of biodiversity may be a difficult concept with which to engage a casual visitor, but everyone can understand food and medicine. Botanical gardens have a tremendous opportunity to be advocates for the environment, to be more than just a beautiful place to visit. Perhaps gardens are now beginning to learn from museums and develop more interesting labels. For example, the Morris Arboretum has made a feature of its ferns and cycads through large colourful labels which relate them to the age of dinosaurs (see Plate 37) and although still a stronghold of the bogstandard (botanical garden standard) label, a self-guided trail leads visitors on a fascinating 'medicinal tree tour'.

Glasshouses as immersion habitats

'Immersion habitats' are now a major feature in many zoos, creating an illusion of a particular environment. Botanical gardens have been doing this for many years in glasshouses and conservatories, growing together trees from tropical regions, or re-creating the image of desert environments. Frequently this has lead to the design of beautiful and technically elaborate structures in gardens; those at Kew and Frankfurt are excellent examples. Visitors move from one precisely created environment to another, temperature and humidity are carefully monitored to give optimum conditions for the plants. It is even possible to design conservatories for hot sunny climates, a process requiring great ingenuity, exemplified by Ambasz's Halsell Conservatory of the San Antonio Botanic Garden in Texas, which enables a self-contained alpine meadow to flourish in the most unlikely and uncompromising situation (Dillon, 1988).

The Victorian Palm House at Kew is a monument to the ingenuity of glasshouse design. Arguably described as 'The greatest glasshouse in the world' (Kew's Temperate House and the Princess of Wales Conservatory are both larger in floor area), its restoration during the 1980s provided the opportunity not only to demonstrate the political and fiscal will to conserve a remarkable building, but also to redefine its role on a conservation theme (Minter,1990). As Prance (1990) notes:

> With strangely prophetic wisdom, when the restoration of the Palm House was planned ten years ago, it was modified to house samples of the two most species-diverse habitats on earth; the tropical rainforest and the coral reef with its associated marine environment; since that time we have become more and more aware of the threats to our ecosystems caused by man's destruction and degradation, be it of the North Sea, the tropical forest, the coral reef or the hedgerow. So, in the present we are privileged to recreate in the Victorian Palm House something of the wonder of the tropical rainforest and of the coral reef. Wonder can lead us to respect and to value such environments. The displays so elegantly housed also prove to us the economic value of many of the plants and organisms, for our lives are touched daily by products of other ecosystems even when we are ignorant of the source.

The concept of combining a marine display into a glasshouse is a novel one, and one which demonstrates the breakdown of traditional barriers – here between botanic garden and aquarium. Kew's rationale for the development was to find a way of exhibiting algae – one of the most important plant groups, yet one usually ignored by botanical gardens in public displays. The former basements of the palm house became the marine displays, which include examples of British coastline species (rocky coast, muddy shore) as well as the tropical marine habitats of coral reef and mangrove swamp (Morris and Riddington, 1990).

Botanical gardens and species conservation

Most larger botanical gardens have had a research function – those that also maintain herbaria have a tradition of involvement in plant taxonomy and systematics. The impact of environmentalism has been to focus research on to conservation issues, to continue to document the world's flora, but also to be involved in practical ways of halting habitat decline and preserving genetic resources. The research function is now a major contribution made by botanical gardens to an understanding of the world's flora, and many gardens are actively involved in plant conservation through breeding and reintroduction programmes. Heslop-Harrison (1976) remarked that:

> In taking up the challenge of developing the living collections for their manifold potential functions in conservation, botanic gardens can find for themselves a new and purposeful role, at a time when economic pressures are growing and their traditional functions in teaching and research are coming under question.

Never has such an effort been more urgently required – it is estimated that some 22,000 plant species are currently endangered; as many as a million of the current complement of animal and plant species could be lost by the year 2000 (Myers, 1983). Raven (1976) estimated that at least 50,000 tropical vascular plants were threatened with extinction by 2000. In Europe, about 2,000 species are classed as rare or threatened; of these, 481 are in cultivation in botanical gardens (Lucas, 1978). In answer to the need for species conservation, the Botanic Gardens Conservation Secretariat launched 'The Botanic Gardens Conservation Strategy' in 1989, with the aim to encourage a co-ordinated approach to conservation based on collections documentation and information exchange. Unfortunately, the Strategy rests on the ability of members to find funds to pay a subscription to support the database – and even in the UK many gardens cannot afford the subscription.

The scale of the loss of plant biodiversity is daunting, so great that even the 1991 Kew Conservation Conference (*From Specimen to Habitat Management*) recognized the inability of botanical gardens to conserve the endangered species of the world's flora. This situation is made apparent when examining Britain's flora – the 48 botanical gardens, though reasonably resourced, cannot conserve even the 51 species which are threatened with extinction – and these from a relatively small (*c.* 2,000 species) flora. Pickering (1992) expands on this situation:

> A list of Rare Vascular Plant Species in Cultivation in the British Isles is kept on an *ad hoc* basis. The Biological Records Centre and the IUCN's Conservation Monitoring Centre produced the fourth edition of this list but were unable to keep it up. A fifth edition was prepared by the Conservation Propagator at the University of Cambridge Botanic Garden. . . . Between the production of the 4th edition in 1978 and the fifth edition in 1983 there was a substantial drop in the number of rare species held in cultivation . . . the sixth edition of the list has been prepared at Chelsea Physic Garden as English Nature no longer fund the Conservation Propagators Post at Cambridge. The post was held by Miss V. Morgan who said that the species were not intrinsically difficult to grow but that the 'Careful and painstaking record-keeping and administration required to maintain a conservation collection as a quality scientific resource has proved too expensive for most botanic gardens in the recent cutbacks. In practice, collections have only survived in gardens where a particular member of staff has bothered to look after them on their own initiative' (Morgan, *pers. comm.*).

If Britain cannot cope, then it is no surprise that botanical gardens in developing countries, (often relics of a colonial past, and with meagre resources), can offer little in terms of conserving species diversity. This is particularly tragic as they are often close to the areas of greatest species loss. Bovey (1991) in a review of African gardens, felt that the only way to ensure the survival of gardens on the continent was for them to seek industrial sponsorship, to develop links with high-profile gardens in Europe, and to relocate to areas of greatest biological interest from their historical urban setting. The need for careful site selection was paramount, and an area of political stability essential. Training (provided by other better resourced gardens) is also a major key to success. These are all reasonable, and highly creditable suggestions, and some of the larger botanic gardens have established links with botanical gardens in Africa as part of major research projects. For example, Missouri Botanical Garden's Research in Madagascar is carried out in association with the Botanical and Zoological Park at Tsimbazaza, and a major objective is to assist and train Malagasy scientists and students. It is this type of support and co-operation which must be seen as the way forward.

It is evident that botanic gardens must concentrate on what is realistic, and in recent years this has lead many botanical gardens to re-evaluate their strategies for living collections (Donnelly, 1989). It is impossible for them to conserve all the endangered elements of the world's flora by maintaining collections of growing plants. Falk (1991) has argued that gardens, with their specialist ability in propagation and growing techniques, should act as a flow-through system, producing large numbers of individuals for re-introduction to damaged habitats. Propagating endangered species is also very costly – the Sainsbury Orchid Project at Kew, dedicated to conserving the ten species of endangered British species, has been made possible only by an endowment of £1 million. The project has succeeded in growing many of the rarer orchids in tissue culture, and some species have been re-introduced into the wild (Pain, 1991). However, as with animal species, there are huge difficulties involved in reintroducing plants to the wild. Maunder (1991) has suggested that there are some 190 reintroduction projects in 18 countries covering 20 plant families; little information is available about these projects, and even defining a successful reintroduction is difficult. For example, Pickering (1992) notes that if the Cafe Maron *Ramosmania heterophylla* propagated at Kew from cuttings of the last tree had succeeded in Rodrigues it would be a sterile clone. Other, more practical difficulties, often face reintroduction programmes – the attempts to re-establish *Sophora toromiro* on Easter Island by the Swedish botanist Bjorn Alden failed in part because of the small number of specimens available, but also due to the ravages of a parasitic worm which destroyed the roots of the young plants. The research effort continues, with attempts to bulk-up populations at the University of Bonn; meanwhile individual plants have been distributed to other botanic gardens (such as Menton, France) to ensure survival of the species, albeit, for the present in cultivation only (Monnier, 1993).

Other approaches are required if botanical gardens are going to achieve conservation aims. For example, Winter (1991) has suggested that attractive plants threatened by over-picking in the wild can be assisted by botanical gardens, which can propagate such species and pass plants and techniques on to the horticultural trade – a role assumed by the National Botanical Institute, Kirstenbosch, South Africa. Another progressive idea is for botanical gardens to manage nature reserves – for example the Jardin Botanico

'Viera y Clavijo' in the Canary Islands has responsibility for managing a number of nature reserves on the islands which house endangered plant species. Wakehurst Place, run by Kew Gardens, includes a fine piece of Wealden woodland which contains several rare English plants (Hepper, 1982).

Gathering information about plant species and their distribution has always been a key role for botanical gardens, and it remains as the cornerstone of conservation. It is imperative fully to document the world's flora, and as pressures on habitats increase, so there is a need to speed up taxonomic studies and complete the inventories of species for every country. In this function the botanical garden is acting exactly as do many of the foremost museums of the world; carrying out scientific collecting, identifying new species, producing species lists, identifying rare and threatened species. Without such data no conservation measures are possible – they are the yardstick from which success or failure can be measured.

There are many examples which might be cited of innovative scientific survey and assessment of floras throughout the world. Inevitably, it is the well-heeled gardens of the west which carry out much of the fieldwork in tropical countries; this pin-points the key difficulty in conserving biodiversity – matching resources and skills for conservation to the countries where the problem is most acute. The situation is the same for museums and botanical gardens. Many gardens combine a commitment to conservation in other countries with the recognition of conservation needs closer to home – for example the Atlanta Botanical Garden announced a formal conservation strategy in 1990, carrying out a plant-collecting trip to Ecuador that year, and establishing a joint programme with the Department of Natural Resources to breed and reintroduce three species of pitcher plants (*Sarracenia*) threatened with extinction in Georgia (Atlanta Botanical Garden,1990).

The Missouri Botanical Garden, a world leader in botanical research, works with local institutions in more than 30 countries to collect and classify the rapidly vanishing plant species of the world. Henry Shaw, founder of the garden, purchased a comprehensive herbarium of 62,000 specimens in 1857; this became the basis for the present-day collection of 4 million specimens of plants, of which some 170,000 were added in 1992 alone through the collecting activities of MBG botanists. The end result of all this activity is the publication of important floras such as *Flora Mesoamericana*, the *Flora of China*, and the *Flora of North America*; the data about plant names and specimen details associated with the publications are stored on the garden's botanical database, TROPICOS. Garden scientists are not only involved in pure taxonomic research, but in innovative projects such as the creation of the rainforest reserve on the Masaolo Peninsula in Madagascar mentioned earlier. A unique aspect of this project is that the local peoples will be taught sustainable agricultural techniques so that they can survive without destroying the forest. The Missouri Botanical Garden is now the national headquarters for The Centre for Plant Conservation, (CPC) a national organization dedicated to the preservation of endangered plants native to the USA. CPC's goal is to prevent the extinction of native species by maintaining living collections, and eventually to reintroduce species to the wild. Another interesting facet of the Missouri garden's work is its Natural Products Program, a link with commercial companies to screen plants for potentially useful chemical compounds, with the search concentrating on compounds for medicinal and agricultural use. One result of this work is a collaborative programme in Cameroon to cultivate an undescribed species of *Ancistrocladus,* collected by garden scientists, which has shown chemical activity against the HIV virus.

Taxonomic research (in museums and botanical gardens) has been conducted at a fairly leisurely pace in the past. As habitat destruction continues to increase, there have been a number of calls for an increase in taxonomic activity – the United States National Academy of Sciences has called for a crash programme of plant taxonomy and for 1,500 taxonomists world-wide (Synge, 1982). The concept of 'quick and dirty' surveys has been raised by a number of eminent scientists in the early 1990s. May (1992) remarked that 'studies need to be quicker and more crude than conventional/traditional methods'. Recent research in Costa Rica has followed this guideline, using 'parataxonomists' to good effect in the survey of rainforest ecosystems (Gamez, 1992). Kew scientists carried out a rapid survey of the species-rich coastal forests of Bahia province, Brazil, to provide a species list for an area which is under threat from logging operations. Several Kew expeditions have been carried out with conservation aims in mind – the designation of some of the islands of the Seychelles as nature reserves was in part due to an assessment of the flora (Synge, 1982).

However, it is all too easy to assume that once a plant is named, and its geographical locality identified, we subsequently know all about its biology – and as a consequence can conserve it. This is a total misconception – plants are part of a complex ecosystem, and collecting and propagating a few specimens of a species is not enough to save the ecosystem itself. More complex biological questions need to be answered; and perhaps equally important is the need to find the resources, political will and scientific expertise to put habitat conservation aims into practice. However, there is a practical way in which lists of plant names can be beneficial in conservation terms – this is by adding names of endangered species to legislation which is enacted at an international or national level. Most laws for the protection of wildlife act through appendices which list vulnerable species; hence Kew, acting as an adviser in Washington in 1973, provided species names which would ensure protection under the Convention on International Trade in Endangered Species (CITES). Similarly, taxonomists have provided species lists for the Berne Convention on the Conservation of European Wildlife and Natural Habitats (1982) and its enactment in Britain, the Wildlife and Countryside Act (1981) and its Amendment (1985). Looking to the future, it seems increasingly possible that the knowledge of taxonomists, whether they be in botanic gardens or museums, will be in great demand to help supply knowledge which will enable practical solutions to be found to conservation problems. These problems are invariably due to man's activities, and it is vital to tailor solutions to human demands, and to geographical areas. Referring to Kew, Lucas (1989) notes:

> The thrust, therefore, of all Kew's tropical expeditions, whether to the arid or humid regions, is now more than ever strongly stimulated by the need, not just to identify the flora, but to see what species are used and for what purpose. We do this in the pursuit of our basic belief that the answer to tropical problems will be found in tropical solutions.

This need to take a more pragmatic approach becomes increasingly important when contemplating the costs of conservation measures. Gee (1990) noted that 'Whilst worldwide concern is being expressed over the rate of species loss, . . . shrinking funds at Kew will mean that new conservation initiatives will have to rely on private money.'

Many botanical gardens seek to preserve genetic diversity through the exchange of seed and the maintenance of seed banks. It is interesting to contemplate that this development was also a symptom of growing environmental awareness – Smith (1982) notes that:

> At the same time as the (Kew) seed unit was being developed concern became widespread across the world at the rapid disappearance of many natural habitats owing to man's direct interference and the consequent threat to the survival of plant species.

Kew began its refrigerated unit in 1969; it transferred to Wakehurst Place in 1973. The principal advantage of seed banks is that a huge quantity of genetic material can be stored in a small space – also, by reducing moisture content of seeds (*c.*5 per cent) and keeping them at low temperatures (-20°C), viability may be extended to hundreds of years. Selecting seed to store is undoubtedly influenced by conservation aims – Kew initially chose to collect from threatened Mediterranean habitats for example, and to build up a collection of rare, endangered or local British species. As Smith (1982) points out:

> Maintaining seeds of wild species in long-term storage can be seen as an act of faith for the future . . . what properties will be sought amongst the collections in the future cannot be predicted but . . . we can be certain that many plants will prove to have valuable and perhaps surprising uses.

These sentiments will be recognized (and echoed) by all curators worldwide – we are all seeking to maintain collections for the future – and who knows what the future will demand of them?

Museum gardens

Creating exciting botanical displays within a museum is extremely difficult, and only a small number of exhibitions have successfully employed living plants. The most notable achievement is that of the National Museum of Natural Science, Ottowa, (Haber *et al.*, 1982) where live plants were used to illustrate a variety of botanical themes as part of a multi-media exhibition; in the UK, Liverpool Museum and the Royal Botanic Garden in Edinburgh have maintained pleasing live plant exhibits within a building. Bearing these difficulties in mind, can museums take a lead from botanic gardens and use their grounds in an effective way to create areas of wildlife interest and promote environmental awareness? Museums with large areas at their disposal are in an enviable position, and can develop areas of considerable interest for the visitor. An excellent example is the ethnobotanical garden sited in the 43-acre grounds of the Sabah Museum, Malaysia, to provide a living store of information about the uses of plants by the indigenous peoples of Sabah. Opened in July 1993, the garden was created as a result of a growing interest in the ability of ethnic groups to recognize and use plants for food, raw materials and medicine. The museum recognized a need to document this knowledge before it was lost, as part of its research programme on the cultures of Sabah. Borneo has one of the richest floras in the world (at least 10,000 species of flowering plants, 95 per cent are found in Sabah), of which 25 per cent are exploited by indigenous groups. The gardens now form the basis for most of the museum's environmental interpretation activities (Guntavid, *pers. comm.*).

Even museums with restricted areas of garden can use them to good effect to re-create natural habitats which can attract wildlife and provide an extra educational facility. The Natural History Museum in Berne, Switzerland, perhaps best known for its magnificent dioramas, has small gardens which have been left to 'go wild', with a resulting blaze of colour which looks good and provides a rich source of nectar for the museum's observation hive of honey-bees. However, persuading the museum authorities that this was a wise development proved difficult – the Keeper of Vertebrates noted 'It was a long story to convince everybody that such a ground is better than a lawn. . . . ' (Lups, *pers. comm.*)

The Hancock Museum in Newcastle upon Tyne, England, a mid-sized natural history museum in an urban setting, faced exactly the same problem when seeking to develop its grounds as a 'Northumbrian Garden'. Opened to the public on its present site in 1884, the museum has a long history of involvement with environmental matters, particularly the documentation of the natural history of the surrounding region. The museum is the property of the Natural History Society of Northumbria, but is administered by the University of Newcastle upon Tyne. In 1988, with the assistance of Northumberland Wildlife Trust, a small formal lawn in the gardens was developed as an experimental hay-meadow by introducing species characteristic of that habitat. The positive reaction to this area from members of the public led to a more ambitious scheme to re-create other habitats – moorland, woodland, grassland and a pond; this was eventually approved, and the work carried out, in 1989. Grant aid was obtained for interpretive signs and a self-guided leaflet which gave information about the habitats and individual species. The pond developed rapidly; damselflies and amphibia were quick to take advantage of the new habitat, as were museum visitors who delighted in the experience. Inevitably, other areas were less successful; it takes a little time to encourage new moorland or woodland in such a setting. Sadly, the department responsible for maintenance of the University of Newcastle grounds were not impressed by the unkempt appearance around the very public face of the museum; they were not prepared to accept the time-scale required for habitat creation. Worse, there was a marked divergence in

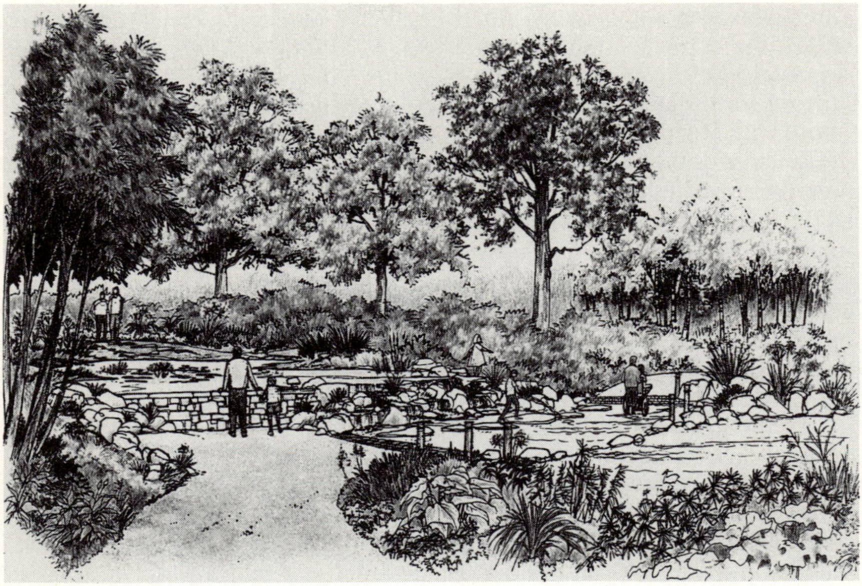

Plate 38 The Natural History Museum, London – an artist's impression of the 'wild gardens'. (Reproduced courtesy of Neil Smith and Adam Loxton.)

opinion regarding the role (and perhaps even the philosophy) of a natural history museum – whereas curatorial staff could see the educational advantages of such a scheme, and see beauty in the native flora, estates staff could only see an unkempt area at a very public spot on the university campus – an area which tarnished their record for good housekeeping and was expensive to maintain and develop. In their book the museum demanded a formal setting. Despite protests, in 1992 the moorland area was bulldozed aside, the pond filled in on safety grounds, and other habitat areas reduced in size. In 1996 part of the scheme still remains, though the initial objective of allowing museum visitors to experience all the major Northumberland plant habitats has been lost. There are obvious lessons to be learned from this experience, not least that of reaching agreement with other colleagues within an organization (whether it be a university or a local authority) about the appropriate site and scale for such developments and the time-scale required to re-create natural habitats. Perhaps with more careful planning and wider consultation, and a phased programme of development, the scheme might have avoided its eventual fate. However, the Hancock experiment must be perceived as a bold and pioneering venture, and there must be many other city-centre museums with formal gardens which could dedicate space to habitat re-creation.

The Natural History Museum in London announced a similar habitat creation programme in 1993, a Wildlife Garden (see Plate 38) which will cover more than an acre of the museum grounds, and re-create a range of British plant and wildlife habitats including ponds, marsh, chalk meadow, bluebell woods and even an 'ancient lane'. The garden, opened in 1995, is seen as a powerful educational resource, an outdoor classroom where schoolchildren can experience the first living exhibition in the museum's history, and where special provision can be made for disabled children or those with learning difficulties. Scientific research will also benefit, as the museum entomologists plan to monitor the colonization of the garden by insects and other invertebrates. Derek Langslow, Chief Executive of English Nature, one of the main sponsors of the garden scheme, has commented that:

> The Museum's plans for the garden are bold and far-sighted. ... The project demonstrates that there is a place for learning about nature in the very heart of our capital, and it will serve as an example for others to follow, not just in London, but throughout the country. (Dobson, 1993.)

Re-creating habitats is not essential of course – even formal garden and parkland can be used for educational activities. In Forest Hill, London, the Horniman Museum has been using its existing gardens in a creative way. The founder of the museum, F. J. Horniman (1835–1906) was the head of the Quaker firm of Horniman's Tea Merchants, and travelled to North America, Japan, Sri Lanka, India and China. A passionate collector, his house at 100 Lordship Lane rapidly became more like a museum, with magnificent collections of art, ethnography and natural history. The demands of the collections became so great that about 1890 Horniman and his family removed to Surrey Mount, a large house sited on high ground behind Lordship Lane, which had extensive gardens. Horniman was a great philanthropist, and encouraged access to his museum, and to his gardens, opening the latter to the public in 1895, when over 90,000 visitors attended; subsequently it was established as a public park. In 1898 the site of 100 Lordship Lane was cleared to enable the building of the museum, designed by C. Harrison Townsend (1851–1928); it was opened to the public in June 1901. E. J. Horniman (1863–1932), son of the founder, offered the museum, 15 acres of gardens, Surrey Mount, the collections and a library of 6,000 volumes as a gift to the people of London (White, 1981).

The gardens themselves are augmented by access to the disused railway line built by the London, Chatham and Dover Railway in 1865 to connect the area of Forest Hill to the re-erected Crystal Palace at Sydenham. This area is now dedicated as a nature reserve, and provides an opportunity for a 'railway trail' to explore ideas of plant recolonization, and to examine the concept of wildlife corridors. There are interesting buildings, including a Dutch Barn collected by Horniman himself, the Coach House and the magnificent Coombe Cliffe Conservatory, lovingly restored and reopened in October 1989. These buildings, and the educational centre (1969), form the focus for many of the environmental activities which take place at the Horniman; there are formal identification trails associated with the Coach House and the Dutch Barn. The range of educational activities offered is extraordinary, but those which have used the gardens specifically include sessions on wildlife gardening, garden design, minibeast safaris, horticultural demonstrations, and natural history photography. In this museum the garden is being used as a tool to relay environmental themes – it is not purely an ornamental setting for a beautiful building.

Wollaton Hall in Nottingham is just this – an imposing building in a graceful parkland setting, an area of tremendous potential to interpret environmental themes which has yet to be realized. It has been argued (Walley, *pers comm*) that this museum established the first nature trail ever to be associated with a museum in the UK (*c.* 1971); there is still a trail in the grounds, and the pond acts as a focus for many visiting school groups, to the detriment of the vegetation. There is much to be said for designing and creating wetland habitats with such use in mind – for example the use of boardwalks around pond margins protects bankside vegetation and has the additional advantage of enabling access to the disabled visitor. The ponds created at Towneley Hall, Burnley are an excellent example (Graham, 1994).

The best botanical gardens are fulfilling an essential role in conservation education, and many lessons can be learned from them by museums. Above all, they demonstrate ways in which natural history museums can use their own grounds and gardens effectively. Some museums have successfully followed this route, even with small gardens in an urban environment; for those with access to large public open space there is much potential which can be realized. Botanical gardens and natural history museums are in the process of redefining their objectives and missions, and it is fascinating to note the convergence of purpose, the new focus of the environment and its conservation giving new goals and new enthusiasms.

References

Alexander, E. P. (1979) *Museums in Motion,* American Association for State and Local History, Nashville.

Anon. (n.d.) *Guide to sculptures,* Morris Arboretum of the University of Pennsylvania, Philadelphia.

Atlanta Botanical Garden (1990) *Conserving our Environment,* Annual Report of the Atlanta Botanical Garden.

Bovey, M. (1991) 'Tropical botanic gardens: an under-utilised resource', Paper presented at the Kew Conference, *From Specimen to Habitat Management,* September.

Dillon, D. (1988) 'Eden under glass', *Museum News,* May–June, 30–32.

Dixon, C. R. (1991) 'Environmental education in Botanic Gardens: a survey of botanic garden staff and the general public', MSc Thesis, University College, London.

Dobson, G. (1993) 'The Wildlife Garden – A living exhibition at The Natural History Museum', NHM Press Release, November.

Donnelly, G. T. (1989) 'The climate is right', *Museum News,* September–October 1989, 68 (5), 45–48.

Falk, D. (1991) 'Integrated conservation techniques: their application to the developing tropical countries', Paper presented at the Kew Conference, *From Specimen to Habitat management,* September.

Foundation Kröller-Müller Museum (1994) *Guidebook to the Kröller-Müller Museum,* Otterlo, Netherlands.

Gamez, R. (1992) 'Costa Rica's National Biodiversity Institute: Linking biodiversity, conservation and development', *International Symposium and First World Congress on the Preservation and Conservation of Natural History Collections, Madrid, 1992,* Vol 3, 363–366.

Gee, H. (1990) 'Fog over London', *Nature,* 347, 18 October, 608.

Graham, M. (1994) 'Environmental education at Towneley Hall Art Gallery and Museum, Burnley', *The Biology Curator,* Issue 1, 16–18.

Haber, E., Dugal, A. and St-Jean, M. (1982) 'The Gallery of Plant Life at the National Museum of Natural Sciences', *CMA Gazette,* 15 (2/3), 8–18.

Hepper, J. N. (1982) *Royal Botanic Gardens Kew – Gardens for Science and Pleasure,* HMSO, London.

Heslop-Harrison, J. (1976) 'Introduction', In Simmons, J. B. *et al.* (eds) *The Conservation of Threatened Plants,* Plenum Press, London.

Hill, A. W. (1915) 'The history and functions of botanic gardens', *Annals of the Missouri Botanic Garden,* Vol 2.

Lucas, G. (1978) 'Organisations and contacts for conservation throughout the world – the practical role of botanic gardens in the conservation of rare and threatened plants'. In Synge, H. and Townsend, H. (eds) *Survival or Extinction,* Bentham Moxon Trust, London.

Lucas, G. (1989) 'Plant collecting and conservation'. In Hepper, F. N. (ed.) *Plant Hunting for Kew,* HMSO, London, 209–213.

Maunder, M. (1991) 'Plant re-introductions – an overview', Paper presented at the Kew Conference, *From Specimen to Habitat Management,* September.

May, R. M. (1992) 'Global change: the need and concern for collecting and preserving', *International Symposium and First World Congress on the Preservation and Conservation of Natural History Collections,* Madrid, 1992, Vol 3, 35–42.

Minter, S. (1990) *The Greatest Glass House – The Rainforests re-created,* HMSO, London.

Monnier, Y. (1993) 'Des plantes alléchantes au Jardin exotique de Menton', *La Lettre de l'OCIM,* No 27, May–June, 3–7.

Morris, P. and Riddington, P. (1990) 'The Marine Display'. In Minter, S. *The Greatest Glasshouse – The Rainforests Re-created,* HMSO, London, 158–166.

Myers, N. (1983) *A Wealth of Wild Species,* Westview Press, Boulder.

Neilson, (1985) 'Botanic gardens: their potential for recreation', *Landscape Research,* 10 (1), 18–21.

Pain S. (1991) 'Saving orchids with Sainsburys', *Kew,* Winter, 22–25.

Pickering, A. T. (1992) 'An evaluation of the roles of Botanic Gardens in recreation and conservation', PhD Thesis, Department of Town and Country Planning, University of Newcastle upon Tyne, Newcastle.

Prance, G. (1990) 'Foreword'. In Minter, S. *The Greatest Glass House – The Rainforests re-created,* HMSO, London.

Raven, P. H. (1976) 'Ethics and attitudes'. In Simmons, J. B. *et al.* (eds) *The Conservation of Threatened Plants,* Plenum Press, London, 155–176.

Raven, P. H. (1992) 'From the Director', *Missouri Botanic Garden Annual Report 1992,* 4–5.

Smith, R. D. (1982) 'The Seed Bank'. In Hepper, F. N. (ed.) *Royal Botanic Gardens, Kew: Gardens for Science and Pleasure,* 167–171.

Synge, H. (1982) 'Kew and Plant Conservation'. In Hepper, F. N. *Royal Botanic Gardens, Kew: Gardens for Science and Pleasure,* 173–179.

White, B. J. (1981) 'The Horniman Museum', *Newsletter of the Dulwich Society,* July, 8–12.

Winter, J. (1991) 'The release of endangered plants into the horticultural trade: conservation or exploitation?', Paper given at the Kew Conference, *From Specimen to Habitat Management*, September.

12. *The environmentally friendly museum*

... any real move towards sustainable development will not be merely environmental; it will mean change in economic, social and political systems. It will require a revolution, a change in civilization from one based on consumption and short-lived competition to one based on conservation and equity between generations.

Stephan Schmidheiny, Looking forward; Our common Enterprise (Introduction to *Environmental Strategies Handbook*, ed. R.V. Kolluru, 1994)

We don't inherit the earth from our ancestors – we borrow it from our children.

Indian proverb

Changes in attitude towards the environment by business and industry has been prompted partially by stricter environmental regulation, but also by a growing realization that wise use of resources is vital for sustainability and efficiency. Continuous pressure from environmentalists has also led organizations to reassess their operating procedures, as have other factors, including the demands of consumers for environmentally friendly products, the need for companies to demonstrate a positive environmental image, and pressures from concerned, better-informed shareholders and employees. The Earth Summit acted as a huge catalyst to encourage this attitudinal change, not least by emphasizing the critical nature of many environmental problems and the short time which remains to reverse them. The environment is now becoming central to the policies and practice of many major companies, with identifiable shifts towards more efficient use of natural resources, (especially energy), an economy based on conservation and long-term investment, and a savings culture. The Business Council for Sustainable Development (BCSD) has coined the term 'eco-efficiency' to describe those organizations which deliver useful goods and services whilst also reducing resource consumption and pollution. Despite the commonly held assumption that business excellence and environmental concerns are mutually exclusive, the BCSD discovered that the opposite was the case. It appeared that trends including customer demand for 'cleaner' products, tougher environmental laws, and the increasingly common policy of banks and insurance companies to favour 'green' companies, give eco-efficient organizations a competitive advantage (Schmidheiny, 1994). Like any other business, museums consume materials and energy and provide products; consequently they can learn much from the management procedures that industry has employed to place a commitment to the environment firmly within their operation – and benefit as a result. Crucial to achieving change is the use of an assessment technique – or more correctly a number of techniques – which collectively are known as environmental audit.

The museum and environmental audit

In an earlier chapter the role of museum scientists in the preparation of environmental impact statements was discussed. Most environmentalists make a firm distinction between this process, the prior evaluation of new developments, and environmental audit, which is confined to an evaluation of the environmental consequences of the existing operations of an organization. Others argue that such a distinction is unworkable, and also that the term 'audit', with its financial connotations is misleading – 'environmental review' has been suggested as a viable alternative, but has yet to be widely adopted. The European Commission's definition of environmental audit (European Commission, 1991) reads as follows:

> A management tool comprising a systematic, documented, periodic and objective evaluation of how well organisations, management and equipment are performing with the aim of contributing to safeguard the environment by facilitating management control of environmental practices, and assessing compliance with company policies, which would include meeting regulatory requirements and standards applicable.

In essence, an environmental audit is an assessment of current performance, but can set targets to allow an organization to achieve environmental excellence. This might be an individual company, a local authority such as a County Council, or one of its component divisions. It is important to recognize that although initially targeted at the production industry, an environmental audit can equally be applied to providers of services, including museums. Although initially utilized as a means of ensuring compliance with the law, (including the Water Act 1989 and the Environmental Protection Act 1990 in the UK for example), environmental auditing should now be seen to be a positive business asset, and even a means of saving money – for example through improved energy efficiency and minimizing waste. Increasingly, governments are examining environmental economics and promoting sustainable development – in the UK the Pearce report (Pearce, 1989) has done much to promote environmental accounting. As a result the business community is beginning to support the concept, with the Confederation of British Industry (CBI) (1991) declaring 'the environment means business'. Despite these words of good intent 'fewer than 50% of companies have developed formal environmental statements, and even fewer environmental audit procedures' (Grayson, 1992). Environmental audit is beginning to be seen as good practice in business, but it is not mandatory in the UK. The European Commission's plans for compulsory environmental audits have also foundered on meeting opposition from industry and the majority of member states.

In the UK, local authorities, many of which run museums, have taken the concept of environmental audit seriously; it has been suggested that this practice is merely an extension of a local government concern for the environment which can be traced back to nineteenth-century improvements in public health and housing. However, Grayson (1992) notes that 'this historic interest has been given an added boost by the growth of environmentalism as a local political issue, and by government demands to demonstrate both efficiency and effectiveness in the use of resources'. The pressures of public opinion and new legislation encouraged local government associations to produce their own guidelines (Association of County Councils *et. al.*, 1990), which suggest that an environmental audit should consist of two parts:

> A State of the Environment Report comprising a comprehensive appraisal of an area's environmental condition, and a regularly updated environmental database which can be compared with local, national or international indicators of environmental excellence.
>
> An Internal Audit comprising the systematic and objective evaluation of the environmental impact of an authority's policies and practices, together with a statement of the actions to be taken to improve matters where necessary. Although the results of the Internal Audit may be made public, it is primarily a document for internal consumption.

The 'State of the Environment Report' relates to environmental resources and conditions within the geographical area of the authority. In such an assessment databases held by museum environmental record centres are of considerable significance, contributing a historical perspective, and museum staff are well qualified to make an input to survey work and data interpretation. This input may be related to biological and geological sites, but could also include archaeological, architectural and historical interests. However, the report must go further than simply assessing the natural and built environment. It should examine wider issues, including air and water quality, waste, noise, energy production and supply, and land use and transport.

In contrast the Internal Audit is an inward-looking process which is aimed at incorporating environmental issues into strategic documents. It seeks to make the environment central to all decision-making and activities rather than peripheral position as in the past. For example, Northumberland County Council's recent audit (Environment Management Ltd, 1992) concluded that:

> there are few defined environmental policies or objectives, and where they do exist, in the current economic climate they are not always given the consideration they deserve, alongside such matters as cost ... a more cohesive, corporate approach could be introduced to environmental matters.

Success in environmental auditing has so far been patchy, with many local authorities doing no more than producing an 'Environment Charter'. It is also evident that where authorities have initiated an environmental audit they have concentrated their efforts on the wider 'state of the environment' issues rather than the internal audit. One survey (Thompson and Therivel, 1991) demonstrated a wide spectrum of practice in local authorities, with suspect, simplified and subjective methodologies leading to statements of variable and sometimes limited quality. Few local authorities have used the comparative matrix recommended by Friends of the Earth (McLaren and Bosworth, 1990), and as a consequence the quality of environmental audit has been poor. Rarely are they carried out by independent specialists; minimal consultation has resulted in reports with little commitment to positive action or to long-term environmental monitoring. Despite this, some authorities in the UK (e.g. Lancashire County Council, 1991), have produced documents, at considerable cost, which demonstrate a real understanding of the issues, and a promise of action.

Local authority museums have a significant role to play in this exercise, both by taking part in the audit, and by informing the public of the findings and proposals for the future. As Grayson (1992) notes:

> In local authorities, as befits democratically-elected organisations, the environmental audit process includes a significant emphasis on responding to the legitimate concerns of the public and interest groups about the decline in the quality of life, especially in urban areas, and on making environmental information more freely available.

There is little evidence to indicate co-operation between museums and their local authority in this field. Few museums have participated in an environmental audit, let alone acted as a vehicle to communicate the results. When local authorities carry out such an audit it is imperative that museums are involved, for they must be audited too, and used to inform the public of the outcome. If museums are to be taken seriously as purveyors of sound ecological knowledge, as organizations which can change visitor perceptions and alter patterns of behaviour for the benefit of the environment, they must themselves be seen to be striving towards 'environmental excellence'.

It is pertinent at this point to ask 'Why is environmental auditing worthwhile?' One of the most important reasons is to change or improve public perception about an organization. The fast-food industry provided one of the best-known examples, when in 1989 McDonald's formed a partnership with the Environmental Defense Fund to carry out a major environmental audit. This partnership guaranteed credibility for the company's environmental programmes (Prince and Denison, 1994; Frause and Colehour, 1994). The major shift in policy which resulted was the decision to eliminate the use of polystyrene packaging, replacing synthetic clamshells with a paper-wrap system which has since proved to be a positive move in terms of consumer support for the company. It signalled the end to a long-running 'McToxics' campaign, and led to good publicity in a wide spectrum of newspapers including the *Washington Post*, *AdWeek* and even *Rolling Stone*, the latter concluding that:

> For years, corporate executives have argued that environmental considerations were a luxury they couldn't afford. McDonald's, nudged and advised by EDF, has

adopted a strategy based on the opposite notion. Environmental action was something it could ill-afford not to do (Gifford, 1991).

McDonald's long-term strategy includes a waste-reduction plan which has the potential to cut the 'waste-stream' in the chain's 8,500 restaurants in the USA by more than 80 per cent, it also includes measures to minimize energy use, and states that 'We will not permit the destruction of rain forests for our beef supply. This policy is strictly enforced and closely monitored' (Prince and Denison, 1994). These operating changes are significant, and have been used in promotional campaigns; according to Frause and Colehour (1994) 'Overall McDonald's has managed to stem the tide of challenges from the environmental community and even to soften consumers' negative perceptions about their environmental performance.' In the longer term McDonald's also envisages saving money as a result of these initiatives. Other companies have also made savings through waste-reduction and pollution-prevention operations, including Chevron which saved $3.8 million in waste disposal costs in one year (Chevron Corporation, 1991), and Du Pont which through reduced consumption of raw materials saved $1 million in one year (McMurray, 1991).

Museums, unlike these giants of commerce and industry, are not perceived by the public as a major threat to the environment. Indeed, in the case of most natural history museums, the reverse is probably the case; they already have a positive image which is promoted through their exhibitions, educational activities and their reputation as a source of identification skills and environmental expertise. They are also, in most cases, small-scale operations, and it might be argued that this is good enough reason for doing nothing. However, the justification for carrying out environmental audits and reviewing the way in which they operate remains the same. Although they start from a higher level of respectability, museums can make significant improvements which will increase their credibility and further enhance their image, and in the longer term reduce costs through energy efficiency schemes and recycling efforts. They can further develop the market in which they already have a considerable advantage. For many museums it is equivalent to 'putting your money where your mouth is' – in other words becoming a green institution through action, rather than simply talking about it. Size is no excuse for lack of action, it is still important to 'think global, act local' and indeed it may be easier to institute changes within a small organization. Museums are in an unparalleled position to capitalize on any significant changes – they usually have an effective publicity machine and are always in the public eye. They have a captive audience to receive messages about their changes in policy and practice, and can persuade others to do likewise.

Carrying out a corporate or comprehensive audit is an expensive and time-consuming process, and probably beyond the scope of most museums. However, there are areas of operation which should be examined, areas where museums can react and respond. The following environmental audits are feasible and desirable if museums are to make progress, and can be carried out individually as part of a phased programme of improvement:

compliance audit – an assessment of compliance with environmental regulations
energy audit – an assessment of energy use within a museum building or a museum site
waste audit – an assessment of the amount of waste created, and what could be recycled
activities audit – an assessment of the environmental impact of specific museum activities
issues audit – an assessment of the museum's impact on a specific environmental problem, which can be carried out at a variety of levels.
life cycle analysis – so-called 'cradle-to-cradle' assessment of the environmental impact of a particular product or service.

It needs to be stressed that these audits are not the preserve of natural history museums alone. All museums and galleries can make a real commitment to the environment by examining their activities and procedures and setting measurable environmental quality targets.

Compliance audit – the museum and environmental legislation

Museums in the USA face a battery of environmental legislation, much of which is peripheral to their main area of activity, but of which they need to be aware. The National Environmental Protection Act 1969/70 (NEPA) is the cornerstone of legislation about the environment, setting out a policy to:

> encourage productive and enjoyable harmony between man and his environment; to promote efforts which will prevent or eliminate damage to the environment and biosphere and stimulate the health and welfare of man; to enrich the understanding of the ecological systems and natural resources important to the nation. . . .

It is the latter section which is especially relevant to the role of museums; it demands environmental impact statements for new developments, which can provide useful contract work for museum scientists. Much of the existing environmental legislation demands scientific investigation, such as a biological assessment, and monitoring procedures – The Endangered Species Act 1973, Fish and Wildlife Co-ordination Act 1934 and Coastal Zone Management Act 1972 for example. Museums may be involved in assessment, but also need to be aware of the status of protected areas if planning research activities or collecting expeditions. Superimposed on such national legislation are those imposed by individual states – New Jersey for example passed a Fresh Water Wetlands Protection Act in 1987 which is a comprehensive Act to promote the conservation of wetland ecosystems (Jain, 1994).

Furthermore, museums need to be aware of the implications of international conventions such as CITES (Convention on International Trade in

Endangered Species) or conventions such as the World Heritage Convention which address the built environment and cultural heritage in addition to sites of outstanding landscape or wildlife value. The scope of the numerous (but mainly ineffective) international conventions which relate to wildlife protection has been described by Lyster (1985).

Environmental policies in Britain are now closely tied to those of the European Community – the long-needed improvements in British legislation in recent years has been largely due to changes initiated in Europe. Further advance will be necessary for European member states to comply with the varied directives which resulted from the Earth Summit, in particular the Biodiversity Convention and Agenda 21 (sustainability) issues. Environmental legislation in Britain prior to Rio has been neatly summarized by Haigh (1990); much of current law is naturally directed towards industry, pollution control measures and maintaining environmental quality and has little or no relevance to museums. There is, however, a significant body of legislation relating to wildlife and the countryside, in particular the National Parks and Access to the Countryside Act 1949 and the Wildlife and Countryside Act 1981; the CITES convention is now enacted primarily through the EC CITES Directive which came into force in 1982, although the UK still continues to use stricter control measures through the Endangered Species (Import and Export) Act of 1976. There are a number of other relevant European directives (relating to birds and their habitats, whales, seals, environmentally sensitive areas) of which museums need to be aware (see Lyster, 1985).

In addition to legislation aimed at protecting wildlife or ecosystems, and maintaining environmental quality, there are a variety of laws relating to health and safety in the work-place – the immediate environment for all curatorial staff. When carrying out environmental audits these issues must also be considered, and are an essential part of any compliance audit, though a separate Health and Safety Audit would also be appropriate. Museums do use potentially dangerous equipment and materials; there are inherent problems in carrying out field work, including exposure to vectors of diseases. Even preparing specimens can be hazardous to health if inappropriate materials or methods are used. Maintaining pest-free museum environments has in the past demanded the use of toxic chemicals, fumigants and pesticides such as dichlorovinylphosphate (DDVP), methyl bromide, naphthalene or paradichlorobenzene (PDB). Even specimens themselves (for example, live or post-mortem material, older specimens which may have been preserved with highly toxic chemicals) pose a risk. As Richards (1994) has remarked:

> The often secure and comfortable environment of a museum may blind curators to its dangers, but the threats are very real. Perhaps with little or no further care, most natural history collections will outlive the curator, but it is possible that some may cause the curator's demise!

Health hazards associated with specimen preservation have been summarized by Linnie (1990). The principal legislation covering these issues in the UK is the Health and Safety at Work Act 1974; legislation referring to maintaining live collections in the UK is embodied in the Zoo Licensing Act 1981 and the Dangerous Wild Animals Act 1976. The use of chemicals in the museum environment is subject to the Control of Substances Hazardous to Health (COSHH) Regulations in the UK, and the Occupational Safety and Health Administration (OSHA) in the USA. Museums must be aware, and respond to these regulations – Richards (1994) gives sound advice for all curators and conservators involved with natural history collections.

Even this brief and superficial review of environmental legislation provides an indication of the importance of a compliance audit as an essential element towards encouraging and maintaining environmental excellence. As noted in Chapter 3, many natural history museums have adopted environmental protection as the cornerstone of their operation and this is frequently written into their mission statements. To support these ethical values and attitudes museums need not only to be aware of the legislation, but also to react positively to it. Awareness, attitudes and procedures relating to environmental legislation must be embodied in all museum policy documents, whether these are individual policies (such as an adopted Health and Safety Policy), more general statements such as Collections Management Plans or strategic documents such as forward plans. There is also an imperative to state which legislation has been examined; all too often Collection and Disposal Policies will include a vague statement of intent along the line of 'the museum will comply with all legislation relating to the protection of wildlife and natural habitats'. This is not sufficient – if any museum is serious about its environmental role, then a compliance audit, carefully carried out, with a final report which lists the environmental legislation which it has examined, and sets realistic objectives, is a necessity.

Energy audit – museums and energy efficiency

> Through efforts to install more energy-efficient light and controls, the Franklin Institute Science Museum lowered its 1992 electric bill $50,000 and its kilowatt hours (kwh) by 5 to 10 percent; in 1991 the Museum of Science in Boston saved $220,000, about 2.7 million kwh. The Smithsonian Institution is slated to receive about $300,000 in credits from its local utility company this year for retrofitting several of their buildings. And when the new Oregon Museum of Science and Industry opened last year, it expected to save more than 230,000 kwh each year, thanks to installing a sophisticated lighting control system (Anon, 1993).

These examples illustrate the very real financial benefits of carrying out an energy audit, and effecting energy conservation measures which by default benefit the environment.

The concept of an energy-saving museum is not new – Malcolm Wells (1971) responded to the upsurge of interest in environmental issues by proposing a design for an 'ecological museum', after pointing out the hypocrisy of museums which claim to support the environment, yet fail to use sewerless toilets, heat-recovery techniques, adequate insulation, or solar power, fail to initiate water conservation or practise recycling. Wells' museum was underground, 'expressing its ecological convictions by originating rain-catching ponds and rooftop plantings, wooded areas and topsoil-building fields of composted trash-garbage-sewage. Underground parking and solar energy traps will be further expressions of this new American land conscience' (Wells, 1971). This may sound a little fanciful (and it is difficult to imagine many architects who would wish to see their creations buried), yet Wells' ideas are centred on saving space and energy. It is gratifying to see that the idea of promoting energy efficiency and environmental awareness in museum design has recently been resurrected, albeit in a small way, at London's innovative Horniman Museum. The 'Centre for Understanding the Environment' (Anon, 1994) will be an 'environmentally-friendly futuristic building which aims to highlight current and future environmental concerns and solutions' (see Plate 39) and will feature many of the design concepts outlined in Wells' plans, including a natural passive ventilation system, reed beds to recycle waste water, and solar panels.

Considerable improvements in technology to enable energy efficiency have taken place in the 25 years following Wells' forward-thinking paper. In the same period, the impact of carbon dioxide and other emissions from energy-generating sources on global warming, and the release of CFCs from a variety of sources (including air-conditioning equipment, refrigeration plant etc.) on the ozone layer have become more fully understood, and regarded as environmental concerns of considerable importance. As a consequence the UK government has set a minimum target of stabilizing emissions of carbon dioxide at 1990 levels by the year 2005. This will necessitate a real move towards energy efficiency in maintaining museum environments (heating, air-conditioning, lighting) for the benefit of both objects and visitors. The wise use of energy has become even more pertinent in the face of shrinking budgets, and is now of significance to all museums; two recent conferences held in the UK – Saving Energy for a Better Environment (November 1991) and Environmental Control in Museums and Galleries (1993) – have addressed the subject (Cassar, 1994).

A recent survey of museums in the UK (Oreszczyn *et al.*, 1994) is less than encouraging, especially when compared to the American museum examples above. Only 10 per cent of museums perceived their energy performance to be good; 50 per cent of respondents had implemented one energy-saving measure during the past five years (usually a modification to lighting), five museums had initiated more than one, but only two had addressed a range of measures including better insulation and good housekeeping. The major factor preventing greater action appeared to be lack of capital, despite the fact that many energy-conservation measures simply involve better housekeeping and usually have a pay-back period of less than four years. Another reason given for lack of action was the misconception that saving energy might affect the museum climate to the detriment of objects, whereas in fact energy efficiency aims to minimize the use of energy whilst maintaining or improving standards. Despite this, energy awareness in museums was generally perceived as having improved during the past five years, with most museums monitoring their energy consumption, and seeking professional advice. The total energy bill for museums in the UK was estimated at £140 million – less than 1 per cent of the UK energy use in

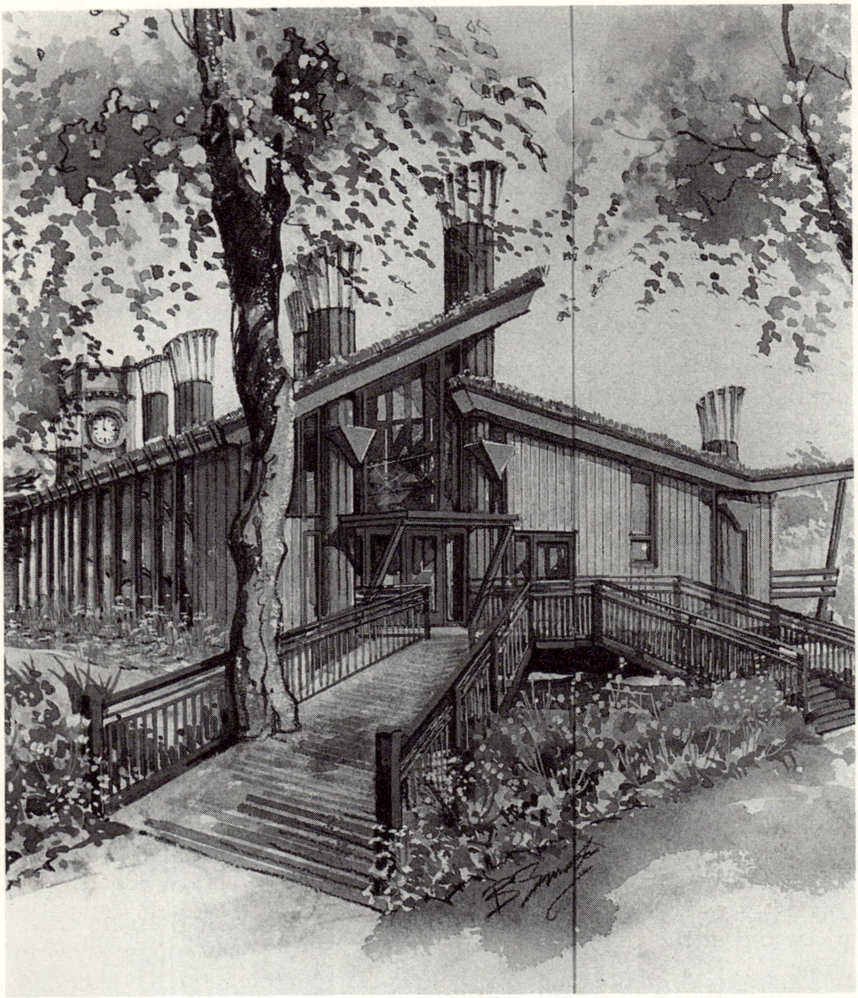

Plate 39 The Centre for Understanding the Environment at the Horniman Museum, London, includes many ecological design features. (Courtesy of The Horniman Museum, Forest Hill, London.)

buildings; of that sum 42 per cent was needed for space (galleries, stores, offices and circulation space) heating, 34 per cent for lighting, 12 per cent for air-conditioning, 5 per cent for water heating, 3 per cent for catering and 4 per cent for other uses. Somewhat surprisingly, older buildings appeared to be more energy efficient. Mixed and fine art collections were above-average consumers, and there was no clear correlation between fuel costs per square metre and total floor area. Clearly museums are a challenge, yet at the same time have considerable potential for energy savings. Above all, a clear strategic approach is needed, with a senior manager taking responsibility for energy-saving, a rolling programme of capital investment, improved housekeeping and careful monitoring of expenditure and energy-use targets.

The imperative is to employ low-energy approaches by selecting efficient equipment, maintaining it, and using it only when necessary. An energy audit is the starting-point for any new approach, and even the most modest of museums, (for example, Whitby Museum – see Pybus and Morris, 1994), having carried out a review, can reduce energy costs through improved lighting, roof insulation, and more efficient heating systems.

Waste audit – recycling in the museum

In its simplest form, this is probably the commonest form of environmental audit undertaken in business, and is conducted to establish the overall amounts of recyclable waste currently being disposed of. Once known, measures can be put in place to ensure that all recyclable materials are salvaged. Such materials can earn money for the organization – for example in the UK the price paid for computer paper was £70–80 per ton, for aluminium cans £450 per ton, and £20 per ton for tin (steel) cans. Initiating recycling programmes for these basic materials is worthwhile, especially if a museum operates a central catering facility or restaurant. In addition, the recycling programme

> is something most employees will feel good about doing in the workplace. Presently, most people practice some form of recycling at home as a way of life and will be anxious to do the same in the workplace. Recycling is a 'feel-good' and easy to perform environmental activity that will help employees feel good about the company they work for (Frause and Colehour, 1994).

It is not simply staff who will react in this way – museum visitors should also be encouraged to recycle – The Natural History Museum in Berne, Switzerland, has strategically placed bins to collect recyclables, and according to its curator, 'we recycle everything that is possible, in the public exhibition areas, and in the rooms behind as well – paper, glass, aluminium, iron, timber. ... ' (Peter Lups, *pers. comm.*). Similarly, Monterey Bay Aquarium has made recycling a priority:

> All materials, including glass, plastic, metal, paper and cardboard are recycled from aquarium business offices, workshops, food service operations and lounges. Desktop recycling containers and other recycling bins are available in all offices. Construction material from special exhibitions are recycled after the exhibits are dismantled. Shredded paper has replaced Styrofoam pellets for shipping fragile goods. ... All aquarium publications are printed on recycled paper ... vendors have been encouraged to minimise the packaging on goods they ship to the aquarium ... paper cups in staff lounges have been replaced with reusable ceramic cups (Monterey Bay Aquarium, 1993).

Waste management can of course be much more comprehensive, examining emissions of all kinds and their release to air, water or land. In a museum context, however, recycling is the only pragmatic form of waste management (other than the disposal of toxic substances) that is essential. A particularly practical example of recycling is the sale of 'Zoo-Doo' (composted elephant and rhinoceros excreta) by Memphis Zoo (C. S, 1992) which is available in a variety of gift packs and sacks emblazoned with the logo 'Just Doo It'. Material which would have been consigned to the nearest landfill site has become, 'a zoo-per idea that's out of the jungle and onto the shelves'. Proof, if it were needed, that recycling is an essential component for the ecologically sound museum, providing usable products and by-products as part of a measured output.

Activities audit – an assessment of specific museum activities

An activities audit can examine the environmental consequences of any aspect of a museum's operation; it might for example seek to assess the consequences of using particular suppliers of goods or services, perhaps assessing their environmental credentials. A major storage project, for example, might require the purchase of a large number of card trays to hold specimens – if this project is to take place, then an activities audit would need to consider the relative merits of trays made of different materials – which tray, and which supplier, is most 'environmentally friendly'? If major building works are due to be undertaken as part of a capital scheme, then it is advisable to write sound environmental practice into specifications and tender documents in order to ensure that recognized environmental quality criteria are not breached, types of substances used are controlled, and that disposal of wastes, including recycling options, are given due consideration.

One area of museum activity which demands particular attention is the museum shop – or any other trading activity. The preference of some individuals for environmentally friendly products is a well-known phenomenon, and Grayson (1992) records that research carried out in 1991 and 1992 suggests that whilst many people remain 'armchair' recyclers, green consumerism has proved remarkably resistant to global recession. The definitive studies of the phenomenon in the USA, the so-called Roper-Johnson surveys of 1990 and 1992 have established the increasing strength of demand for 'green' products, with 'truly green consumers' accounting for 25 per cent of the survey sample in 1992 – an increase of 3 per cent from 1990, as well as indicating that individuals are also exhibiting higher levels of environmental consciousness in their actual behaviour (Frankel and Coddington, 1994). Other surveys support these findings – A *Wall Street Journal*/NBC poll carried out in July 1991 reported that 46 per cent of individuals surveyed had purchased a product during the past six months because the product or manufacturer had a good reputation for protecting the environment. Interestingly 53 per cent of respondents refused to buy a product specifically because of environmental concerns about the manufacturer or product. Over half the interviewees had bought more expensive products because they were better for the environment (cited in Prince and Denison, 1994, p 254). The success of 'natural products' companies, including The Body Shop and Earth's Best indicates the potential of a strong environmental commitment and reputation.

As a consequence of these changes the museum shop is faced with both an opportunity and a real dilemma. The opportunity is to capitalize on a new market by specializing in environmentally friendly products – becoming a major focus for goods which do not exploit the environment, and which encourage an understanding of it. The dilemma is whether this can be achieved, and still ensure a viable trading operation; most museum shops exist primarily to generate additional income, rather than provide a service. A further factor also needs to be considered in natural history museums, where the pressure to avoid selling harmful products is paramount. Despite the environmental messages which these museums, aquaria and zoological gardens provide through their displays and educational activities, they are so frequently let down by their trading operations; many still continue to sell marine curios such as corals and molluscan shells for example. And what about seemingly harmless products? – most natural history museum shops are populated by huge colonies of plastic dinosaurs, rubber sharks and octopuses – but at what environmental cost? What environmental degradation is caused, what quantity of toxins and wastes are produced, how much energy expended, to enable the world's museums to sell plastic stegosaurs? One can only suspect that it is considerable.

Mahan (1979), when considering a code of ethics for natural history museums, analysed questionnaire returns from 121 museum shops in the USA, and noted that:

> A surprising number of American museums still stock such material as individual artefacts, fossils, objects made of bones of animals as well as insects (some of which are endangered), and other material that one could question from their standpoint of ethical propriety.

In descending order of priority these were:

1. Items sold that encourage exploitation, harm to endangered or threatened species, or destruction of scientific sites:
   - mineral specimens
   - fossils
   - shellfish and corals
   - insect-collecting equipment
   - bones of animals (including ivory and whalebone)
   - original archaeological material
   - feathers and fur
   - preserved insects
   - miscellaneous items
2. Sale of de-accessioned specimens.
3. Sale of books that encourage collecting, keeping wild animals in captivity; books that are not accurate scientifically or strongly biased.
4. Sales of items that show disrespect for minority ethnic groups.
5. Sale of natural history art and anthropological replicas not identified as such, and replicas of poor quality.
6. Items which suggest competition with commercial, non-museum, gift shops – items which are irrelevant to the nature of the museum itself.

This list gives some useful guidance for reviewing the items offered for sale. Inevitably there must be compromise in the museum shop; it must make a profit, it must provide what the customer wants, yet at the same time an activities audit of trading should be carried out in order to ensure that no products are sold which contravene museum environmental policy, and that suppliers are environmentally responsible. This is especially important if the museum is to become involved in environmental marketing.

Issues audit – museums and specific environmental concerns

Issues audits are usually carried out by large commercial organizations to determine how their industrial processes affect major environmental concerns such as loss of biodiversity and natural resources, acidification, global warming, loss of the ozone layer or other pollution effects. Museums, being small, service organizations, have little real impact when compared to these industrial giants, yet it is still important for them to consider ways in which they might reduce their environmental impact. An issues audit is really unnecessary in most museums, but some simple procedures could be adopted, including:

biodiversity
- adopt a selective collecting policy

natural products
- do not use peat in the museum garden
- use only recycled paper
- product specifications – choose the least environmentally damaging
- use no hardwoods from tropical countries

pollution
- ensure fleet vehicles are correctly maintained
- change to greenfreeze technology
- encourage cycle/public transport use by staff
- be selective when purchasing cleaning products and equipment
- avoid any products containing CFCs
- avoid the use of pesticides and herbicides
- dispose of hazardous wastes responsibly
- phase out halon extinguishers.

Life-cycle analysis

Life-cycle analysis or life-cycle assessment has been defined (Henn and Fava, 1994) as 'an evolving management approach for reducing the impact of a product, package or activity upon human health and the environment'. With (perhaps predictable) origins in the 1960s, current ideas about life-cycle analysis are an amalgam of two separate approaches, first product life-cycle assessment, a methodology designed to reduce environmental releases and waste, especially energy, and secondly, life-cycle cost analysis, a purely economic device. It is used in industry to examine the environmental cost in each stage in the life of the product from obtaining the raw materials to manufacture, distribution, use and eventual disposal or recycling. It is

The changing attitudes & behaviour of humankind, partly influenced by environmentalism

'Hydrocarbon Man'	Ecological human beings
Modified emphasis →	
Male dominance	Female ascendancy
Specialization	Multidisciplinary holism
Mechanics	Organics
Quantity	Quality
Analysis of parts (reductionism)	Synthesis (symbiosis synergism)
Form (What is it?)	Function (What does it do and why?)
Linear	Nonlinear, cycles
Measurable internal costs	Total life cycle costs (internal and external)
Authoritarian hierarchies	Interactive networks
Individualism	Communitarianism
Wasteful affluence	Economic sufficiency
Abundance forever	Carrying capacity concerns
Pricing base on internal costs	Full-cost pricing
Shareholder accounting	Stakeholder accounting
Closed systems	Open systems
Materialistic self-interest	Enlightened self-interest
Exploitation	Conservation
Individual rights	Individual & collective responsibilities
Confrontation	Cooperation
Exclusion	Inclusion
Cure	Prevention
Short-term	Longer-term
Rising material expectations	Recognised physical limits
Independence	Interdependence
Collective military security	Collective economic security
Physical growth	Sustainable development
Gross national product	Gross national well-being
Survival of the fittest competition	Competitor alliances for mutual survival
Volume of production output	Quality of input, throughput, and output

Figure 12.1 The changing attitudes and behaviour of humankind, partly influenced by environmentalism. (After Henn and Fava 1994.)

'described and applied in different ways in different industries for different purposes with mixed results and widely varying degrees of enthusiasm' but according to Henn and Fava (1994) is 'only one manifestation of a great transformation currently underway from what might be called Hydrocarbon Man ... living in the mechanistic Machine Age, towards Ecological Human Beings moving into the emerging Age of Integrated Systems.' This shift marks a move to sustainability, the only concept which can ensure the long-term viability of the Earth's resources and human survival. (See Figure 12.1)

Life-cycle analysis is essentially a type of systems analysis; systems analysts have recognized that there are many similarities in the ways in which ecosystems and technological systems function – for example both have pathways of energy, materials and information flow, feedback loops, can store resources and use outside sources of energy. There are identifiable differences of course; natural ecosystems are directly solar-powered and self-maintaining, whilst technosystems use solar power indirectly and require

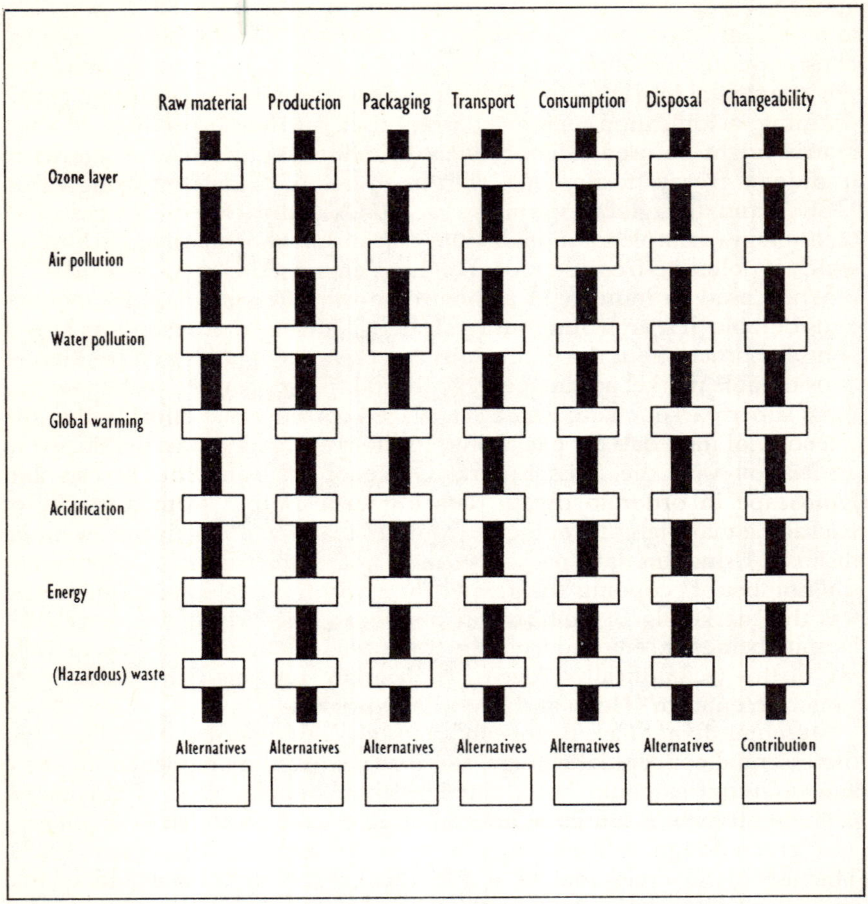

Figure 12.2 The Milion-Matrix life-cycle assessment provides a useful means of determining the environmental impact of carrying out museum activities. The matrix could be used to establish (for example) the environmental costs of establishing, maintaining and closing an exhibition. (Based on the Matrix devised by DUIJF Consultancy, Vught, The Netherlands.)

maintenance. The parallels are sufficient, however, to encourage this new way of thinking about the environmental costs of any operation – with continued emphasis on recycling and reuse it is possible to think in terms of cycles – or a cradle-to-cradle process rather than a cradle-to-grave one.

An example of the use of life-cycle analysis in a museum context might be a decision regarding the selection and purchase of a new piece of equipment – perhaps a new refrigerator. A simple Dutch rating system known as the Milion-Matrix method (designed by DUIJF Consultancy – cited in Henn and Fava, 1994) subjectively scores environmental issues for each stage in the life cycle of the product on a scale of 1–5 (see Figure 12.2). Along one axis might be raw materials, production, packaging, transport, consumption/operation, and disposal, along the other axis lie predicted environmental impacts – ozone layer, air pollution, water pollution, global warming, acidification, energy consumption, and waste products. Such a matrix might be used to good effect to determine the relative merits of traditional freezer technology (with its source of ozone-destroying CFCs, HCFCs and the global warming gas HFC) compared to 'greenfreeze' technology (a simple solution of water, salt and alcohol marketed as Flo-Ice) which is pollution-free and scores high on energy efficiency.

When used by industry in a committed way, life-cycle analysis requires considerable research time and substantial financial investment. It is a complex process, but the end results can place companies in a position of considerable market advantage. A good example is provided by Migros, the Swiss supermarket chain, which has placed considerable emphasis on the selection of materials for packaging. Its life-cycle analysis was developed in association with the Swiss Federal Office of Environment, Forests and Landscape in order to obtain data for waste output from a variety of packaging materials at every stage from raw material extraction to eventual disposal. Using this data, packages can be awarded 'ecopoints' – the higher the number of ecopoints the greater the environmental impact; one result was the marketing of milk in soft plastic pouches which are recyclable. Despite being far removed from the traditional container, effective marketing of this (scientifically proven) 'ecofriendly' product led to ready consumer acceptance (Henn and Fava, 1994, pp 605–8).

Museums do not have the resources to carry out research at this level, but the essential conceptual change, the new way of thinking which life-cycle analysis requires could be adopted to their benefit. One of the major processes for which museums are responsible is the creation of exhibitions, and 'green design' is beginning to achieve recognition (Mackenzie, 1993). The use of life-cycle analysis within this creative process would enable museums to assess the environmental impact of the various components of an exhibition and the processes and activities related to it, from initial planning through to design, construction, usage and eventual closure. What are the initial inputs in terms of raw materials and energy? Are we assured that raw materials, especially timber, is not of rainforest origin? If live animals are to be used, can we guarantee that they have been captive-bred rather than collected in the wild? What are the environmental costs of the lighting system? How much energy will be consumed in heating the exhibition space? How much of the exhibition has been designed to be recycled? Are any of the materials reusable? These questions rarely, if ever, feature in any discussion of exhibition design, yet they are important questions relating to the life cycle of an exhibition. The answers may, in some areas, lack the scientific objectivity of data one might expect for an industrial process, but they are still questions which need to be asked. How often has the end point of an exhibition been the return of specimens and objects to store, and the consignment of the fabric – the timber, metalwork and fibreglass – to the nearest landfill site?

Guidelines for the environmentally responsible museum – achieving change

Following the *Exxon Valdez* oil spill in 1989, a coalition of commercial and environmental organizations formed Project CERES (Coalition for Environmentally Responsible Economies) to develop a code of conduct to allow organizations to demonstrate their environmental awareness and responsibility. The code includes several important prerequisites for sustainable development (Henderson, 1991). Known as the 'Valdez Principles' they are primarily aimed at major commercial enterprises, but are equally applicable to museums. I have adapted the Principles to reflect the nature of museums; the following version sets ten major environmental targets:

- minimize all activities that may cause environmental damage to the biosphere
- conserve non-renewable natural resources through efficient use and careful planning
- minimize the creation of waste; dispose of hazardous waste materials in a responsible manner
- make wise use of energy; if possible use environmentally safe and sustainable energy sources to meet operating requirements
- respect the surrounding environment and local communities
- sell products that minimize environmental impact
- take compensatory action for environmental damage caused by operating activities
- make environmental policy central to the organization; make a senior manager responsible for environmental policy
- carry out an annual environmental audit to evaluate progress in implementing these principles and in complying with all applicable laws and regulations
- inform the public about your achievements.

To put these objectives into practice, to ensure excellence, the model proposed by Greeno (1994) is one which has much merit. His 'five-star' arrangement of interrelated tasks comprises *positioning, learning, communications, personal involvement* and *catalyzing progress. Positioning* is the initial assessment – where does your museum stand at the moment in terms of the environment, and where do you want it to be in the future? The assessment needs to be honest, to set realistic and measurable goals and to be aware of external forces such as environmental legislation and the demands of museum visitors. It is an initial environmental audit by any other name. Setting measurable goals is important – in line with the maxim 'what gets measured gets managed' – but environmental performance measurement is still undeniably in its infancy, even in major companies. Devising performance measures is a part of *learning*; it essentially involves monitoring progress and researching how environmental performance might be measured and assessed. However, the learning sequence equally implies a growing corporate understanding of the significance of environmental performance, as well as the education of individual members of museum staff regarding environmental issues and the museum's mission. This is important for the next phases – *communication* and *personal involvement.* Formal (e.g. Board Meetings, staff meetings, annual reports) and informal mechanisms (e.g. open discussion, press releases) are important in *communication* to ensure that members of staff, board members and members of the public are aware of the efforts, and progress that the museum is making. The *personal involvement* and commitment of senior staff is essential, especially where strategic change is required to fulfil environmental goals. *Catalyzing progress* demands improved monitoring and a commitment to change, and a total understanding of the issues by top management.

Environmental marketing – a new way to attract museum visitors?

The attitudinal changes which encapsulate the growing development of environmental responsibility has encouraged the phenomenon known as 'environmental marketing', which places the environment in a prominent place in marketing products and services. Frankel and Coddington (1994) have provided a two-part definition of environmental marketing:

> 1. Environmental marketing consists of the set of core principles and practices that enable companies to put consumers' and others' concerns about the environment to their financial advantage, in an ethically appropriate manner.
> 2. Environmental marketing also defines a changed attitude about, and new context for, making marketing decisions – one in which the environment is a pervasive consideration.

The limiting clause in section one is significant – in other words environmental marketing is not about exploiting genuine concerns of customers for the environment to enable company profits. Section two indicates the importance of placing the environment at the heart of the marketing process.

There are of course inherent dangers in promoting a 'green' image – it has proved tempting for many companies to employ an 'environmental' sales pitch, claiming a commitment to environmental quality, or promoting an 'environmentally friendly' product. The assumption of a low level of sophistication on the part of customers, and that claims would not be investigated, has led to a significant number of lawsuits in the USA against companies making environmental claims for products as diverse as waste bags, hair spray and coffee filters. Premature or inappropriate advertising, and misleading and inaccurate labelling has resulted in a backlash against companies whose aggressive marketing and sales strategies precede environmental reality. Particularly culpable are glossy, corporate advertising campaigns, which are often so superficial that they are meaningless, and consequently open to criticism. Environmental groups, always looking for such bigotry, have exposed such claims mercilessly, and even official agencies in the USA have clamped down on excessive claims. By the end of 1992 the Federal Trade Commission (FTC) had filed 16 cases against manufacturers for misleading environmental advertising and labelling (Frause and Colehour, 1994). The need for environmental advertising standards is paramount, with consistent definitions for terms such as recycled, recyclable, compostable, biodegradable and the like. Interestingly, California introduced its Truth in Environmental Advertising Act in 1990 which aims to set such standards – other states (Indiana, Rhode Island, New York) have since followed suit. Product certification programmes run by creditable third-party organizations, are becoming more commonplace. They assess the environmental performance of products and packaging; Germany has its successful Blue Angel programme whilst Canada has issued EcoLogo licences to 120 companies and 700 products.

Environmental marketing is still in a developmental phase and it is apparent that there are many hurdles to be overcome before any organization launches a 'green' marketing campaign. It must be aware of expected standards and definitions, and be able to prove its claims beyond doubt. Despite this, there is little doubt that environmental marketing is gaining in significance, and without it any organization can lose its competitive position and its credibility. Successful environmental marketing campaigns, such as those carried out by Procter & Gamble and Dwight & Church (cited in Frankel and Coddington, 1994) have provided short-term financial benefits, helped to set longer-term goals and have benefited the environment.

What are the implications of environmental marketing for museums? Although many museums, through their displays and activities, present an environmentally friendly image, they could undoubtedly do more. The supposition that natural history museums, or any museum or other heritage attraction is managed with the environment in mind is a total fallacy. Very few organizations have yet to undertake an environmental audit, let alone respond to the findings. If museums are prepared to put the environment at the centre of their business, then they already are in an advantageous position. The museum world has seen a marked shift towards the employment of marketing and promotional staff during the 1980s and 1990s, primarily due to the need to increase visitor numbers and income in response to declining levels of core-funding. Museums are now expected to run as businesses. They are reliant on multiple sources of funding, and are competing in a world where there is an ever-increasing number of visitor attractions within the heritage sector and elsewhere. Museums need to gain an edge in order to compete effectively for funds from sponsors to attract visitors and then to encourage repeat visits – they now have the marketing expertise to do this. If, to gain an advantage, they wish to develop an environmental marketing strategy, they have the added benefit, especially in natural history museums, of resident environmental expertise – a remarkable combination with which to develop an environmental marketing strategy.

However, the lessons learned by other organizations when promoting an environmental image or product should be heeded, in particular the need to be totally honest. It is also essential that a successful environmental marketing strategy is backed by genuine corporate commitment, including a training element to ensure that all staff are aware of the strategy. Environmental marketing needs to sell soft, not hard, in other words it is essential to go beyond image enhancement and explain the changes or actions that the museum has made. There is also a need to continue to educate the consumer – in this case, the museum visitor – and museums are in an advantageous position to do so when compared to most businesses; equally, if they have carried out sufficient market research, they will understand their visitor profile and be able to target the market niche correctly.

Postscript – an environmental agenda for natural history museums

This book has examined the ways in which natural history museums have responded to environmental issues. It is apparent that they have always recognized the fragility and beauty of the natural environment, but that their activities are now more closely geared to environmental conservation. It is also evident that changes in policy have accelerated since the 1960s and the growth of environmentalism. Conservation in all its guises is now regarded by many major natural history museums as the major focus for their scientific, curatorial and educational activities. This shift, which is pragmatic as well as philosophical, has been motivated in part by environmental concerns, in particular those relating to biodiversity, but also by the need for natural history museums to reposition themselves within the heritage and conservation sectors. Up until the 1960s they had remained locked into roles which had been defined in the nineteenth century. Research, collecting and educational activities lacked real focus, and generally, natural history museums were seen as outmoded institutions – they were buildings housing stuffed, dusty and faded animals. Environmentalism provided the impetus for change. This was not simply a new direction for individual natural historians who were aware of environmental degradation and loss of species, but a sea change in the way in which whole institutions perceived themselves. A parallel change has taken place in related organizations too, notably zoological and botanical gardens, providing a new justification for their activities. The environment is also an important promotional mechanism, not only to the public but also to governing bodies, funding agencies and potential sponsors; environmental concerns usually receive sympathetic responses.

How might natural history museums respond further to environmental conservation? Despite the considerable advances of the past thirty years, there is no reason for complacency. Indeed, the environment is subject to ever-increasing threats, and museums need to continue to respond appropriately, not only through active research and educational programmes, but also by implementing in-house conservation measures in line with the Valdez Principles. Developing a broad environmental strategy is paramount. The key components of such a strategy might be summarized as:

- *capitalize on existing environmental activities within the museum including* biodiversity/systematics research

environmental recording activities
environmental education programmes
the environmental content of existing exhibits
environmental consultancy activities

- *promote changes in policy which are made as a result of environmental audits, which could include*
  the adoption of an environmental plan
  the museum's commitment to environmental legislation
- *promote changes in museum activities and processes, including*
  energy-saving and recycling programmes initiated in the museum
  new displays relating to key environmental issues
  new educational programmes addressing environmental issues
  new directions for trading activities – an environmental approach.

These key issues provide a checklist for action for natural history museums. By continuing to focus on environmental concerns they can be seen to be organizations which matter, centres where important issues can be researched, debated and promoted. Most museum stakeholders, including the museum visitor, the scientific community, the museum's governing body, and employees have a major interest in environmental issues and recognize the need to continue to promote them. To do so the museum needs to be innovative and unafraid of implementing change, but first it must recognize environmental conservation as its *raison d'être*, and be committed to the concept of sustainability. It must be hoped that the important roles played by natural history museums, especially in relation to environmental education and biodiversity issues, will gradually be recognized, to their benefit.

References

Anon. (1993) 'Science museums lighting the way to energy efficiency', *ASTC Newsletter*, 21 (3), 8.

Anon. (1994) 'Developments: The Horniman Museum', *Museums Journal*, 94 (11), 8.

Association of County Councils, Association of District Councils and Association of Metropolitan Authorities (1990) *Environmental Practice in Local Government*, 236 pp.

C. S. (1992) 'End product', *Museum News*, 71 (4), 16.

Cassar, M. (ed.) (1994) *Museums, Environment, Energy*, HMSO, London.

Chevron Corporation (1991) *Save Money and Reduce Toxics*, Company brochure, San Francisco.

Confederation of British Industry (1991) *Environmental Technology; Competing in a Growing Market*, CBI, London, 33 pp.

Environment Management Limited (1992) *Summary Report of Environmental Audit for Northumberland County Council*, Middlesbrough, 47 pp.

European Commission (1991) *Draft Proposals for a Council Regulation Establishing a Community Scheme for the Evaluation and Improvement of Environmental Performance in Certain Activities and the Provision of Relevant Information to the Public (Eco-Audit)*, Brussels, 44 pp, (Document XI/83/93, Revision 4).

Frankel, J. and Coddington, W. (1994) 'Environmental marketing'. In Kolluru, R. V. (ed.) *Environmental Strategies Handbook: A Guide to Effective Policies and Practice*, McGraw-Hill Inc., New York, 643–677.

Frause, B. and Colehour, J. (1994) *The Environmental Marketing Imperative: Strategies for Transforming Environmental Commitment into a Competitive Advantage*, Probus Publishing Company, Chicago.

Gifford, B. (1991) 'The greening of the Golden Arches', *Rolling Stone*, 22 August, 34.

Grayson, L. (1992) *Environmental Auditing: A Guide to Best Practice in the UK and Europe*, British Library Science Reference and Information Service and Technical Communications, Oxford.

Greeno, J. L. (1994) 'Corporate environmental excellence and stewardship'. In Kolluru, R. V. (ed.) *Environmental Strategies Handbook: A Guide to Effective Policies and Practice*, McGraw-Hill Inc, New York, 43–64.

Haigh, N. (1990) *EEC Environmental Policy and Britain*, Longman, Harlow.

Henderson, H. (1991) *Paradigms in Progress; Life Beyond Economics*, Knowledge Systems, Indianapolis, IN.

Henn, C. L. and Fava, J. A. (1994) 'Lifecycle analysis and resource management'. In Kolluru, R. V. (ed.) *Environmental Strategies Handbook: A Guide to Effective Policies and Practice*, McGraw-Hill Inc, New York, 541–641.

Jain, R. K. (1994) 'Regulatory framework; United States, Canada and Mexico'. In Kolluru, R. V. (ed.) *Environmental Strategies Handbook: A Guide to Effective Policies and Practice*, McGraw-Hill Inc, New York, 67–117.

Lancashire County Council (1991) *Lancashire: A Green Audit: A First State of the Environment Report*, Lancashire CC, Planning Department, Preston, 352 pp.

Linnie, M. J. (1990) 'Pest control in museums: the use of chemicals and associated health problems', *Museum Management and Curatorship*, 9, 234–239.

Lyster, S. (1985) *International Wildlife Law*, Grotius, Cambridge.

Mackenzie, D. (1993) 'The role of museum designers in contributing to a better environment', *Museum Design*, Issue 3, 2.

Mahan, H. D. (1979) 'Ethics in science museums', *ICOM Natural History Committee Newsletter* No 5, Paper/texte No 8.

McLaren, D. and Bosworth, T. (1990) *Environmental Audits of Local Authorities: Terms of Reference*, Friends of the Earth, London, 6 pp.

McMurray, S. (1991) 'Chemical firms find that it pays to reduce pollution at the source', *Wall Street Journal*, June 11, A1, A6.

Monterey Bay Aquarium (1993) *Conservation Fact Sheet*, Monterey Bay Aquarium, Monterey, California.

Oreszczyn, T., Mullany, T. and Ni Riain, C. (1994) 'A survey of energy use in museums and galleries'. In Cassar, M. (ed.) *Museums, Environment, Energy*, HMSO, London, 17–38.

Pearce, D. (1989) *Sustainable Development: The Implications of Sustainable Development for Resource Accounting, Project Appraisal and Integrative Environmental Policy*, London Environmental Economics Centre, London, 192 pp.

Prince, J. and Denison, R. (1994) 'Developing an environmental action plan for business: environment as a standard operating practice – a case study'. In Kolluru, R. V. (ed.) *Environmental Strategies Handbook: A Guide to Effective Policies and Practice*, McGraw-Hill Inc., New York, 239–258.

Pybus, D. and Morris, J. M. (1994) 'Environmental improvements and energy efficiency in Whitby Museum'. In Cassar, M. (ed.) *Museums, Environment, Energy*, HMSO, London, 97–114.

Richards, J. P. (1994) 'Health and Safety in Natural History Museums'. In Stansfield, G., Mathias, J. and Reid, G. (eds) *Manual of Natural History Curatorship*, HMSO, London, 213–231.

Schmidheiny, S. (1994) 'Looking forward: our common enterprise'. In Kolluru, R. V. (ed.) *Environmental Policies Handbook: A Guide to Effective Policies and Practices*, McGraw-Hill Inc., New York, 1–8.

Thompson, S. and Therivel, R. (eds) (1991) *Environmental Auditing*, Oxford Polytechnic Schools of Biological and Molecular Sciences and Planning, Oxford, Working Paper 130, 69 pp.

Wells, M. B. (1971) 'Museums and the ecological imperative'. In The Environmental Committee of AAM (eds) *Museums and the Environment: A Handbook for Education*, American Association of Museums, Arkville Press, 177–182.

Index